Beyond Cosmology to Post-Cosmology

Beyond Cosmology to Post-Cosmology

A Preface to a New Theory of Different Worlds

Peter Baofu

Viva Books

New Delhi | Mumbai | Chennai | Kolkata | Bengaluru | Hyderabad | Kochi | Guwahati

First Indian Edition 2012

VIVA BOOKS PRIVATE LIMITED

- 4737/23 Ansari Road, Daryaganj, New Delhi 110 002
  E-mail: vivadelhi@vivagroupindia.net, Tel. 42242200
- 76, Service Industries, Shirvane, Sector 1, Nerul, Navi Mumbai 400 706
  E-mail: vivamumbai@vivagroupindia.net, Tel. 27721273, 27721274
- Jamals Fazal Chambers, 26 Greams Road, Chennai 600 006
  E-mail: vivachennai@vivagroupindia.net, Tel. 28290304, 28294241
- B-103, Jindal Towers, 21/1A/3 Darga Road, Kolkata 700 017
  E-mail: vivakolkata@vivagroupindia.net, Tel. 22836381, 22816713
- 7, Sovereign Park Apartments, 56-58, K. R. Road, Basavanagudi, Bangalore 560 004
  E-mail: vivabangalore@vivagroupindia.net, Tel. 26607409, 26607410
- 101-102 Mughal Marc Apartments, 3-4-637 to 641, Narayanguda, Hyderabad 500 029
  E-mail: vivahyderabad@vivagroupindia.net, Tel. 27564481, 27564482
- Beevi Towers, First Floor, SRM Road, Kaloor, Kochi 682 018, Kerala
  E-mail: vivakochi@vivagroupindia.net, Tel: 0484-2403055, 2403056
- 232, GNB Road, Beside UCO Bank, Silpukhuri, Guwahati 781 003
  E-mail: vivaguwahati@vivagroupindia.net, Tel: 0361-2666386

www.vivagroupindia.com

Published by arrangement with
Cambridge International Science Publishing
7, Meadow Walk, Great Abington
Cambridge CB1 6AZ, UK

ISBN : 978-81-309-1993-5

Published by Vinod Vasishtha for Viva Books Private Limited, 4737/23, Ansari Road, Daryaganj, New Delhi-110 002. Printed and bound by Syndicate Binders, Noida.

To Those in Future Post-Cosmology

BOOKS ALSO BY PETER BAOFU

Category I: Social Sciences and Related Fields

- 32. *The Future of Post-Human War and Peace* (2010)
- 31. *The Future of Post-Human Law* (2010)
- 30. *The Future of Post-Human Organization* (2009)
- 29. *The Future of Post-Human Mass Media* (2009)
- 28. *The Rise of Authoritarian Liberal Democracy* (2007)
- 27. *Beyond the World of Titans, and the Remaking of World Order* (2007)
- 26. *Beyond Capitalism to Post-Capitalism* (2005)
- 25. Volume 1: *Beyond Democracy to Post-Democracy* (2004)
- 24. Volume 2: *Beyond Democracy to Post-Democracy* (2004)
- 23. *The Future of Capitalism and Democracy* (2002)

Category II: Natural Sciences and Related Fields

- 22. *The Future of Post-Human Personality* (2010)
- 21. *The Future of Post-Human Sexuality* (2010)
- 20. *The Future of Post-Human Urban Planning* (2009)
- 19. *The Future of Post-Human Engineering* (2009)
- 18. *The Future of Post-Human Unconsciousness* (2008)
- 17. *The Future of Complexity* (2007)
- 16. *Beyond Nature and Nurture* (2006)
- 15. *The Future of Post-Human Space-Time* (2006)
- 14. *The Future of Post-Human Consciousness* (2004)

Category III: Formal Sciences and Related Fields

- 13. *The Future of Post-Human Formal Science* (2010)
- 12. *The Future of Post-Human Language* (2009)
- 11. *The Future of Post-Human Geometry* (2009)
- 10. *The Future of Post-Human Mathematical Logic* (2008)
- 9. *The Future of Information Architecture* (2008)

Category IV: The Humanities and Related Fields

- 8. *Beyond Ethics to Post-Ethics* (2010)
- 7. *The Future of Post-Human Martial Arts* (2009)
- 6. *The Future of Post-Human Creative Thinking* (2009)
- 5. *The Future of Post-Human Knowledge* (2008)
- 4. *The Future of Aesthetic Experience* (2007)
- 3. *Beyond Civilization to Post-Civilization* (2006)
- 2. Volume 1: *The Future of Human Civilization* (2000)
- 1. Volume 2: *The Future of Human Civilization* (2000)

Contents

Part Three: The Speculative Ends

Part Four: Conclusion

TABLES

Category III: Visions on the Mind

Category IV: Visions on History

Category IX: Visions on Culture

FOREWORD

Thinkers have long been fascinated about the nature of the universe—with greater or lesser success—as measured by the number of adherents to their arguments.

Dr. Peter Baofu in a rousing tour-de-force bundles and summarizes their arguments, subjects them to a systematic analysis in the context of ideas ranging from the Ancient Greeks to the present day, and then offers a new perspective.

Dr. Baofu's contribution to cosmology comes at a time when nations are struggling with issues of nuclear annihilation that threaten to greatly reduce the developmental progress of this cosmos. It would behoove the perpetrators of such threats to shift their attention to a more positive agenda that promises a higher level of existence for humankind.

Sylvan Von Burg
George Washington University

ACKNOWLEDGMENTS

As this is also true for my previous books, I have written to challenge conventional wisdom and to propose a new way of thinking.

This book challenges the existing views in cosmology and to offer a new theory to understand its future fate.

It thus comes as no surprise that this book receives no external funding nor help from any formal organization or institution, because of its political incorrectness.

My only reward is the joy of discovering something not thought of before in history.

In addition, I greatly appreciate the foreword by Sylvan von Burg at George Washington University.

In any event, and as always, I bear the sole responsibility for the ideas presented in this book.

ABBREVIATIONS

ALD = Peter Baofu. 2007. *The Rise of Authoritarian Liberal Democracy: A Preface to a New Theory of Comparative Political Systems.* Cambridge, England: Cambridge Scholars Publishing, Ltd.

BCIV = Peter Baofu. 2006. *Beyond Civilization to Post-Civilization: Conceiving a Better Model of Life Settlement to Supersede Civilization.* NY: Peter Lang Publishing, Inc.

BCPC = Peter Baofu. 2005. *Beyond Capitalism to Post-Capitalism: Conceiving a Better Model of Wealth Acquisition to Supersede Capitalism.* NY: The Edwin Mellen Press.

BCOS = Peter Baofu. 2010. *Beyond Cosmology to Post-Cosmology: A Preface to a New Theory of Different Worlds.* Cambridge, England: Cambridge International Science Publishing, Ltd.

BDPD1 = Peter Baofu. 2004. Volume 1. *Beyond Democracy to Post-Democracy: Conceiving a Better Model of Governance to Supersede Democracy.* NY: The Edwin Mellen Press.

BDPD2 = Peter Baofu. 2004. Volume 2. *Beyond Democracy to Post-Democracy: Conceiving a Better Model of Governance to Supersede Democracy.* NY: The Edwin Mellen Press.

BEPE = Peter Baofu. 2010. *Beyond Ethics to Post-Ethics: A Preface to a New Theory of Morality and Immorality.* Charlotte, NC: Infomration Age Publishing.

BNN = Peter Baofu. 2006. *Beyond Nature and Nurture: Conceivng a Better Way to Understand Genes and Memes.* Cambridge, England: Cambridge Scholars Publishing, Ltd.

BWT = Peter Baofu. 2007. *Beyond the World of Titans, and the Renaking of World Order: A Preface to a New Logic of Empire-Building.* Cambridge, England: Cambridge Scholars Publishing, Ltd.

FAE = Peter Baofu. 2007. *The Future of Aesthetic Experience: Conceiving a Better Way to Understand Beauty, Ugliness and the Rest.* Cambridge, England: Cambridge Scholars Publishing, Ltd.

FC = Peter Baofu. 2007. *The Future of Complexity: Conceiving a Better Way to Understand Order and Chaos.* London, United Kingdom: World Scientific Publishing Co.

FCD = Peter Baofu. 2002. *The Future of Capitalism and Democracy.* MD: The University Press of America.

FHC1 = Peter Baofu. 2000. Volume 1. *The Future of Human Civilization.* NY: The Edwin Mellen Press.

FHC2 = Peter Baofu. 2000. Volume 2. *The Future of Human Civilization.* NY: The Edwin Mellen Press.

FIA = Peter Baofu. 2008. *The Future of Information Architecture: Conceiving a Better Way to Understand Taxonomy, Network, and Intelligence.* Oxford, England: Chandos Publishing (Oxford) Limited.

FPHC = Peter Baofu. 2004. *The Future of Post-Human Consciousness.* NY: The Edwin Mellen Press.

FPHCT = Peter Baofu. 2009. *The Future of Post-Human Creative Thinking: A Preface to a New Theory of Invention and Innovation.* Cambridge, England: Cambridge Scholars Publishing, Ltd.

FPHE = Peter Baofu. 2009. *The Future of Post-Human Engineering: A Preface to a New Theory of Technology.* Cambridge, England: Cambridge Scholars Publishing, Ltd.

FPHFS = Peter Baofu. 2010. *The Future of Post-Human Formal Science: A Preface to a New Theory of Abstraction and Application.* Cambridge, England: Cambridge Scholars Publishing, Ltd.

FPHG = Peter Baofu. 2009. *The Future of Post-Human Geometry: A Preface to a New Theory of Infinity, Symmetry, and Dimensionality.* Cambridge, England: Cambridge Scholars Publishing, Ltd.

FPHK = Peter Baofu. 2008. *The Future of Post-Human Knowledge: A Preface to a New Theory of Methodology and Ontology.* Oxford, England: Chandos Publishing (Oxford) Limited.

FPHL = Peter Baofu. 2009. *The Future of Post-Human Language: A Preface to a New Theory of Structure, Context, and Learning.* Cambridge, England: Cambridge Scholars Publishing, Ltd.

FPHLAW = Peter Baofu. 2010. *The Future of Post-Human Law: A Preface to a New Theory of Necessity, Contingency, and Justice.* Cambridge, England: Cambridge Scholars Publishing, Ltd.

FPHMA = Peter Baofu. 2009. *The Future of Post-Human Martial Arts: A Preface to a New Theory of the Body and Spirit of Warriors.* Cambridge, England: Cambridge Scholars Publishing, Ltd.

FPHML = Peter Baofu. 2008. *The Future of Post-Human Mathematical Logic: A Preface to a New Theory of Rationality.* Cambridge, England: Cambridge Scholars Publishing, Ltd.

FPHMM = Peter Baofu. 2009. *The Future of Post-Human Mass Media: A Preface to a New Theory of Technology.* Cambridge, England: Cambridge Scholars Publishing, Ltd.

FPHO = Peter Baofu. 2009. *The Future of Post-Human Organization: A Preface to a New Theory of Communication, Decision-Making, and Leadership.*.Cambridge, England: Cambridge Scholars Publishing, Ltd.

FPHP = Peter Baofu. 2010. *The Future of Post-Human Personality: A Preface to a New Theory of Normality and Abnormality.* Cambridge, England: Cambridge International Science Publishing, Ltd.

FPHS = Peter Baofu. 2010. *The Future of Post Human Sexuality: A Preface to a New Theory of the Body and Spirit of Love-Makers.* Cambridge, England: Cambridge Scholars Publishing, Ltd.

FPHST = Peter Baofu. 2006. *The Future of Post-Human Space-Time: Conceivng a Better Way to Understand Space and Time.* New York: Peter Lang Publishing, Inc.

FPHU = Peter Baofu. 2008. *The Future of Post-Human Unconsciousness: A Preface to a New Theory of Anomalous Experience.* Cambridge, England: Cambridge Scholars Publishing, Ltd.

FPHUP = Peter Baofu. 2009. *The Future of Post-Human Urban Planning: A Preface to a New Theory of Density, Void, and Sustainability.* Cambridge, England: Cambridge Scholars Publishing, Ltd.

FPHWP = Peter Baofu. 2010. *The Future of Post-Human War and Peace: A Preface to a New Theory of Aggression and Pacificity.* Cambridge, England: Cambridge Scholars Publishing, Ltd.

• PART ONE •

Introduction

CHAPTER 1
INTRODUCTION—THE AWE OF THE UNIVERSE

> [The quantum theory of gravity for a unified theory of physics] would be the ultimate triumph of human reason—for then we would know the mind of God.
> —Stephen Hawking (1988:175)

The Euphoric Framing of a Worldview

What makes a unified theory of physics about the universe so extraordinarily promising that "it would be the ultimate triumph of human reason—for then we would know the mind of God"? (S. Hawking 1988: 175)

The euphoric framing of this worldview is all the more questionable, especially because this claim comes from a world-renowned physicist who later changed his mind and proposed something else instead: "Some people will be very disappointed if there is not an ultimate theory that can be formulated as a finite number of principles. I used to belong to that camp. I have changed my mind. I'm now glad that our search for understanding will never come to an end and that we will always have the challenge of new discovery." (S. Hawking 2005)

Later, in 2010, Hawking changed his mind again, when he wrote in *The Grand Design* that "it is not necessary to invoke God" at all to understand the working of the universe.

So, which camp is correct in this debate on understanding the universe? Contrary to the two opposing camps, the understanding of the universe is neither possible nor desirable to the extent that its respective defenders would like us to believe.

Of course, this is not to say that cosmology is a worthless field of study, or its other fields of study (like metaphysics, theology, astrology, and esotericism) should be dismissed. Needless to say, neither of these two extreme views is reasonable either.

The important point to remember is that this book provides an alternative (better) way to understand the nature of the universe, especially in relation to its contested beginnings and speculative ends—while learning from different approaches in the literature but without favoring any one of them (nor integrating them, since they are not necessarily compatible with each other). Thus, this book offers a new theory to transcend the existing approaches in the literature on cosmology in a way not conceived before.

If successful, this seminal project is to fundamentally change the way that we think about cosmology, from the combined perspectives of the mind, nature, society, and culture, with enormous implications for the human future and what I originally called its "post-human" fate.

Cosmology vs. Cosmogony

A good point of departure is to define the term "cosmology" and its relation to a different term "cosmogony"—although the two have much in common.

The term "cosmology" is from "Greek κοσμολογία" (with "κόσμος, kosmos" to mean "cosmos, the world," or "the universe"; and "-λογία, -logia" to mean "study") and thus refers to "the study of the universe in its totality, and by extension, humanity's place in it." (WK 2010 & 2010a)

Although the term "cosmology" is of recent origin (as it was "first used in 1730 in Christian Wolff's *Cosmologia Generalis*"), the "study of the universe has a long history involving science, philosophy, esotericism, and religion." (WK 2010)

By contrast, another term "cosmogony" (or "cosmogeny") is from "Greek κοσμογονία (or κοσμογενία)" (with "κόσμος" to mean "cosmos, the world," or "the universe"; and "γί(γ)νομαι/γέγονα" to mean "to be born, come about") and thus refers to "any theory concerning the coming into existence or origin of the universe, or about how reality came to be." (WK 2010a)

Therefore, "cosmogony" differs from "cosmology," because "cosmology…studies the universe at large and throughout its existence" and "technically does not inquire directly into the source of its origins," as in "cosmogony." (WK 2010a) An opposite of "cosmogony" is "eschatology" (which refers to the study of the speculative ends of the world instead).

This distinction is true, even though it is not always followed by those who use the two terms (and can confuse them occasionally). For instance, "the cosmological argument from theology regarding the existence of God is technically an appeal to cosmogonical rather than cosmological ideas." (WK 2010a)

This distinction is all the more important for science, because, for instance, "physical cosmology is the science that attempts to explain all observations relevant to the development and characteristics of the universe as a whole. Questions regarding why the universe behaves in such a way [as in cosmogony] have been described by physicists and cosmologists as being extra-scientific, though speculations are made from a variety of perspectives that include extrapolation of scientific theories to untested regimes and philosophical or religious ideas." (WK 2010a)

This distinction does not disappear, even though some efforts have been made to naturalize cosmogony, so as to make it more scientific—because of two inherent limitations. The first inherent limitation is "based in the philosophy of science and the epistemological constraints of science itself, especially with regards to whether scientific inquiry can ask questions of 'why' the universe exists." (WK 2010a)

And the second inherent limitation is "more pragmatic" in nature, because "there is no physical model that can explain the earliest moments of the universe's existence (Planck time) because of a lack of a consistent theory of quantum gravity." (WK 2010a)

However, the study of the universe in cosmology cannot be complete without some consideration of cosmogony (and, for that matter, eschatology) too, especially in regard to the profound questions concerning the contested origins (or beginnings) and the speculative ends of the universe.

For this reason, this book will also consider cosmogony (and eschatology) in the context of cosmology.

The Different Conceptions of the Universe

This distinction between cosmology and cosmogony then leads to another clarification, this time in regard to what is exactly meant by the term "universe" in the first place—regardless of whether it is used in the context of the study of the universe (as in cosmology) or of the origins of the universe (as in cosmogony).

The term "universe" is "from the Old French word *Univers*" and "from the Latin word *universum*" (with "un, uni" to mean "one" and "vorsum, versum" to mean "something rotated, rolled, changed") and therefore refers to "everything rotated as one" or "everything rotated by one." (WK 2001b; OED 1971)

Even then, however, there are numerous conceptions about what "everything rotated as one" or "everything rotated by one" is supposed to mean.

For instance, hereafter are four different conceptions of the universe, for illustration—namely, (a) the universe as "reality and probability," (b) the universe as "reality," (c) the universe as "connected space-time," and (d) the universe as "observable reality." (WK 2010b)

The Universe as "Reality and Probability"

The first conception of the universe to be illustated here has to do with the broadest definition of the universe as "reality and probability." (WK 2010b)

For instance, "the medieval philosopher Johannes Scotus Eriugena...defined it [the universe] as simply everything: everything that exists and everything that does not exist. Time is not considered in Eriugena's definition; thus, his definition includes everything that exists, has existed and will exist, as well as everything that does not exist, has never existed and will never exist." (WK 2010b)

In modern times, "[t]his all-embracing definition...reappears in quantum physics, perhaps most obviously in the path-integral formulation of [Richard] Feynman [1965]. According to that formulation, the probability amplitudes for the various outcomes of an experiment given a perfectly defined initial state of the system are determined by summing over all possible paths by which the system could progress from the initial to final state. Naturally, an experiment can have only one outcome; in other words, only one possible outcome is made real in this Universe...." (WK 2010b)

So, "[i]n this well-defined mathematical sense, even that which does not exist (all possible paths) can influence that which does finally exist (the experimental measurement)....This conception of the Universe embracing both the existent and the non-existent loosely parallels the Buddhist doctrines of shunyata and interdependent development of reality, and Gottfried Leibniz's more modern concepts of contingency and the identity of indiscernibles." (WK 2010b)

The Universe as "Reality"

The second conception of the universe is less broad and has to do with the universe as "reality" only. (WK 2010b)

In this conception, "the Universe is defined as everything that exists, has existed, and will exist. According to this definition and our present understanding, the Universe consists of three elements," as show below.

- "Space and Time"—"collectively known as space-time or the vacuum" (WK 2010b)
- "Matter and Energy"—"matter and various forms of energy and momentum occupying space-time" (WK 2010b)
- "Physical Laws"—"the physical laws that govern the first two"

In fact, "these three elements of the Universe (spacetime, matter-energy, and physical law) correspond [very] roughly to the ideas of Aristotle. In his book *The Physics*..., Aristotle divided το παν (everything) into three roughly analogous elements: matter (the stuff of which the Universe is made), form (the arrangement of that matter in space) and change (how matter is created, destroyed or altered in its properties, and similarly, how form is altered). Physical laws were conceived as the rules governing the properties of matter, form and their changes." (WK 2010b)

The Universe as "Connected Space-Time"

The third conception of the universe has a different interpretation of space-time, in terms of "disconnected space-times," with "each existing but unable to interact with one another." (WK 2010b)

Or different put, "[a]n easily visualized metaphor is a group of separate soap bubbles, in which observers living on one soap bubble cannot interact with those on other soap bubbles, even in principle. According to one common terminology, each 'soap bubble' of space-time is denoted as a universe, whereas our particular space-time is denoted as the Universe, just as we call our moon the Moon. The entire collection of these separate space-times is denoted as the multiverse." (WK 2010b; G. Ellis 2004)

By definition, "the other unconnected universes may have different dimensionalities and topologies of space-time, different forms of matter and energy, and different physical laws and physical constants, although such possibilities are currently speculative." (WK 2010b)

The Universe as "Observable Reality"

And the fourth conception of the universe is much less broad, by comparison, and treats the universe as "observable reality" only. (WK 2010b)

In this sense, "the Universe is everything within our connected space-time that could have a chance to interact with us and vice versa. According to the general theory of relativity, some regions of space may never interact with ours even in the lifetime of the Universe, due to the finite speed of light and the ongoing expansion of space." (WK 2010b)

As an example, "radio messages sent from Earth may never reach some regions of space, even if the Universe would live forever; space may expand faster than light can traverse it. It is worth emphasizing that those distant regions of space are taken to exist and be part of reality as much as we are; yet we can never interact with them. The spatial region within which we can affect and be affected is denoted as the observable universe." (WK 2010b)

Also, "the observable universe depends on the location of the observer. By traveling, an observer can come into contact with a greater region of space-time than an observer who remains still, so that the observable universe for the former is larger than for the latter. Nevertheless, even the most rapid traveler may not be able to interact with all of space." (WK 2010b)

In general, this conception of the universe is often taken "to mean the universe observable from our vantage point in the Milky Way Galaxy." (WK 2010b)

Some Deep Questions in Cosmology

With these four conceptions of the universe in mind—a good question to ask then concerns the main challenges in the field of cosmology.

In other words, what are the deep questions which have challenged the best and the brightest minds in cosmology through the ages?

Consider the following five questions, for illustration only.

- Ex: "What is the origin of the Universe? What is its first cause? Is its existence necessary?" (WK 2010)
- Ex: "What are the ultimate material components of the Universe?" (WK 2010)
- Ex: "What is the ultimate reason for the existence of the Universe? Does the cosmos have a purpose?" (WK 2010)
- Ex: "What is the purpose of Consciousness in the Cosmos? Does Reason explain how we know what we know in totality of the Cosmos? Does Cosmological Reasoning reveal Metaphysical truths?" (WK 2010)

- Ex: What is the ultimate fate of the universe? Will the universe have an end?

Surely, there can be many other questions, but these five are among the most often cited and are deemed sufficient for the present purpose of illustation.

The Theoretical Debate

The answers to these five deep questions in cosmology depend, of course, on the different theoretical approaches in the literature that a scholar uses.

For illustration, there are five main theoretical approaches in the debate on the nature of the universe to be summarized hereafter, and they can be called, in the absence of better words, namely, (a) the *scientific* argument, (b) the *religious* argument, (c) the *metaphysical* argument, (d) the *esoteric* argument, and (e) the *post-cosmological* argument (which is my original contribution to the debate)—to be analyzed hereafter, in that order.

Lest misunderstanding occurs, two clarifications should be made here, as this is something that I often did in my previous books—*almost verbatim*.

Firstly, there can be other arguments (and theories within each) besides the first four here, that is, the "scientific," the "metaphysical," the "religious" argument, and the "esoteric" argument (as indicated above). The advantage to select these four here has to do with their illustrative representation of the diverse theories in the literature (which are deemed sufficient for the purpose at hand in this book).

And secondly, the four arguments are not necessarily mutually exclusive. For instance, those who advocate the "religious" argument can also consider the "scientific" argument, although they do not focus on the latter.

And the reverse is also true, in that those who make the "scientific" argument can also consider the "religious" argument, although they do not focus on the latter.

For instance, the physicist Frank Tipler (1994) made good use of theology in his book *The Physics of Immortality* (1994:154 & ch.4), when he "tried to synthesize theology with physics through the Omega Point Theory. This theory assumes that the universe is closed and that intelligence life...will gain greater and greater control of all matter and energy sources in the universe before their control becomes total at the Omega Point." (P. Baofu 2000a: Ch.7)

In other words, their disagreement is often one in degree, not in kind.

The Scientific Argument

With these two clarifications in mind—the first major theoretical approach is called, in the absence of better words, the *scientific* argument, which focuses, relatively speaking, more on the role of science (like physics and astrophysics, in particular) to provide physical explanations on the nature of the universe (like its beginning and evolution).

In this sense, the word "scientific" in the title of the argument is suggestive, because it refers to the use of science (like physics and astrophysics) for the understanding of the universe.

Those who make use of physics and astrophysics to study the universe are known as "physical cosmologists"—and this approach to cosmology is called "physical cosmology." (WK 2010)

However, there is no consensus even among physical cosmologists on various issues concerning the nature of the universe, except that they all agree on the "scientific" argument as the best way to study them.

Celestial Mechanics

In the early days of physical cosmology, the field of study "was what is now known as celestial mechanics, the study of the heavens. The Greek philosophers Aristarchus of Samos, Aristotle, and Ptolemy proposed different cosmological theories." (WK 2010)

For instance, "the geocentric Ptolemaic system [that the Earth is the center of the universe] was the accepted theory to explain the motion of the heavens until Nicolaus Copernicus, and subsequently Johannes Kepler and Galileo Galilei, proposed a heliocentric system in the 16th century [that the Earth is not the center of the universe but revolves, together with other planets, around the Sun]. This is known as one of the most famous examples of epistemological rupture in physical cosmology." (WK 2010 & 2010c)

Later, Isaac Newton, with his 1687 publication of *Principia Mathematica*, finally solved "the problem of the motion of the heavens....Newton provided a physical mechanism for Kepler's laws, and his law of universal gravitation allowed the anomalies in previous systems, caused by gravitational interaction between the planets, to be resolved. A fundamental difference between Newton's cosmology and those preceding it was the Copernican principle that the bodies on earth obey the same physical laws as all the celestial bodies. This was a crucial philosophical advance in physical cosmology." (WK 2010)

The Theory of General Relativity

Physical cosmology took another major step after the publication of the 1917 paper titled "Cosmological Considerations of the General Theory of Relativity" by Albert Einstein, who showed "how the geometry of space and time is influenced by whatever matter is present," such that space can curve and time can slow down, for instance. (WK 2010 & 2010e)

This pathbreaking work of "general relativity prompted cosmogonists such as Willem de Sitter, Karl Schwarzschild and Arthur Eddington to explore the astronomical consequences of the theory, which enhanced the growing ability of astronomers to study very distant objects. Prior to this (and for some time afterwards), physicists assumed that the Universe was static and unchanging. In parallel to this dynamic approach to cosmology, a debate was unfolding regarding the nature of the cosmos itself." (WK 2010)

The Big Bang Theory

Then came later models of the universe to further explore "the possibility that the cosmological constant introduced by Einstein in his 1917 paper may result in an expanding universe, depending on its value." (WK 2010)

The term "cosmological constant" here is "usually denoted by the Greek capital letter lambda: Λ" and "was proposed by Albert Einstein as a modification of his original theory of general relativity to achieve a stationary universe," such that the universe neither contracts nor expands. (WK 2010f)

The reasoning here has to do with "the density parameter, Ω [Omega Lambda]," which is defined as "the ratio of the actual (or observed) density ρ to the critical density ρc of the Friedmann universe. The critical density is the watershed between an expanding and a contracting Universe." (WK 2010f & 2010g)

So, if the actual (or observed) density ρ is more than the critical density ρc, the universe will contract because of the larger gravitational pressure. But if the actual (or observed) density ρ is less than the critical density ρc, the universe will expand because of the lack of sufficient gravitational pressure to cause the universe to collapse. And if the actual (or observed) density ρ is equal to the critical density ρc, the universe will neither contract nor expand, which is exactly what Einstein preferred and is known as his "stationary universe."

Unfortunately for Einstein—in 1927, "the big bang theory was proposed by the Belgian priest Georges Lemaître...which was subsequently corroborated by Edwin Hubble's discovery of the red shift in 1929 and later by the discovery of the cosmic microwave background radiation by Arno Penzias and Robert Woodrow Wilson in 1964. These findings were a first step to rule out some of many alternative physical cosmologies," including Einstein's stationary universe. (WK 2010)

More specifically, in favor of the big bang theory, "the critical density is estimated to be approximately five atoms (of monatomic hydrogen) per cubic metre, whereas the average density of the Universe is believed to be 0.2 atoms per cubic metre. Therefore, the Universe will expand forever." (WK 2010g; M. Rees 2000)

With this new finding, Einstein had to change his mind and abandoned his idea of "the cosmological constant"; he even "called it the 'biggest blunder' of his life...However, the cosmological constant remained a subject of theoretical and empirical interest." (WK 2010f)

More recently, "observations made by the COBE and WMAP satellites observing this background radiation have effectively, in many scientists' eyes, transformed cosmology from a highly speculative science into a predictive science, as these observations matched predictions made by a theory called Cosmic inflation, which is a modification of the standard big bang theory. This has led many to refer to modern times as the 'Golden age of cosmology.'" (WK 2010)

But this "Golden age of cosmology" is not without its critics. The successes of physical cosmology have invited some serious criticisms; and hereafter are three most important ones.

The first critique is that the big bang theory assumes some fundamental hypothetical entities which are not yet proved empirically.

More specifically, "the big bang today relies on a growing number of hypothetical entities, things that we have never observed—inflation, dark matter and dark energy are the most prominent examples. Without them, there would be a fatal contradiction between the observations made by astronomers and the predictions of the big bang theory....It would, at the least, raise serious questions about the validity of the underlying theory." (NS 2004)

Secondly, the popular acceptance of the big bang theory has also overshadowed non-standard physical cosmologies in the study of the universe, which is not necessarily a healthy state in cosmology.

For instance, "the big bang is not the only framework available for understanding the history of the universe. Plasma cosmology and the steady-state model both hypothesize an evolving universe without beginning or end. These and other alternative approaches can also explain the basic phenomena of the cosmos, including the abundances of light elements, the generation of large-scale structure, the cosmic background radiation, and how the redshift of far-away galaxies increases with distance. They have even predicted new phenomena that were subsequently observed, something the big bang has failed to do." (NS 2004)

Of course, "supporters of the big bang theory may retort that these theories do not explain every cosmological observation. But that is scarcely surprising, as their development has been severely hampered by a complete lack of funding. Indeed, such questions and alternatives cannot even now be freely discussed and examined. An open exchange of ideas is lacking in most mainstream conferences. Whereas Richard Feynman could say that 'science is the culture of doubt,' in cosmology today doubt and dissent are not tolerated, and young scientists learn to remain silent if they have something negative to say about the standard big bang model." (NS 2004)

Thus, for the critics, this state of intolerance in physical cosmology "makes unbiased discussion and research impossible.'" (WK 2010h; NS 2004)

And thirdly, the big bang theory is concerned solely with the vantage point of science (like physics and astrophysics) in the process of understanding the universe, but there are other approaches to cosmology which have been ignored in physical cosmology, and good examples are "religious cosmology," "metaphysical cosmology," and "esoteric cosmology" (as will be elaborated below).

The Religious Argument

The second theoretical approach is called, in the absence of better words, the *religious* argument, which focuses, relatively speaking of course, more on the role of religion (like pantheism and creationism) to provide relligious explanations on the nature of the universe (like the relationships between the universe and God).

In this sense, the word "religious" in the title of the argument is suggestive, because it refers to the use of religion (like creationism) for the understanding of the universe.

Those who make use of religion to study the universe are known as "religious cosmologists"—and this approach to cosmology is called "religious cosmology." (WK 2010)

The term "religious cosmology" thus refers to some alternative "ways of explaining the history and evolution of the universe based, at least in part, on the acceptance of principles that cannot be justified by accepted scientific arguments" (like the theories in physical cosmology as already discussed above). (WK 2010m)

Even then, like the researchers in "physical cosmology"—there is also no consensus even among "religious cosmologists" on various issues concerning the nature of the universe, except that they all agree on the "religious" argument as the best way to study them, in a way different from those in physical cosmology, for instance.

Monism

A good starting point concerns a speculative view known as "monism," which is often mixed with both philosophical and religious ideas. So, this sub-section on the religious argument overlaps a bit with the next one on the metaphysical argument.

With this caveat in mind—the term "monism" here refers to the view that "the universe is really just one thing, despite its many appearances and diversities," and this view, when mixed with religion, suggests that "there is one God, with many manifestations in different religions," in the world. (WK 2010i)

Monism, in the context of its ideal types, however, can come in three major (different) kinds.

- "Idealism, Phenomenalism, or Mentalistic Monism"—this kind of monism "holds that only mind is real." (WK 2010i)
- "Neutral Monism"—this kind of monism "holds that both the mental and the physical can be reduced to some sort of third substance, or energy." (WK 2010i)
- "Physicalism or Materialism"—this kind of monism "holds that only the physical is real, and that the mental or spiritual can be reduced to the physical." (WK 2010i)

Of course, there can be "other positions [which] are hard to pigeonhole into the above categories" in a strict and clear way. (Wk 2010i)

In any event, the monist idea that there is "a monotheistic God that is omnipotent and all-pervading" is quite appealing to a lot of earthly religions, such as some versions of Judaism, Christianity, Islam, Buddhism, and Hinduism. (WK 2010i)

Pantheism

A closely related view, known as "pantheism," is "from the Ancient Greek: πᾶν (pan) meaning 'All' and θεός (theos) meaning 'God'" and therefore literally means "All is God." (WK 2010j)

Pantheism thus refers to the view that "the Universe (Nature) and God are identical, or that the Universe…is the only thing deserving the deepest kind of reverence." (WK 2010j)

Because pantheism holds that "the Universe" is "the sum total of all that was, is and shall be…, rather than…a transcendent, mental, personal or creator entity," it therefore does not accept "a personal or creator god. Although there are divergences within Pantheism, the central ideas found in almost all versions are the Universe as an all-encompassing unity and the 'sacredness' of Nature." (WK 2010j)

In general, pantheism can come in three major (different) kinds, as shown below. (WK 2010j; P. Harrison 1999)

- "Monist Physicalist or Naturalistic Pantheism"—this kind of pantheism "holds that there is only one type of substance, and that substance is physical, i.e., energy and matter. Historically this version was held by Stoics such as Zeno of Citium or Marcus Aurelius, and in modern times by John Toland, Ernst Haeckel, D.H. Lawrence and Paul Harrison….In this version, the term god—if used at all—is basically a synonym for Nature or Universe, seen from the point of view of reverence." (WK 2010j)
- "Monist Idealist Pantheism"—this kind of pantheism "holds that there is only one type of substance, and that substance is mental or spiritual. Ultimate reality consists of a single consciousness. This version is uncommon in modern times but is represented in the Eleatic school of Xenophanes and Parmenides and by Consciousness-Only schools of Buddhism." (WK 2010j)
- "Dualist Pantheism"—this kind of pantheism "holds that there are two major types of substance, physical and mental/spiritual. Dualistic pantheism is very diverse, and may include beliefs in reincarnation, cosmic consciousness, and paranormal connections across Nature. It is represented most widely today in literal versions of Paganism." (WK 2010j)

Therefore, there is a strong relationship (or similarity) between "monism" and "pantheism," because "almost all pantheists are monists" (which means that only "some pantheists may…be not-monists," in accepting two different substances in the universe). (WK 2010j)

Panentheism

Another related religious view is "panentheism," which is "from Greek πᾶν (pân)" to mean "all," "ἐν (en)" to mean "in"; and "θεός (theós)" to mean "God"; so the whole term literally means "all-in-God." (WK 2010k) By contrast, "pantheism" means "all is God." (WK 2010j)

"Pantentheism" is thus "a belief system which posits that God exists and interpenetrates every part of nature, and timelessly extends beyond as well. Panentheism is distinguished from pantheism, which holds that God is synonymous with the material universe." (WK 2010k)

In other words, "in pantheism, 'God is the whole'; in panentheism, 'The whole is in God.' This means that the Universe in the first formulation is practically the Whole itself, but in the second the universe and God are not ontologically equivalent. In panentheism, God is not necessarily viewed as the creator or demiurge, but the eternal animating force behind the universe, with the universe as nothing more than the manifest part of God." (WK 2010k)

So, in panentheism, "the cosmos exists within God, who in turn 'pervades' or is 'in' the cosmos. While pantheism asserts that God and the universe are coextensive, panentheism claims that God is greater than the universe and that the universe is contained within God." (WK 2010k; E. Fahlbusch 1999)

Creationism

Another closely related view is "creationism," which is more radical, with "the religious belief that humanity, life, the Earth, and the universe were created in some form by a supernatural being or beings. The term is usually used to refer to the denial, on religious grounds, that biological processes (most commonly, evolution) can adequately account for the history, diversity, and complexity of life on earth—the creation-evolution controversy." (WK 2010l)

In other words, creationism considers God as the creator, whereas panentheism does not make this necessary claim.

Good examples of creationism are the belief systems in Jewish and Christian sects, which are "usually based on a literal reading of the Genesis creation myth." (WK 2010l; R. Numbers 2010)

However, for the critics, religious cosmology, be it based on monism, pantheism, panentheism, or creationism, has its serious problems.

The most serious challenge to religious cosmology by questioning its claims has to do with the rise of science (as shown in the previous sub-section on "scientific cosmology"), which, especially "since the end of the 19th century," has much decreased the "belief in creationism" and other religious explanations, because "scientific theories have been presented that support more naturalistic explanations for the universe and for life." (WK 2010l)

For instance, "both creation science and intelligent design have been characterized as pseudoscience by the mainstream scientific community. The most notable disputes concern the evolution of living organisms, the idea of common descent, the geological history of the Earth, the formation of the solar system and the origin of the universe." (WK 2010l; NCSE 2008; IAP 2006; AAAS 2006)

The Metaphysical Argument

The third theoretical approach is called, in the absence of better words, the *metaphysical* argument, which focuses, relatively speaking of course, more on the role of philosophy (like metaphysics, in particular) to provide philosophical explanations on the nature of the universe (like its orgins and the ultimater material components of the universe, for instance).

In this sense, the word "metaphysical" in the title of the argument is suggestive, because it refers to the use of philosophy (like metaphysics) for the understanding of the universe.

Those who make use of philosophy (especially metaphysics) to study the universe are known as "metaphysical cosmologists"—and this approach to cosmology is called "metaphysical cosmology." (WK 2010)

Metaphysical cosmology "deals with the world as the totality of space, time and all phenomena. Historically, it has had quite a broad scope, and in many cases was founded in religion. The ancient Greeks did not draw a distinction between this use and their model for the cosmos." (WK 2010)

In this sense, "metaphysical cosmology" and "religious cosmology" (as already elabroated in the previous sub-section) has much in common.

However, there is still some fundamental difference between the two, because "in modern use," metaphysical cosmology "is distinguished from religious cosmology in that it approaches these questions using philosophical methods (e.g. dialectics)." (WK 2010)

In addition, unlike "physical cosmology"—metaphysical cosmology "addresses questions about the Universe which are beyond the scope of science." (WK 2010)

However, like those in "physical cosmology" and "religious cosmology"—there is also no consensus even among "metaphysical cosmologists" on various issues concerning the nature of the universe, except that they all agree on the "metaphysical" argument as the best way to study them.

Mechanism

A good way to introduce the metaphysical argument in more detail has to do with the metaphysical view known as "mechanism," which refers to "the theory that all natural phenomena [in the universe] can be explained by laws of nature. It's the opposite of vitalism, which claims organisms have 'vital forces' which aren't physical." (WK 2010n)

In metaphysics, "mechanism…comes in two different flavors. They are both doctrines of metaphysics, but they are different in scope and ambitions: the first [universal mechanism] is a global doctrine about nature; the second [anthropic mechanism] is a local doctrine about human beings and their minds." (WK 2010n)

In the present context concerning the nature of the universe, let's look more closely at the first flavor, namely, "universal mechanism."

The term "universal mechanism" refers to the view that "the universe is best understood as a completely mechanical system—that is, a system composed entirely of matter in motion under a complete and regular system of laws of nature. The mechanists understood the achievements of the scientific revolution to show that every phenomenon in the universe could eventually be explained in terms of 'mechanical' laws: that is, in terms of natural laws governing the motion and collision of matter. It follows that mechanism is a form of thoroughgoing determinism: if all phenomena can be explained entirely through the motion of matter under physical laws, then just as surely as the gears of a clock completely determine that it will strike 2:00 an hour after it strikes 1:00, all phenomena are completely determined by the properties of that matter and the operations of those natural laws." (WK 2010n)

In fact, "the determinism implied by universal mechanism is even stronger than clockwork. Whereas the mechanism of a clock may cease to work predictably as its parts break down, the 'parts' of the system in universal mechanism are nothing less than everything in the universe—anything that they 'broke down' into would still be a part of the universe, and so would still be subject to the mechanistic laws of nature." (WK 2010n)

Isaac Newton also contributed to the acceptance of "mechanism" because his "work in mechanics seemed to successfully explain the motion of everything in heaven and in earth according to the operation of a single mechanical principle." (WK 2010n)

But mechanism later sufferred from two major problems.

The first problem with mechanism is that Newton's work in mechanics (like his idea of "universal gravitation," for instance) "was something of a disappointment to the older cadre of mechanists, since mechanism originally sought to explain all phenomena entirely in terms of the motion and collision of material bodies, whereas Newton's principle of gravitation required action at a distance." (WK 2010n) However, this "action at a distance" requires an explanation, and later in the 20th century, Albert Einstein dethroned Newtonian mechanics with his alternative theory of relativity to explain gravity in relation to the curvature of space.

And the second problem with mechanism is that its determnism was later challenged by "quantum mechanics," especially in relation to the uncertainty principle by Werner Heisenburg in the 1920's.

Dynamism

Unlike "mechanism"—a related metaphysical veiw is known as "dynamism," which was "conceived by Gottfried Leibniz (1646–1716) and developed into a full system of cosmology. Dynamism in metaphysical cosmology explains the material world in terms of active, pointlike forces, with no extension but with action at a distance. Dynamism describes that which exists as simple elements, or for Leibniz, monads, and groups of elements which have only the essence of forces." (WK 2010o)

In this sense, "dynamism" differs from "mechanism," because the former "was developed as a reaction against the passive view of matter in philosophical mechanism. Interaction between elements takes place without contact, through modes or even harmonics of motion, yielding all phenomena in the Universe." (WK 2010o)

In modern times, "elements of Dynamism also developed into process philosophy, via Alfred North Whitehead and others, as well as systems theory via Ludwig von Bertalanffy and William Ross Ashby. Immanuel Kant was another philosopher who helped developed the theory of dynamism. More recently Colin Stott argues that many of the problems of traditional Western philosophy arise from an unsustainably passive view of both matter and consciousness." (WK 2010o)

But, for the critics, an interesting question to ask here is whether or not, or to what extent, the metaphysical view of "dynamism" for change (or the ontology of becoming) commits the opposite extreme, just as "mechanism" has its own version of extremity for constancy (or the ontology of being).

In *The Future of Post-Human Consciousness* (or *FPHC*), I already went to great lengths to show why this choice between the two is a false dilemma.

Atomism

Another metaphysical view is known as "atomism," which was "developed by Leucippus and his student Democritus in the fifth century BC. These atomists theorized that the natural world consists of two fundamental and opposite, indivisible bodies—atoms and void (void is mere nothing, or the body's negation)." (WK 2010p)

According to this view, "[a]toms are intrinsically unchangeable and move about the void combining into different clusters….Atoms are reality's very small, indestructible building blocks." (WK 2010p)

In this context, it is no wonder that "the word atomism derives from the ancient Greek adjective *atomos*, which literally meant 'uncuttable,'" with "*a*" for "not" and "*tomos*" for "cuttable," because t*omos* is "a conjugate of the Greek verb *temnein* (to cut)." (WK 2010p)

But there are serious problems with "atomism." Consider two of them below, for illustration.

Firstly, "the particles that chemists and physicists of the early 19th century thought were indivisible, and therefore identified with the uncuttable *a-toms* of long tradition, were found in the 20th century to be composed of even smaller entities: electrons, neutrons, and protons. Further experiments showed that protons and neutrons are made of fundamental quarks. These particles at present show no experimental evidence of size or substructure." (WK 2010p)

Nowadays, some physicists even think that, "as regards quarks, electrons, and other fundamental leptons are concerned, the possibility that they too are composed of smaller particles cannot be ruled out." (WK 2010p)

And secondly, the void is not really so empty as assumed in atomism. Nowadays, it is well known that that "'empty' space is not really empty all the way, because it can be full of energy (e.g., random quantum fluctuations at the sub-atomic level, and, for that matter, dark energy in the universe) and matter (e.g., different versions of sub-atomic particles, and, for that matter, dark matter in the universe)," as I already explained this in *FPHUP* (2009c: 250). (F. Wilczek, Frank 2008)

The Esoteric Argument

The fourth theoretical approach is called, in the absence of better words, the *esoteric* argument, which focuses, relatively speaking of course, more on the role of esotericism (and/or occultism) to provide esoteric explanations on the nature of the universe (like its planes of existence, emanation, and involution, for instance).

In this sense, the word "esoteric" in the title of the argument is suggestive, because it refers to the "psycho-spiritual" understanding of the universe in a way which is "difficult to understand" and "limited to a small circle" (of practitioners). (WK 2010q; MWD 2010)

Those who make use of esotericism (and/or occultism) to study the universe are known as "esoteric cosmologists"—and this approach to cosmology is called "esoteric cosmology." (WK 2010)

In general, esoteric cosmology refers to "an intrinsic part of an esoteric or occult system of thought. It almost always deals with at least some of the following themes: emanation, involution, evolution, epigenesis, planes of existence or higher worlds (and their emanation and the connections between them), hierarchies of spiritual beings, cosmic cycles (e.g., cosmic year, Yuga), yogic or spiritual disciplines and techniques of self-transformation, and references to mystical and altered states of consciousness." (WK 2010q)

In this sense, "esoteric cosmology" has much in common with "religious cosmology" and "metaphysical cosmology," because they are all concerned with issues like "the origin, purpose, and destiny of the universe and of consciousness and the nature of existence. For this reason it is sometimes difficult to distinguish where religion or philosophy end and esotericism or occultism begins." (WK 2010q)

Yet, esoteric cosmology differs from both religious cosmology and metaphysical cosmology, in that "esoteric cosmology is distinguished from religion in its more sophisticated construction and reliance on intellectual understanding rather than faith, and from philosophy [metaphysics] in its emphasis on techniques of psycho-spiritual transformation." (WK 2010q)

However, like those in "physical cosmology," "religious cosmology," and "metaphysical cosmology"—there is also no consensus even among "esoteric cosmologists" on various issues concerning the nature of the universe, except that they all agree on the "esoteric" argument as the best way to study them.

Consider three examples of esotericism in action, for illustration, in what follows.

Emanationism

A good example of the esoteric argument has to do with "emanationism," which is "from the Latin 'emanare' meaning 'to flow from'" and thus refers to "the mode [in the universe] by which all things are derived from the First Reality, or Principle. All things are derived from the first reality or perfect God by steps of degradation to lesser degrees of the first reality or God, and at every step the emanating beings are less pure, less perfect, less divine." (WK 2010r)

In this sense, "emanationism is a transcendent principle from which everything is derived, and is opposed to both Creationism (wherein the universe is created by a sentient God who is separate from creation) and materialism (which posits no underlying subjective and/or ontological nature behind phenomena, being immanent)." (WK 2010r)

In emanationism, the First Reality or "the Absolute, its nature and its activity, must be inseparably one thing only, namely will, such that the nature and activity of the Absolute is both one and the same (again, will)....In addition, agnosis, or the lack of Subjective gnosis, is a primordial privation which must be corrected before a metaphysical 'Oneing' [as argued by Plotinus] can occur. Through this process, the transcendent yet immanent will of individuals is made self-reflexive by recollecting back further and further. Eventually it will reach that nature, the Noetic (and real) self, which is antecedent to the phenomenal, corporeal self. The ontologically transcendent yet immanent Self is seen as being one's unactualized nature, and this nature will remain unactualized until contemplation is brought to fruition, thereby bringing into actuality what had been merely potential." (WK 2010r)

For the critics, however, this idea of emanationism is so esoteric (difficult to comprehend) that it raises more profound questions than answer them. Consider two of them below, for illustration.

Firstly, how exactly can an individual self be both ontologically "immanent" and "transcendent"? And what is the mechanism which can link the two worlds of ontology for an individual to move from one to the other? To be esoteric by itself is not a proof of the possibility of its claims.

And secondly, why should it necessarily be desirable for an individual self to "contemplate" until that point that it can be reunited with the First Reality again? Or why should this "actuality" be desirable in the first place? To be esoteric by itself is also not a proof of the desirability of its claims.

Involutionism

A closely related esoteric argument concerns the idea of "involution" which refers to the process in which, "[w]hen the universe attains a stage of sufficient density, the individual spirit is able to descend and participate in the evolution. Involution thus refers to the incarnation of spirit in an already established matter, the necessary prerequiste of evolution:" (WK 2010s)

Thus, both "evolution" and "involution" are "part of a complex sequence of cosmic cycles, called Round." (WK 2010s)

Let me call this system of thought *involutionism*, in the absence of better words (just as there is emanationism for the process of emanation).

In involution, there is "that period of time devoted to the attainment of self-consciousness and the building of the vehicles through which the spirit in man manifests....Its purpose is to slowly carry life lower and deeper into denser and denser matter for the building of forms, till the nadir of materiality is reached. From that point, life begins to ascend into higher Worlds. This succeeding period of existence, during which the individual human being develops self-consciousness into divine omniscience, is called 'spiritual evolution.'" (WK 2010s)

For the critics, however, the idea of "involutionism" fares no better when compared with "emantionism," since the same problems exist, albeit in a different way.

The first problem is, Why does the cosmic evolution necessarily have a process for involution in the first place? This is an inherent assumption in involutionism which remains to be shown.

And the second problem is, Why is it necessarily desirable for life to develop and ascend into the so-called "higher Worlds"? Again, this is an inherent normative assumption in involutiionism which remains to be justified.

The Idea of the Planes of Existence

Still another fascinating esoteric argument involves the idea of different "planes of existence" in the universe, where the word "plane" refers not only to the tradittional idea of the "physical" plane in everyday life but also, more importantly, to "a subtle state of consciousness that transcends the known physical universe." (WK 2010t)

This idea of different "planes of existence" can be found "in religious, and esoteric teachings—e.g., Vedanta (Advaita Vedanta), shamanism, Hermeticism, Neoplatonism, Gnosticism, Kashmir Shaivism, Sant Mat/Surat Shabd Yoga, Sufism, Druze, Kabbalah, Theosophy, Anthroposophy, Rosicrucianism (Esoteric Christian), Eckankar, Ascended Master Teachings, etc.—which propound the idea of a whole series of subtle planes or worlds or dimensions which, from a center, interpenetrate themselves and the physical planet in which we live, the solar systems, and all the physical structures of the universe." (WK 2010t)

In esoteric cosmology, "most cosmologists suggest that there are both positive and negative planes. They indicate that an Earth-departed soul is propelled towards the plane that corresponds to the level of its merits or demerits." (WK 2010t)

For instance, "this understanding can be found in the secular or popular Christianity, where in afterlife, a departed soul finds itself in either heaven or hell. Other traditions, including the inner or esoteric Christianity, suggest that there are finer differences between various positive and negative planes, and that each plane beyond Earth plane becomes more and more subtle. It is commonly believed that positive planes are inhabited by subtle bodies doing righteous deeds, whereas negative planes are inhabited by subtle bodies doing unrighteous deeds." (WK 2010t)

Again, for the critics, the idea of different "planes of existence" does not do away the persistent problems in esoteric cosmology. Consider two examples below, for illustration.

Firstly, esoteric cosmologists assume something which cannot be comprehended.

For instance, these different "planes of existence" originate, "at the dawn of the universe's manifestation, in The Supreme Being Who sent out—from the unmanifested Absolute *beyond comprehension* [italic added]—the dynamic force of creative energy…into the abyss of space." (WK 2010t)

But how can something which is "beyond comprehension" be attributed to have originated the "planes of existence" in the evolution of the universe?

And secondly, these different "planes of existence" are not necessarily desirable, in that they are inherently discriminative, in that there are "positive" planes for "subtle bodies doing righteous deeds" and "negative" planes for "subtle bodies doing unrighteous deeds." (WK 2010t)

The Post-Cosmological Argument

And fifthly, unlike the previous four arguments (viz., the scientific, the religious, the metaphysical, and the esoteric)—the fifth one to be introduced here can be called, in the absence of better words again, the *post-cosmological* argument, which learns from all other theoretical approaches (which include the four previous arguments and also other examples as will be introduced in the rest of this book), without, however, favoring any of them, nor trying to integrate them (as they are not necessarily compatible with each other, as the introduction of the theoretical debate in the previous sub-sections has shown).

The Theory of Post-Cosmology

My *post-cosmological* argument can be more precisely called *the theory of post-cosmology*, since it is to propose a new theory of cosmology.

The theory of post-cosmology has four distinctive features to remember.

Firstly, my theory makes good use of all theoretical approaches in the literature on cosmology, be they about the scientific argument, the religious argument, the metaphysical argument, or the esoteric argument—especially from the most comprehensive combined perspectives of the mind, nature, society, and culture.

Secondly, just like many other theories of mine in my previous books, my theory here does not heavily favor any specific theory over others in the literature, nor trying to integrate them (as they are not necessarily compatible with each other).

Thirdly, my theory treats the issue of cosmology in the critical context of the so-called "contested" beginnings and "speculative" ends—especially in the larger dialectic context of transcending them unto post-cosmology.

And fourthly, it contains fifteen major theses, namely, (a) the first thesis: the absoluteness-relativeness principle, (b) the second thesis: the predictability-unpredictability principle, (c) the third thesis: the explicability-inexplicability principle, (d) the fourth thesis: the preciseness-vagueness principle, (e) the fifth thesis: the simpleness-complicatedness principle, (f) the sixth thesis: the openness-hiddenness principle, (g) the seventh thesis: the denseness-emptiness principle, (h) the eighth thesis: the slowness-quickness principle, (i) the nineth thesis: the expansion-contraction principle, (j) the tenth thesis: the convention-novelty principle, (k) the eleventh thesis: the evolution-transformation principle, (l) the twelveth thesis: the symmetry-asymmetry principle, (m) the thirteenth thesis: the regression-progression principle, (n) the fourteenth thesis: the sameness-difference principle, and (o) the fifteenth thesis: the post-human odyssey—to be elaborated in the rest of the book, with a summary in the concluding chapter.

Of course, as this is something that I often emphasized in my previous books, other principles (besides the fourteen as cited above) are also relevant, but these fourteen are the most relevant in the current case study (in terms of the number of citations of each principle in the book).

Even then, in some cases, the difference between any two given principles, for instance, in terms of the number of citations in a book, is rather small, so the reason in those cases is more aesthetic (than otherwise), because it looks nicer to list only fourteen theses for fourteen principles (than twenty-one theses for all of the twenty-one principles) in the Table of Contents.

This is true, even if different studies of the same kind can yield different views about the degree of relevance for each principle, depending on the specific nature of a research in question, needless to say. So, if a different author analyzes the same subject matter in a different way, the relevance of the principles will be different.

With this clarification in mind—the seminal project here, if successful, will fundamentally change the way that we think about cosmology, from the combined perspectives of the mind, nature, society, and culture, with enormous implications for the human future and what I originally called its "post-human" fate.

Theory and Meta-Theory

The *post-cosmological* argument, or *the theory of post-cosmology*, presupposes some kind of meta-theory, in special relation to methodology and ontology—just as all theories, either explicitly or implicitly, require their own versions of meta-theory.

In my case, as this is something that I often explained in my previous books (almost *verbatim*, with only some updated revisions for each new book), I already proposed my distinctive approach to methodology (known as "sophisticated methodological holism") and ontology (known as "existential dialectics").

The next two sections are to introduce my meta-theory, or existential dialectics and sophisticated methodological holism, respectively.

A Unified Theory of Everything

In the end, my distinctive approach to meta-theory serves as a foundation to unify all domains of knowledge for an overarching understanding of all forms of reality (by way of some ontological principles at the meta-theoretical level and the comprehensive perspectives of the mind, nature, society, and culture at the theoretical level—as will be explained in the next two sections).

This is so, without commiting the sins of reductionism and reverse-reductionism (as often seen in many holistic approaches, with the fad of systems approach as a most recent notorious example, as already critically explained in *The Future of Post-Human Formal Science*, or *FPHFS*), since my distinctive approach makes good use of different schools of thought without favoring any of them nor trying to integrate them (as they are not necessarily compatible with each other), so as to include (or adjust for) heterogeneity, conflict, subjectivity, and complexity, for example.

In this sense, the word "unified" does not have to mean an integrative approach, in a narrow sense, nor a systems approach, in a broad one, both of which I have critically rebuked (in my previous books), while learning from them.

This overarching project (consisting of numerous theories of mine, as will be summarized in the next section), thus, constitutes my grand unifed theory of everything in the world and beyond.

The Logic of Existential Dialectics

To start, the ontology of existential dialectics can be summarized in this section—to be repeated, *often verbatim* (with only a few updated revisions), from my previous works—as this is something that I often do in each new book of mine as an introduction to some background information for the convenience of the readers.

After all, this book, like all others of mine, is in conversation with all previous ones of mine, for the final aim to converge all of them into a larger grand project on the future of intelligent life, both here on earth and there in deep space unto multiverses.

The summary can be organized in relation to (a) the conception of existential dialectics (or its ontics), (b) the syntax of existential dialectics (or its ontomethodology and ontologic), (c) the semantics of existential dialectics (or its ontosemantics), and (d) the pragmatics of existential dialectics (or its ontopragmatics), respectively hereafter—again *often verbatim*.

This means that, for those readers who had read some of my previous books, the summary below is really not much new (*almost verbatim*, with only some updated revisions).

But for those who have never read my previous books, it is a good review of them.

The Conception of Existential Dialectics (or Its Ontics)

This ontology was originally designated as "existential dialectics" in *Beyond Capitalism to Post-Capitalism* (herein abbreviated as *BCPC*), although it was already analyzed in other books of mine like the 2 volumes of *The Future of Human Civilization* (*FHC*), *The Future of Capitalism and Democracy* (*FCD*), *The Future of Post-Human Consciousness* (*FPHC*), and the 2 volumes of *Beyond Democracy to Post-Democracy* (*BDPD*).

Lest any confusion occurs, it is important to stress at the outset that the word 'existential' in "existential dialectics" has nothing to do with Existentialism, which I rebuked in *FHC*, *FCD*, and also *FPHC*. Instead, the word here merely refers to the existence of intelligent life (both primitive and advanced) in a broad sense.

Some Basic Concepts

The conception of existential dialectics (or its ontics) makes use of different concepts (like "sets," "elements," "relations," "operations," "functions," "truth values," "axioms," "postulates," and "principles"—as shown in *Table 4.3*), which are important for the understanding of any logic of ontology.

That clarified—existential dialectics, as a language of ontology, can be analyzed in different ways that a language, as an analogy, is often analyzed, namely, in relation to phonology ("the study of patterns of a language's sounds"), phonetics ("the study of the physical aspects of sounds of human language"), morphonology ("the study of the internal structure of words"), syntax ("the study of how words combine to form grammatical sentences"), semantics ("the study of the meaning of words [lexical semantics] and fixed word combinations [phraseology]"), and pragmatics ("the study of how utterances are used…literally, figuratively, or otherwise…in communicative acts"). (WK 2007)

But since my theory of existential dialectics makes use of the English language for communication here, it does not propose a new way to make the patterns of a language's sounds (as in phonology), to study the physical aspects of sounds (as in phonetics), or even to strive for a new organization of the internal structure of words for the English language (as in morphonology).

For this reason, the language of existential dialectics to be analyzed hereafter is not concerned with the phonological, phonetic, and morphonological aspects of the English language. Instead, the analysis will explore the syntax, semantics, and pragmatics of existential dialectics as a language of ontology to improve our understanding of reality.

The Ontic Dispute

But first, there is an ontic dispute to be addressed, which is related to what I called "the dilemma of ontology" as introduced earlier in *Sec. 1.4* of *The Future of Post-Human Knowledge* (*FPHK*).

A good point of departure concerns what I want to call *the ontic dispute*, on how to understand the very nature of ontology.

Traditionally, scholars in the field used to study the conception of ontology in terms of its entities (e.g., material objects, abstract numbers), properties (e.g., duration, plurality), and relationships (e.g., causation).

This is a good starting point, but it suffers from different problems. Consider three examples of bias below, which are quite common in the literature, and the recent version (of treating concepts and principles in ontology on the basis of "plurality," "dynamism," "duration," "interaction," "life," "consciousness," and "volition") by Reginald Firehammer (2005) only constitutes a latest addition to this old (impoverishing) tradition.

Firstly, the classification of entities, properties, and relationships is too material-centric, in often focusing more on the existence of material entities than on non-material ones.

Secondly, the classification is too anthropo-centric, in often favoring the factor of human agency in depicting reality.

And lastly, the classification is too confused among the three categories relating to each other, in often lumping them together without adequate consideration of their intricated interrelationships (both in relation to each and in relation to all).

A good alternative is precisely what I want to offer in this debate, that is, a new classification of ontological entities, properties, and their relationships in terms of *method*, *structure*, *process*, *agency*, and *outcome* instead. Let me explicate what these words mean below.

In relation to "method"—it concerns how ontology is to be studied. In this sense, the word 'method' here should not be confused with the different usage of the same term 'method' as a technique of investigation in the context of methodology (as already introduced in *Sec. 1.2* of *FPHK*).

In relation to "structure"—it concerns what the general patterns of entities, properties, and relationships are.

In relation to "process"—it concerns how these general patterns change over time.

In relation to "agency"—it concerns how agency affects any change of these general patterns over time (without putting, however, too much emphasis on the role of human agency).

And in relation to "outcome"—it concerns what the outputs of this change over time are.

The Syntax of Existential Dialectics (or Its Ontomethodology and Ontologic)

The syntax of existential dialectics, analogously speaking, refers here to the structure of an ontology which can be studied by way of combining ideas into complex relationships like ontological principles to depict reality in the world—and can be analyzed in seven sub-sections, namely, (a) the selection criteria, (b) further clarifications on classification, (c) ontomethodology, (d) ontologic, (e) the principles as short cuts, (f) the principles as family resemblances, and (g) the dialectic constraints imposed by the principles, respectively hereafter.

The Selection Criteria

To start, not any pair of relationship can be chosen as an ontological principle in existential dialectics; otherwise, there could be billions of them under the sky.

Four illustrations below suffice to clarify this point.

(a) Firstly, an ontological principle cannot *overlap* with other principles, in that it is too closely related with other ones.

Of course, one cannot totally rule out any relationship between two entities, especially in a complex world of everything being linked to everything else—but the comparison here is relative, not absolute.

In this sense, the flexibility-inflexibility pair cannot be an ontological principle, since it has something closely in common with other ones (like the simpleness-complicatedness principle, for instance).

As an illustration, something which is flexible in interpretation is more likely to allow multiple viewpoints and thus makes the overall picture more complicated than otherwise (simple).

(b) Secondly, an ontological principle cannot be *redundant* in relation to other principles, in that it somehow duplicates other principles.

In this sense, for instance, the directness-indirectness pair cannot be an ontological principle, because it somehow duplicates other ontological principles (like the preciseness-vagueness principle, for instance).

As an illustration, something which is direct means to be right to the point and does not need to go to an unnecessarily long loop—so it is less vague, in being clearer about the thing in question.

(c) Thirdly, an ontological principle cannot be *derived* from other ones—as if it is a child-parent relationship.

In this sense, the convenience-inconvenience pair cannot be an ontological principle, because it can be derived from other ones (like the slowness-quickness principle, for instance).

For instance, something which is convenient already implies that it can be obtained quickly (rather than slowly)—all things considered.

(d) And fourthly, an ontological principle cannot be *trivial* (or *parochial*)—as if it is only one component competing with other ones for inclusion in a set.

In this sense, the consistency-inconsistency pair cannot be an ontological principle, because its validity is limited (or parochial), as it can be easily challenged and replaced by opposing major (not minor) theories (e.g., the correspondence theory of truth, the pragmatic theory of truth, the semantic theory of truth, and whatnot), not just in accordance to the consistency theory of truth (for consistency).

In this sense, the consistency argument is only one among different opposing arguments in relation to the logic of what constitutes "truth."

Although these examples are not exhaustive, they illustrative the selectiveness of any ontological principle to be existentially dialectic.

Further Clarifications on Classification

Lest any misunderstanding carelessly occurs, some further clarifications on classification should be made here (and summarized in *Table 1.8*).

(a) Firstly, the total number of ontological principles is unknown, to be discovered later, as our knowledge of the world becomes more advanced.

Consequently, the principles as introduced in my books are not exhaustive, with new ones being added, whenever more of them are discovered in later research.

At least, future generations can pick up where I leave off and continue the discovery.

In this sense, my work should be treated as a pioneering effort for the development of a systematic, comprehensive analysis of a new general ontology for the future of knowledge.

(b) Secondly, the principles are not rigidly classified, as they can be reclassified in a different way.

For instance, the preciseness-vagueness principle is classified under the category of "structure" but can be reclassified under the category of "method," although in so doing, it has a different meaning in the context of method.

The same logic applies to the same-difference principle under the category of "outcome," which can be reclassified under the category of "structure," although in so doing, once more, it has a different meaning in the context of structure.

(c) Thirdly, the principles are to be understood together, not that each prinicple is to be treated independently of others. After all, the principles are mutually constraining, in that they work together as a whole.

(d) Fourthly, the principles are relevant to all subject matters, but some principles are, relatively speaking of course, more useful to some subject matters than others under certain conditions—as implied in the symmetry-asymmetry principle.

(e) And fifthly, the principles constitute only two levels of analysis, this time, at the ontological and methodological levels—while other levels of analysis (from the perspectives of the mind, nature, society, and culture) are also needed, in order to understand reality in its totality (as explained in the section on "sophisticated methodological holism," so as to avoid the dual dangers of reductionism and reverse-reductionism).

Ontomethodology

With these selection criteria and clarifications in mind—the syntax of existential dialectics can be divided into two main smaller sections for analysis, since they are related to each other in the context of some fundamental principles, that is, (a) ontomethodology and (b) ontologic, to be addressed hereafter, respectively.

In the context of ontomethodology, some ontological principles are proposed here in relation to the nature of methodology, which however is tied up with the general issue of ontology in the context of knowledge.

In this sense, ontomethodology (in the context of these ontological principles in the category of method) are related with ontologic in general (in the context of ontological principles in general).

Some good instances of ontological principles involving ontomethodology include the formalness-informalness principle, the predictability-unpredictability principle, the explicability-inexplicability principle, the partiality-totality principle, and the absoluteness-relativeness principle, which will be introduced in conjunction with ontological principles in the ontologic of existential dialectics in general hereafter.

Ontologic

With this clarification of the relationship between ontomethodology and ontologic in mind (which is comparable to the relationship between methodology and ontology as described in *Sec. 1.3* of *FPHK*)—the first three principles in existential dialectics were introduced in *BCPC*. And more principles were later added to the logic in my subsequent books (like *FC*, *FAE*, and others).

More specifically, in *BCPC*, I proposed three principles (based on previous works of mine, not just something out of the blue) for the ontological logic, namely, (a) the regression-progression principle on the direction of history, (b) the symmetry-asymmetry principle on the relationships among existents, and (c) the change-constancy principle on the alteration of things—or in a more elegant term, the dynamics of space-time (in *FC*).

In *The Future of Complexity* (*FC*), I added three more principles, on top of the three aforementioned, namely, (d) the partiality-totality principle on the relationships between the parts and the whole, (e) the order-chaos principle on the pattern of things, and (f) the predictability-unpredictability principle on the occurrence of things—as part of the ontological logic of existential dialectics.

In *The Future of Aesthetic Experience* (or simply *FAE*), another principle is added, namely, (g) the evolution-transformation principle (on the multiple kinds of agency).

Also in *FAE*, I then reclassified the 7 principles into four categories, namely, (i) in relation to method, for the partiality-totality principle and the predictability-unpredictability principle, (ii) in relation to process, for the change-constancy principle and the order-chaos principle, (iii) in relation to agency, for the symmetry-asymmetry principle and the evolution-transformation principle, and (iv) in relation to outcome, for the regression-progression principle.

In *The Rise of Authoritarian Liberal Democracy* (*ALD*), I added two more principles, namely, (h) the softness-hardness principle on the force of agency (which is to be classified under the category about agency) and (i) the same-difference principle on the metamorphosis of change (which is to be classified under the category about outcome).

In *The Future of Information Architecture* (*FIA*), I introduce three additional principles, that is, (j) the simpleness-complicadness principle on the interconnection among things, (k) the preciseness-vagueness principle on the refinement of things, and (l) the slowness-quickness principle on the speed of change—with the first two in relation to structure and the third in relation to process.

In *The Future of Post-Human Unconsciousness* (*FPHU*), I added three more principles, that is, (m) the openness-hidden principle on the detection of things, in relation to structure, (n) the explicability-inexplicability principle on the underlying mechanisms of things, in relation to method, and (o) the expansion-contraction principle on the growth of things, in relation to process.

In *The Future of Post-Human Knowledge* (*FPHK*), I added one more principle, that is, (p) the absoluteness-relativeness principle on the multiplicity of entities, in relation to method.

In *The Future of Post-Human Mathematical Logic* (*FPHML*), I added one more principle, that is, (q) the formalness-informalness principle on the formal requirements of systems, in relation to method.

In *The Future of Post-Human Engineering* (*FPHE*), I added one more principle, that is, (r) the theory-praxis principle on the duality of knowledge, in relation to agency.

In *The Future of Post-Human Creative Thinking* (*FPHCT*), I added one more principle, that is, (s) the convention-novelty principle on the nature of creative thinking, in relation to agency.

In *The Future of Post-Human Geometry* (*FPHG*) here, I added one more principle, that is, (t) the finiteness-transfiniteness principle on the nature of numbers, in relation to structure.

And in *The Future of Post-Human Urban Planning* (*FPHUP*), I added one more principle, that is, (u) the denseness-emptiness principle on the distribution of entities in space, in relation to structure.

With this update, there are twenty-one principles so far in existential dialectics, that is, five principles for the category in relation to method (viz., the formalness-informalness principle, the partiality-totality principle, the predictability-unpredictability principle, the explicability-inexplicability principle, and the absoluteness-relativeness principle), five principles for the category in relation to structure (viz., the finiteness-transfiniteness principle, the simpleness-complicatedness principle, the preciseness-vagueness principle, the openness-hiddenness principle, and the denseness-emptiness principle), four principles for the category in relation to process (viz., the change-constancy principle, the slowness-quickness principle, the order-chaos principle, and the expansion-contraction principle), five principles for the category in relation to agency (viz., the theory-praxis principle, the convention-novelty principle, the symmetry-asymmetry principle, the evolution-transformation principle, and the softness-hardness principle), and two principles for the category in relation to outcome (viz., the regression-progression principle and the same-difference principle).

There are thus twenty-one principles in existential dialectics—so far (as summarized in *Table 4.4*).

The Principles as Short Cuts

Yet, these principles should be treated with caution, lest misunderstanding occurs, since they do not constitute rigid dualities (or dichotomies).

The reason is that each pair in an ontological principle consists of two opposites, which are, however, merely short cuts both for multiple variations and degrees, as well as for different interactions with multiple other entities unlike them.

In this light, each pair can end up having hundred (or even thousand, if not more) different versions, which interact with hundred (or even thousand, if not more) other entities—as summarized in *Table 4.5*.

There are two clarifications here.

Firstly, this conception of shortcuts is not mutually exclusive nor absolute, in that the opposites can come in all shapes and sizes, with different degrees. For instance, by analogy, just as there are different degrees of the two colors "white" and "black"—there are likewise different degrees of the opposites in each ontological pair, to the extent that there can be multiple entities (not only two) interacting with each other in each pair.

And secondly, this conception of shortcuts do not ignore other possible entities in interacting with the two opposites (with their different versions) in each pair. For instance, by analogy, there are not only the two colors "white" and "black" as opposites, since there are other colors too besides them like "yellow," "green," "purple," or else. The same logic can be applied to each ontological pair, in that they also interact with other entities, not with only two of them (with their different versions).

Therefore, with these two clarifications in mind—each pair in an ontological principle serves only as an abbreviation for something more complicated and, therefore, although it contains two opposites, it should not be confused as a duality (dualism).

Existential dialectics rejects any dualism (or dichotomy) as too rigid and instead allows the multiplicity of entities, to the extent that between the two opposites in each pair exist many other alternatives to choose from. They are named in that short form for aesthetic elegance only, instead of listing all possible entities between the two opposites in the title (which would be too cumbersome and sound extremely awkward).

To be dialectic is to go beyond any rigid dichotomy and transcend into something different altogether in the long haul. One may be tempted to call the dialectic logic here with a different name like existential "multilectics" (instead of "dialectics"), but this naming is incorrect (or even misleading), for the two reasons aforestated.

After all, all the possible entities between any two opposites in an ontological pair are still about the two opposites, although each ontological pair come in all shapes and sizes (with different degrees) and do not exist by themselves but also interact with other entities unlike them. The virtue of revealing an ontological pair is to show how they relate within themselves (in multiple versions) and also interact with others unlike them (also in multiple versions).

This will be clear in a later sub-section titled "the dialectic constraints imposed by the principles."

The Principles as Family Resemblances

In addition, each ontological principle is generic, with some other comparable ontological pairs to be put in the same family (like a *family resemblance*)—as summarized in *Table 4.6*.

Ludwig Wittgenstein (1953) once suggested the idea of "family resemblance" in explaining different games classified under the same family called *games*.

Why should, for instance, playing football and chess as playing "games," when it is well understood that football is not the same as chess? The answer is that, although each game is different and has different rules, many of them (though not all) share, more or less, some commonalites (e.g., scoring as necessary for winning).

And this is so, even though some games share more than some others in any given selection of critieria, and no two games are exactly identical. So, Wittgenstein's point here is that there is no essential core which is common to all games, and the best that one can look for is some characteristics which are common to many (but not all) games. (A. Biletzki 2006)

(a) Appropriate Family Resemblances

In the context of existential dialectics, I want to distinguish two types of family resemblances, and, in the absence of better words, let me call them (a) *appropriate* family resemblances and (b) *inappropriate* family resemblances.

In regard to appropriate family resemblances—good examples include some of the pairs as already explained in an earlier section on selection criteria, like the flexibility-inflexibility pair, which can be put in the family resemblance of the simpleness-complicatedness principle, although the two pairs are not exactly identical. Likewise, the directness-indirectness pair can be put in the family resemblance of the preciseness-vagueness principle, although, again, the two pairs are not exactly identical.

In this sense, which specific pair in a family should be used to designate the name of the family can be at times a bit arbitrary, but with good reason.

For illustration, in the context of method, the partiality-totality principle can take the different form like individualisticness-holisticness, just as the explicability-inexplicability principle can take the different form like underlyingness-regularness—although each two pairs are not exactly identical and have slightly different meanings and usages.

A more comprehensive listing of this family resemblance for all other principles is shown in *Table 4.6* on the syntax of existential dialectics.

In fact, this section on family resemblances overlaps with the semantics of existential dialectics, which will be introduced later in the next section (and summarized in *Table 4.11*).

(b) Inappropriate Family Resemblances

However, there are some family resemblances which are not appropriate, and let me distinugish two of them by calling them, in the absence of better words, (b1) *inadequate* family resemblances and (b2) *compound* family resemblances (as summarized in *Table 4.10*).

(b1) In regard to "inadequate" family resemblances—there are some pairs which are not adequate, because they fail the test of selection criteria (as already pointed out in an earlier section), especially in relation to the criterion of not being *trivial* (or *parochial*)—as if it is only one component competing with other ones for inclusion in a set.

For instance, the *consistency-inconsistency* pair is inadequate, because it is only one component competing with other ones for inclusion in the larger set of the formalness-informalenss principle (like the *completeness-incompleteness* pair and the *soundness-unsoundness* pair, as discussed in my book titled *The Future of Post-Human Mathematical Logic*, or *FPHML* in short).

The same is true for the *cognition-noncognition* pair, which overlaps with both the completeness-incompleteness pair and the soundness-unsoundness pair in the larger set of the formalness-informalenss principle.

(b2) In regard to "compound" (or non-atomic) family resemblances—there are some pairs which are not appropriate for a different reason, because they fail the test of selection criteria (as already pointed out in an earlier section), especially in relation to the criterion of not *overlapping* with other principles, in that it is too closely related with other ones.

For instance, the *normality-abnormality* pair is compound (and thus inappropriate), because it overlaps with several other principles, like the symmetry-asymmetry principle, the density-emptiness principle, the convention-novelty principle, and the same-difference principle, for instance (as already discussed in *FPHP*).

The normality-abnormality pair overlaps with the symmetry-asymmetry principle, because being "abnormal" (say, for those with mental illnesses) is often treated not equally by others who are "normal." So, the relationship reveals the symmetry-asymmetry principle.

The normality-abnormality pair also overlaps with the density-emptiness principle, because being "abnormal" (say, for those with mental illnesses) is often not as dense in the overall (not local) distribution of a population as those who are "normal." So, the relationship reveals the density-emptiness principle.

The normality-abnormality pair also overlaps with the convention-novelty principle, because being "abnormal" (say, for those with mental illnesses) often exhibits non-conventional (non-conformist) ideas and behaviors when contrasted with those who are "normal." So, the relationship reveals the convention-novelty principle.

The normality-abnormality pair also overlaps with the same-difference principle, because being "abnormal" (say, for those with mental illnesses) shares some similarities in outcome when compared with those who are "normal" (in that both contribute to the control and regulation of human thoughts and behaviors in society, albeit in different ways). Yet, being "abnormal" also has something different from being "normal," in that abnormal thoughts and behaviors follow a different pattern when contrasted with those which are normal. So, the relationship reveals the same-difference principle.

These illustrations of indequate and compound family resemblances show how difficult it is to come up with an ontological principle in existential dialectics out of the blue, because many of them violate the selection criteria in one way or the other. It is all too easy to join two opposing words and simply make them into a pair, but it is all too difficult to pass the test of the selection criteria to be qualified as an ontological principle in existential dialectics.

Other examples of compound (and inadequate) family resemblances are shown in *Table 4.10* on "The Syntax of Existential Dialectics VII: Types of Inappropriate Family Resemblances."

The Dialectic Constraints Imposed by the Principles

The principles, as they constitute the syntax of existential dialectics, are dialectic in character, such that, when they are applied, they impose dialectic constraints on how reality is to be understood.

Consider, say, the symmetry-asymmetry principle as an illustration here, in order to summarize two main characters of the dialectic constraints in question (as summarized in *Table 4.7*).

Firstly, to be dialectic here is to go beyond the narrow dichotomies (and, for that matter, any rigid multi-dimensional classificatory scheme), be they about "self" vs. "world," "freedom" vs. "unfreedom," "barbarity" vs. "civilization," "individuality" vs. "communality," and so on.

One way to do so (to go beyond) is to consider them all in terms of co-existence (without favoring one over the rest). For instance, my theory of "post-civilization" (to be summarized later in the section on the pragmatics of existential dialectics) is to go beyond barbarity and civilization in terms of understanding barbarity and civilization as being co-existent.

And the same logic can be said in relation to my theories of "post-democracy," "post-capitalsm," and others (also to be introduced later in the section on the pragmatics of existential dialectics), in regard to freedom vs. unfreedom, equality vs. inequality, communality vs. individuality, spirituality vs. materiality, and so on.

But to consider them all (in the dichotomies—and, for that matter, in any rigid multi-dimensional classifactory scheme) as co-existent is not the same as to imply that the opposing categories in any classificatory scheme are all equal, since, in accordance to the symmetry-asymmetry principle (as an illustration here), if they are equal in terms of being considered as co-existent, they are asymmetric in terms of being unequal in dialectic interaction (e.g., X can be more dominant than Y in case A, or Y is more dominant than X in case B), although in some special cases, they can be relatively equal (e.g., X and Y are relatively equal in case C).

For this reason, there are different versions of "post-democracy" and "post-capitalism" in my theories. As an illustration, in version I of the theory of post-democracy, freedom is more dominant than equality, whereas in version II of the theory of post-democracy, equality is more so than freedom.

But this "X more than Y" has to be understood in the context of dialectic logic (not in symbolic logic, as conventionally understood), in that both "X" and "Y" are important in post-democracy (in the context of dialectic logic), but in an asymmetry way.

By contrast, in symbolic logic, it often favors one over the other—be it in regard to privileging freedom over equality (in Fascism), favoring freedom relatively more than equality (in Liberal Democracy), or favoring equality relatively more than freedom (in Socialist Democracy). In regard to the latter two cases (about Liberal Democracy and Socialist Democracy), the difference between dialectic logic and symbolic logic can be one in degree, not in kind—in this sense, albeit not in other senses.

The same reasoning can be said about the relatonships between individuality and communality, between spirituality and materiality, and between formal legalism and informal legalism in the different versions of my theory of "post-capitalism."

Secondly, to be dialectic is to go beyond the narrow dichotomies (and, for that matter, any rigid multi-dimensional classificatory scheme) in another way, this time, in a transcendent way, that is, in exploring other possibilities or even other issues not considered within the narrow confines of narrow dichotomies (and, for that matter, any rigid multi-dimensional classificatory scheme).

As an analogy, in this second meaning, to go beyond the narrow dichotomy of "black" vs. "white" is not just to choose both "black" and "white" (as in the first meaning) but also to explore other options (e.g., "green," "purple," "blue," etc.—and, alternatively, "shade," "line," "curve," etc.).

By the same reasoning, to go beyond "democracy" is to transcend democracy (as in version III of my theory of "post-democracy") and to explore other possibilties of lifeforms (e.g., "floating consciousness," "hyper-spatial consciousness," etc., to live beyond the narrow obsession with freedom and equality).

This dialectic character of the principles in existential dialectics has important implications for the pragmatics of existential dialectics (as will be clear shortly, in the section on the pragmatics of existential dialectics).

The Semantics of Existential Dialectics (or Its Ontosemantics)

The syntax of existential dialectics so understood in terms of ontological principles only gives us the structure of ontology in the world, in an abstract (general) sense. These principles by themselves do not tell us the specific meanings in a given context.

In order to grasp the specific meanings of the principles in a given context, it is necessary to study the semantics of existential dialectics (as summarized in *Table 4.11*).

And the analysis of the ontological principles as family resemblances in the previous section on syntax is only a starting point (and thus overlaps a bit with the semantics of existential dialectics here).

For this reason, I have often gone to great lengths in my previous books on different subjects to explain the specific meanings of the principles when applied in different contexts.

Consider two examples below for illustration, which are not exhaustive.

In *FPHST*, for instance, I used the first three principles (i.e., the change-constancy principle, the regression-progression principle, and the symmetry-asymmetry principle) to propose "the perspectival theory of space-time," for a better way to understand space and time—especially, though not exclusively, in relation to future post-human history (as summarized in *Table 1.3*, *Table 1.4*, *Table 1.5*, and *Table 1.6*).

In so doing, I had to introduce concepts and theories specific to the field of physics and other related fields (e.g., "absolute space" and "absolute time" in "classical mechanics" and "relative space-time" in "the theory of relativity").

And in *BNN*, I also exploited the three principles to propose the "transcendent" approach to the study of genes and memes as a new way to understand the interaction between nature and nurture.

In so doing, I had to explore concepts and theories in the world of evolutionary biology (e.g., "mutation," "variation," "adaptation," "selection," and "inheritance" in classical evolutionary theory) and neural biology (e.g., "chromosome," "gene," "DNA," "RNA," "protein," "neuron," "neural network," and "behavior").

Other examples, besdies these two here, are given in other books of mine on different topics.

Therefore, this semantic dimension of existential dialectics is important, since, in each of my books when the ontological principles are used, each principle becomes relevant in a different meaning when contextualized within a given case study in question.

In this book, for instance, the principles are relevant in relation to the subject matter of cosmology (as summarized in the concluding chapter), and they can be shown in all shapes and sizes (since they are also to be treated as "short cuts," as indicated before), without, however, being reduced to "specific ontology" (as opposed to "general ontology"), as explained in *Table 4.9*.

The Pragmatics of Existential Dialectics (or Its Ontopragmatics)

Even the semantics of existential dialectics is not enough to understand reality, since semantics by itself does not tell us the nature of intentionality and interests in relation to the matrix of power formations and conflicting interests when existential dialectics as an ontology is applied.

Surely, one cannot completely separate the pragmatics of existential dialectics from its semantics (and for that matter, its syntax), but the point here is to give each a more elaborated analysis.

For this reason, the pragmatics of existential dialectics needs to be studied too, in special relation to the inclusion of power formation and conflicting interests in the application of the principles of existential dialectics (as summarized in *Table 4.12*).

It is thus no wonder that all my previous books have shown in different ways how and why these ontological principles and their theoretical applications can reveal the future world to come in a way that humans have never known, in a totally different battleground for competing human (and later post-human) interests by myriad groups fighting for their own versions of hegemony.

Perhaps a summary of my previous works in relation to existential dialectics and their contributions to my new theories as proposed over the years is deemed revealing here, for the further understanding of the pragmatics of existential dialectics (in close relation to the semantics of existential dialectics too).

That stressed—the following summary in four sub-sections is something that I often do in my previous books too, as a way to introduce the intellectual background of my ideas.

So, for those readers who had read my previous works, the following sub-sections serve as a reminder, as they are repeated hereafter *often verbatim* (with only some updated changes, with the rest kept intact).

These sub-sections are, namely, (a) the two-way street connecting theory with meta-theory, (b) the theoretical application of existential dialectics, (c) direct and indirect applications of existential dialectics, and (d) multiple applications of existential dialectics—to be summaried herefafter, respectively.

The Two-Way Street Connecting Theory and Meta-Theory

However, lest any misunderstanding occurs, it is important to remember that the pragmatics of existential dialectics is not a one-way street (that is, using the ontological principles for theoretical insights in praxis) but a two-way one, namely, (a) from meta-theory to theory, and (b) from theory to meta-theory.

(a) On one side of the street, the ontological principles can inspire some theoretical insights in praxis, that is, in relation to some specific fields.

(b) On the other (opposing) side of the street, however, the study of a subject matter in the specific fields in turn reveals some more hitherto unknown ontological principles to be discovered and identified.

For this reason, three new principles were added in *FC*, one in *FAE*, two in *ALD*, and three in *FIA*, on top of the original three in *BCPC*, for example (as there are more)—after some research on the specific subject matters.

Both theory and meta-theory enrich each other in all my works over the years.

The Theoretical Application of Existential Dialectics

This two-way street connecting theory and meta-theory has contributed to my original construction of 43 specific theories in 33 books (so far, up to now, in early 2010) on numerous fields in all domains of knowledge, which can be classified in terms of specific "meta-theories" (about epistemology, methodology, logic, ontology, etc.) and specific "theories" (about the mind, nature, culture, society, and the rest—as will be further explained later in the sub-section on "the holistic organization of an inquiry" in the section about "sophisticated methodological holism").

The final aim of my numerous new specific theories (on both theory and meta-theory) is to converge all of them (as proposed in my numerous books) into a grand project about the future of intelligent life, both here on earth and in deep space.

In fact, all these theories of mine serve as a foundation to unify all domains of knowledge for an overarching understanding of all forms of reality (by way of some ontological principles and the comprehensive perspectives of the mind, nature, society, and culture), while allowing for heterogeneity, conflict, subjectivity, and complexity, for example.

With this grand project in mind—the summary of my specific theories on meta-theory are given in this section (on "existential dialectics") and the next (on "sophisticated methodological holism"), whereas the specific theories on theory will be summarized in the following categories of (a) society, (b) culture, (c) the mind, (d) nature, and (e) the rest, as shown below, in that order.

(a) Society

The category of "society" here can be divided into four sub-categories, i.e., (a1) social organizations, (a2) social institutions, (a3) social structure, and (a4) social systems.

(a1) Social Organizations

In relation to social organizations, I proposed a new thesis on the ambivalent nature of virtual organizations in the future, known as "the theory of e-civic alienation" in *FCD* (and summarized in *Table 4.19*).

And in *FPHO*, another theory called "the combinational theory of organization" is worked out, although it was first proposed in Ch.6 of *ALD*.

(a2) Social Institutions

There are different social institutions, be they about (a1.1) politics, (a2.2) economics, (a2.3) military affairs, and (a2.4) mass media, for instance.

(a2.1) In relation to politics, I offered different visions on (a2.1.1) political theory, (a2.1.2) comparative politics, and (a2.1.3) international relations, to cite three main ones.

(a2.1.1) In the context of political theory, my vision of a path-breaking political system in future times concerns the different forms of "post-democracy" to supersede democracy unto the post-human age.

My vision of post-democracy was called "the theory of post-democracy" (as summarized in *Table 4.22*, *Table 4.23*, and *Table 4.24*), whereas *Table 4.25* explains the distinctions among democracy, non-democracy, and post-democracy.

And *Table 4.26* stresses the multiple causes of the emergence of post-democracy, while *Table 4.27* clarifies some possible misunderstandings in regard to post-capitalism and post-democracy.

(a2.1.2) In the context of comparative politics, I also showed, in *ALD*, how and why authoritarianism can be an advanced form of liberal democracy (just as it has its own counterpart in non-liberal democracy), or what I originally called "authoritarian liberal democracy" (as summarized in *Table 4.21*), before it is to be superceded by "post-democracy."

(a2.1.3) In the context of international relations (both at the institutional and structural levels), I provided, in *BWT*, "the theory of cyclical progression of empire-building," for a better way to understand the logic of empire-building on earth and beyond—with the rise of what I originally envisioned as "hyper-empires" (like "the Chinese Union" and "the Indian Union"), "meso-empires" (like "the European Union" and "the North American Union") and "micro-empires" (like "the Middle Eastern Union" and "the Latin American Union") in the coming "post-post-Cold War era" as a visionary illustration, as summarized in *Table 4.20.*

In addition, in *FPHWP*, I proposed "the multifaceted theory of war and peace" to understand war and peace, in a small sense—or aggression and pacificity, in a broad sense.

(a2.2) In relation to economics, I also offered another vision, this time to understand the future forms of economic systems, that is, the different forms of "post-capitalism" to supersede capitalism unto the post-human age.

My vision of post-capitalism was referred to (in *FCD*, *BDPD* and *BCPC*) as "the theory of post-capitalism" (as summarized in *Table 4.28*, *Table 4.29*, *Table 4.30*, *Table 4.31*, and *Table 4.32*). And *Table 4.33* clarifies the differences among capitalism, non-capitalism, and post-capitalism, whereas *Table 4.34* shows multiple causes of the emergence of post-capitalism.

This vision on "post-capitalism" constitutes the third theoretical application of existential dialectics, this time, in that there is no wealth without poverty (or the wealth/poverty dialectics).

(a2.3) In relation to military affairs, I also proposed "the multifaceted theory of war and peace" (in *FCD*, although it was first so labeled in *FPHK*) for a new way to understand the causes of war—and, for that matter, the rise and fall of different forms of warfare over time.

This theory was further expanded in *FPHWP*, where I suggested a new way to understand war and peace, in a small sense—or aggression and pacificity, in a broad sense.

(a2.4) In relation to mass media, I also proposed "the totalistic theory of communication" in *The Future of Post-Human Mass Media* (or in short, *FPHMM*).

(a3) Social Structure

In relation to social structure, I proposed two additional theories, that is, "the theory of the cyclical progression of hegemony" in *FCD* (and also in *FPHC* and *BDPD*), and "the theory of the cyclical progression of empire-building" (in *BWT*, as already indicated above)—with important implications for the symmetry-asymmetry princple in existential dialectics, just to cite one example.

Also, in *BNN*, I also proposed "the theory of contrastive advantages" to understand the persistence of social structure by gender, race, ethnicity, or any other groupings.

(a4) Social Systems

In relation to social systems, there are several sub-categories for illustration, which can be about (a4.1) urbanization, (a4.2) technology, (a4.3) demography, (a4.4) the environment, and (a4.5) social trends as a whole.

(a4.1) In relation to urbanization, I proposed "the contingent theory of urban planning" in *FPHUP* for a new way to understand density and void.

(a4.2) In relation to technology, I worked out "the ambivalent theory of technology" in *FPHE* for a new way to understand the nature of technology.

Some other books (like *FPHMM*, *FAE*, *FCD*, and *FHC*) also deal with technology, and a good example concerns "the synthetic theory of information architecture" in *FIA* on taxonomy, network, and intelligence, in conjunction with the rise of technology.

(a4.3) In relation to demography, I proposed, first in *FHC* (and later in other books like *FCD*, *FPHC*, *FPHU*, etc.), "the theory of post-humanity," in relation to the "post-human" age at some distant point of "after-postmodernity," long after human extinction, to be eventually superseded by "post-humans" of various forms.

Some good candidates of the post-human lifeforms include, for instance, "thinking robots," "thinking machines," "cyborgs," "genetically altered superior beings," "floating consciousness," "hyper-spatial consciousness," "unfolding unconsciousness," "the hyper-sexual body," and "the hyper-martial body." This post-human vision of mine was first originally worked out in *FHC* and further elaborated in both *FCD*, *FPHC*, *FPHU*, *FPHS*, and *FPHMA*, for instance.

Lest any confusion occurs, I need to emphasize two clarifications here about the term "post-human" as a neologism in my works.

Firstly, the word "post-human" here should *not* be confused with another term which looks similar but has a totally different meaning in the literature of postmodernism, namely, "post-humanism"—which constitutes a critique of "humanism" as tradionally understood (especially, though not exclusively, in relation to the idea of progress in science and reason in the Enlightenment project). (WK 2008)

And secondly, the word "post-human" here should also *not* be confused with a similar term which is used to champion the ideology of technology for the future co-existence between humans and cyborgs in "trans-humanism." (WK 2008a)

Instead, my term "post-human" in relation to "posthuman-ism" also rejects "transhumanism" (especially, though not exclusively, in relation to the promises of technology) and refers to something else altogether, that is, the future extinction of humans and its post-human successors in deep space and beyond unto multiverses.

Later, in *BEPE*, I went further and provided a more comprehensive critique of "transhumanism," together with other issues.

With these two clarifications in mind (as summarized in *Table 1.31*)—in the end, a most fundamental question about intelligent life now has an answer, in that, if asked, "What is the future of human civilization?"—my answer in *FCD* (89) is thus: "As addressed in Ch.7 of *FHC*, a later epoch of the age of after-postmodernity (that is, at some point further away from after-postmodernity) will begin, as what I called the 'post-human' history (with the term 'post-human' originally used in my doctoral dissertation at M.I.T., which was finished in November 1995, under the title *After Postmodernity*, still available at M.I.T. library, and was later revised and published as *FHC*). The post-human history will be such that humans are nothing in the end, other than what culture, society, and nature (with some luck) have shaped them into, to be eventually superseded by post-humans (e.g., cyborgs, thinking machines, genetically altered superior beings, and others), if humans are not destroyed long before then."

(a4.4) In relation to the environment, I also addressed the issue of sustainability, especially in *FHC*, *FCD*, and *FPHUP*. For instance, in *FPHUP*, I specifically proposed a new theory, "the contingent theory of urban planning," on the need to go beyond the contemporoary obsession with sustainability, in the context of not only the environment but also other issues.

(a4.5) In relation to social trends as a whole, I proposed in Ch.9 of *FCD* (367-8), that "civilizational history will continue into the following cyclical progression of expansion, before it is to be superseded (solely as a high probability, since humans might be destroyed sooner either by themselves or in a gigantic natural calamity) by posthumans at some distant point in after-postmodernity (which I already discussed in *FHC*)" unto multiverses (different constellations of universes):

Local → Regional → Global →Solar →
Galactic → Clustery…→ Multiversal

In *BDPD*, this thesis of mine was referred to as "the theory of the cyclical progression of system integration and fragmentation," at the systemic level—in close relation to another thesis analyzed in Ch.9 of *FCD*, which was called, in the absence of better words, "the multifaceted theory of war and peace" in *FPHK* (as already indicated above).

In *FC*, I also proposed "the dialectic theory of complexity" to account for the factor of uncertainty, which can have a major role to play in the world.

(b) Culture

The category of "culture" here can also be divided into different sub-categories like (b1) aesthetics, (b2) religion, (b3) popular (mass) culture, (b4), logic, (b5) civilization as a whole, (b6) ethics, (b7) epistemology (methodology), (b8) metaphysics (ontology), and (b9) ideology.

However, (b7) epistemology (methodology), (b8) metaphysics (ontology), and (b9) ideology are already discussed elsewhere, be they about "meta-theory" (e.g., methodology and ontology under the categories of "existential dialectics" here and "sophisticated methodological holism" in the next section) or "theory" (e.g., ideology under the categories of "social institutions," as described earlier, and also of "civilization as a whole," as will be addressed shortly).

(b1) Aesthetics

In relation to aesthetics, I proposed "the transformative theory of aesthetic experience" in *FAE* about some great transformations of aesthetic experience in future history.

(b2) Religion

In relation to religion, I proposed "the comparative theory of religion" in *FHC* (and also in *FCD*, although it was first so labeled in *FPHK*).

(b3) Popular (Mass) Culture

In relation to popular culture, I proposed, in *FHC*, a new way to understand mass culture in terms of some fundamental dilemmas, which is called, in the absence of better words, "the dualistic theory of mass culture" in *FPHO*.

(b4) Logic

In relation to logic, I also worked out, in *FPHML*, "the contrastive theory of rationality" for a new way to understand rationality and knowledge (in the context of mathematical logic).

(b5) Civilization as a Whole

In relation to civilization as a whole, I also argued, in *FHC*, no freedom without unfreedom (as shown in *Table 4.13*) and no equality without inequality (as shown in *Table 4.14*), especially in relation to the seven dimensions of life existence first worked out in *FHC* (i.e., the technological, the everyday, the true, the holy, the sublime/beautiful, the good, and the just).

Even more interestingly, in *BDPD* and later in *BCIV*, I further revealed a theoretical refinement of existential dialectics, this time, in arguing that there is no civilization without barbarity, with human civilization to be eventually superseded by what I originally analyzed as "post-human post-civilization" (which should not be confused with "post-human civilization"), in the context of the freedom/unfreedom and equality/inequality dialectics.

In the final analysis, civilization cannot live without barbarity and has to learn to co-exist with it in ever new ways. Preposteorus as this may seem to many contemporaries—it is no more imperative to preserve civilization than necessary to destroy barbarity, and the ideal of civilization is essentially bankrupt, to be eventually superseded by "post-civilization" (as summarized in *Tables 4.35-4.43*).

This freedom/unfreedom and equality/inequality dialectics holds in pre-moderny, modernity, postmodernity, and what I orignally called "after postmodernity" in *FHC* and *FCD*—especially, though not exclusively, in the context of the emergence of "post-humans."

There are two specific theories about "after-postmodernity" here. The first theory about "after-postmodernity" is about the trinity of pre-modernity, modernity, postmodernity, and after-postmodernity at the cultural level—and was called, in *BDPD*, "the theory of the trinity of modernity to its after-postmodern counterpart," although it was first analyzed in *FHC* and later in *FCD* and also *BCIV* (as summarized in *Table 1.34*, *Table 1.35*, *Table 1.36*, and *Table 1.37* and will be elaborated further in the later sub-section on "the rest" about future history).

The second theory about "after-postmodernity" is about pre-modernity, modernity, postmodernity, and after-postmodernity at the historical level. For instance, a conclusion in *FHC*, which is shocking indeed for many contemporaries, is none other than that "[t]he post-human history will therefore mark the end of human history as we know it and, for that matter, the end of human dominance and, practically speaking, the end of humans as well. The entire history of human civilization, from its beginning to the end, can be summarized by four words, linked by three arrows (as already discussed in *FHC*)":

Pre-Modernity → Modernity →
Post-Modernity → After-Postmodernity

In *BDPD*, this thesis of mine was specifically called "the theory of the evolution from pre-modernity to after-postmodernity," at the historical level—as shown in *Table 1.34*, *Table 1.35*, *Table 1.36*, and *Table 1.37* (and will be elaborated further in the later sub-section on "the rest" about future history).

Therefore, "[t]he end of humanity in the coming human extinction is the beginning of post-humanity. To say an untimely farewell to humanity is to foretell the future welcome of post-humanity." (P. Baofu 2002: 89) This thesis of mine was known in *BDPD* as "the theory of post-humanity," at the systemic level—as already indicated in (a4.3) above.

(b6) Ethics

And in relation to ethics, I also proposed, in *BEPE*, "the theory of post-ethics" for a new way to understand morality and immorality.

(c) The Mind

This important vision about "post-humans" (as summarized above) brings us to the category of "the mind" here, which, as is often understood in everyday's usage, refers to the "brain" and the "body" (together with their interactions) and has different levels of consciousness which can be both "spiritual" and non-spiritual.

Therefore, the "mind" already presupposes the "brain," "body," and "spirit" (or "soul"), so that one does not fall into the intellectual trap of the "mind-body dualism" as debunked in modern (and especially, postmodern) philosophy (which I already went to great lengths to explain in *The Future of Human Civilization*).

That said—the category of "the mind" can be divided into two sub-categories, namely, in relation to (c1) biology and (c2) psychology.

(c1) Biology

In relation to biology, there are two concerns here, namely, (c1.1) genes and intelligence and (c1.2) the human body.

(c1.1) In relation to genes and intelligence, I worked out "the theory of contrastive advantages" (which was originally proposed in *FCD* and further worked out in *BNN*, as summarized in *Table 1.24*), to show the nature-nurture interactions of multiple levels in action (e.g., the biological, the psychological, the structural, the systemic, the cultural, and others) for humans and post-humans on earth and beyond, to the extent that different groups, be they on the basis of race, gender, ethnicity, class, age, or else, are not equal, on average, and have never been, nor will they be, even if there can be exceptions.

(c1.2) In relation to the human body, for instance, in *FPHMA*, I suggested "the expansive-contractive theory of martial arts," in which two great future transformations of the martial body were envisioned, that is, both "with the hyper-martial body" and "without the body" (as summarized in *Table 1.32*)—although this issue can also be classified under the category of "culture."

And in *FPHS*, I proposed "the theory of virtual sexuality," in which three great future transformations of the sexual body were envisioned, that is, "virtual pleasure," "the hyper-sexual body," and "the post-sexual floating mind" (as summarized in *Table 1.33*).

(c2) Psychology

In relation to psypchology (which also overlaps with biology too, however), I also proposed some greater transformations of the mind to come in the future, especially though not exclusively in the post-human age.

These greater transformations of the mind in psychology can be further sub-divided into different categories, namely, (c2.1) consciousness, (c2.2) unconsciousness (together with subconsciousness), (c2.3) creativity, (c2.4) learning, (c2.5) aggression (and pacificity), and (c2.6) personality.

(c2.1) In relation to consciousness, I analyzed, in *FPHC*, the nature of consciousness in relation to the conceptual dimensions of consciousness (as shown in *Table 1.12*), the theoretical levels of consciousness (as shown in *Table 1.13*), the thematic issues of consciousness (as shown in *Table 1.14*), the different dimensions of human existence (viz., having-ness, belonging-ness, and being-ness) in relation to consciousness (as shown in *Table 1.15*, *Table 1.16*, *Table 1.17*, and *Table 1.18*).

In *BCPC*, the nature of consciousness was further explored in relation to cognition, emotion, and behavior, which can be in consciousness and other mental states (e.g., subconsciousness and unconsciousness), in the context of existential dialectics, as shown in *Table 1.19*, *Table 1.20*, *Table 1.21*, *Table 1.22,* and *Table 1.23.*

In fact, in *FCD*, I already proposed three different theories in relation to the limits of cognition, emotion, and behavior, namely, "the theory of cognitive partiality," "the theory of emotional non-neutrality," and "the theory of behavioral alteration."

Then, I argued that the existence of human consciousness will be superseded one day with "floating consciousness" (as first proposed in *FCD*) and "hyper spatial consciousness" (as first proposed in *FPHC*) as a climax of evolution in consciousness, after the future extinction of human consciousness (as shown below):

Primordial consciousness → Human consciousness →
Post-human consciousness (with floating consciousness and hyper-spatial consciousness as a climax in the evolution of consciousness)

In *BDPD*, these latest theses of mine were called "the theory of floating consciousness" and "the theory of hyper-spatial consciousness," both at the cosmological and psychological levels (as summarized in *Table 1.26* and *Table 1.25*).

(c2.2) In relation to unconsciousness, for instance, in *FPHU*, the unconsious mind will undergo tremendous change, to the rise of what I originally called "unfolding unconsciousess" in relation to anomalous experience, as a climax of evolution in unconsciousness, in conjunction with the conscious mind (both human now and post-human later), especially at the psychological level, as shown below:

Primordial unconsciousness → Human unconsciousness →
Post-human unconsciousness (with unfolding unconsciousness
as a climax in the evolution of unconsciousness)

In *FPHU*, this thesis of mine is called "the theory of unfolding unconsciousness" in a small sense, or "the unfolding theory of anomalous experience," in a larger one, as shown in *Table 1.27* and *Table 1.28*.

(c2.3) In relation to creativity, for instance, in *FPHCT*, I also proposed "the comprehensive theory of creative thinking" to explore the role of creative thinking, in relation to technology and other factors, together with my original proposal of important creative techniques and traits (as shown in *Table 1.29*) and its double nature of desirability and undesirability (as shown in *Table 1.30*).

(c2.4) In relation to learning, for instance, in *FPHL*, I worked out "the multilogical theory of learning" for a new way to understand the nature of structure and context in language, in relation to learning and its future development in the context of the mind.

(c2.5) In relation to aggression (and pacificity), I proposed "the multifaceted theory of war and peace" in *FPHWP* (as mentioned above).

(c2.6) And in relation to personality, I worked out "the contrarian theory of personality" in *FPHP* for a new way to understand normality and abnormality.

(d) Nature

The category of "nature" here refers to the state of nature, which can be divided into some sub-categories like (d1) physics, (d2) cosmology, (d3) biology, and (d4) geography.

But since biology in (d3) overlaps with biology in (c1)—only phyiscs, cosmology, and geography are addressed here.

(d1) Physics

In relation to physics, I also proposed "the perspectival theory of space-time" in *FPHST* (based on *FPHC*) for a new way to understand space and time, as shown in *Table 1.3*, *Table 1.4*, *Table 1.5*, and *Table 1.7*.

In *FPHG* (based on *FPHC* and *FPHST*), I proposed "the selective theory of geometry" for a new way to understand geometry, with implications for the understanding of space-time in the context of infinity, symmetry, and dimensionality for future lifeforms that our world has never known (as shown in *Table 1.6*, *Table 1.7*, *Table 1.8; Table 1.9*, *Table 1.10*, and *Table 1.11*).

And in *FC*, I proposed "the dialectic theory of complexity" for a new way to understand order and chaos in the state of nature and beyond.

(d2) Cosmology

In relation to cosmology, my theories on physics (as indicated above) already have important implications for understanding the nature and the future of the universe (which was also addressed in *FHC*, among other works aforcited).

I also proposed "the theory of hyper-spatial consciousness" in *FPHC* (and also in *FPHG*) for a new way to understand the future evolution of the mind to enter into different dimensions of space-time that our current world has never known.

In addition, in *BCOS*, I proposed "the theory of post-cosmology" for a new way to understand the beginnings and ends of the cosmos and beyond.

(d4) Geography

And in relation to geography, I also proposed "the theory of the geopower of nature" in *ALD* for a different way to understand the relationships between geography and other entities (like the mind, culture, and society).

(e) The Rest

The category of "the rest" here refers to what cannot be strictly classified in any of the previous categories (i.e., society, culture, the mind, and nature). A good illustration of "the rest" refers to that which is, namely, (e1) historical.

(e1) Historical

Precisely here, I proposed "the theory of the evolution from pre-modernity to after-postmodernity" (as first worked out in *FIIC* and later in *FCD* and *FPHC*), which overlaps—as a reminder from (b5)—with the issue concerning "civilization as a whole" in (b5).

For instance, in both *FCD* and *FPHC*, I worked out the structure of "post-human civilization" in terms of the trinity of after-postmodernity (i.e., "free-spirited after-postmodernity," "post-capitalist after-postmodernity," and "hegemonic after-postmodernity").

Both conceptually and theoretically, the trinity of after-postmodernity is a sequential extension to the trinity of modernity (i.e., "free-spirited modernity," "capitalist modernity,"and "hegemonic modernity") and the trinity of postmodernity (i.e., "free-spirited postmodernity," "capitalist postmodernity," and "hegemonic postmodernity") as first proposed in *FHC*.

And the trinity of pre-modernity (i.e., "pre-free-spirited pre-modernity," "pre-capitalist pre-modernity" and "hegemonic pre-modernity") was later conceived in *BCIV* to complete the historical set from pre-modernity to after-postmodernity.

In *BDPD*, this thesis about the trinity of pre-modernity, modernity, postmodernity, and after-postmodernity was collectively known as "the theory of the trinity of modernity to its after-postmodern counterpart," at the cultural level—as already so mentioned in (b5) earlier, and summarized in *Table 1.34*, *Table 1.35*, *Table 1.36*, and *Table 1.37*.

At the structural level, all these trinities are subject to the existential constraints (e.g., the freedom/unfreedom and equality/inequality dialectics in the context of "the cyclical progression of hegemony"), be the historical epoch in pre-modernity, modernity, postmodernity, or after-postmodernity in future times. Each of the historical epochs has its ever new ways of coming to terms with the ever new (different) mixtures of freedom/unfreedom and equality/inequality.

This is importantly so, not because, as is falsely assumed in conventional wisdom, one certain way is superior (or better) than another in terms of achieving more freedom and less unfreedom, or more equality with less inequality.

On the contrary, indeed, in each of the historical epochs, each increase of unfreedom greets each freedom achieved, just as each increase of inequality welcomes each equality achieved, albeit in ever new (different) ways. In *BDPD*, this thesis of mine was labeled as "the theory of the cyclical progression of hegemony," at the structural level, though it was first analyzed in *FCD*.

In *BDPD*, more theoretical applications of existential dialectics were further examined, in relation to five main features, in the context of the duality of oppression, namely, (a) that each freedom/equality achieved is also each unfreedom/inequality created, (b) that the subsequent oppressiveness is dualistic, both by the Same against the Others and itself and by the Others against the Same and themselves, (c) that both oppression and self-oppression can be achieved by way of downgrading differences (between the Same and the Others) and of accentuating them, (d) that the relationships are relatively asymmetric among them but relatively symmetric within them, even when the Same can be relatively asymmetric towards itself in self-oppression, and the Others can be likewise towards themselves, and (e) that symmetry and asymmetry change over time, with ever new players, new causes, and new forms, be the locality here on Earth or in deep space unto multiverses—as summarized in *Table.4.13*, *Table 4.14*, *Table 4.15*, and *Table 4.16*.

The same logic also holds both in relation to wealth and poverty (as addressed in *BCPC* and summarized in *Table 4.17* on the wealth/poverty dialectics) and in relation to civilization and barbarity (as addressed in *BCIV* and summarized in *Table 4.18* on the civilization/barbarity dialectics).

In *BDPD*, this thesis on existential dialectics was labeled as "the theory of existential dialectics."

Direct and Indirect Applications of Existential Dialectics

A different way to appreciate the usefulness of existential dialectics is by way of the analysis of its (a) direct and (b) indirect applications.

(a) In direct applications, on the one hand, the logic of existential dialectics can shed some theoretical insights on diverse phenomena in the world, and good instances are the usage of the principles of existential dialectics for the theoretical insights on the freedom/unfreedom dialectics, the equality/inequality dialectics, and the wealth/poverty dialectics—as introduced earlier in (b5) and (e) in the previous sub-section.

For instance, my books like *FPHST*, *BNN*, *FAE*, *FC*, *FIA*, *FPHU*, *FPHK*, *FPHML*, *FPHE*, *FPHMM*, *FPHCT*, *FPHG*, *FPHUP*, *FPHL*, *FPHO*, *FPHMA*, *FPHS*, *FPHLAW*, *FPHFS*, *FPHWP*, *FPHP*, *BEPE*, and *BCOS* use the principles to reveal some theoretical insights on the perspectives of space and time (as in *FPHST*), of nature and nurture (as in *BNN*), of beauty and ugliness (as in *FAE*), of order and chaos (as in *FC*), of taxonomy and network (as in *FIA*), of the unconscious mind (as in *FPHU*), of the nature of knowledge (as in *FPHK*), of mathematical logic (as in *FPHML*), of engineering (as in *FPHE*), of mass media (as in *FPHMM*), of creative thinking (as in *FPHCT*), of geometry (as in *FPHG*), of urban planning (as in *FPHUP*), of language (as in *FPHL*), of organization (as in *FPHO*), of martial arts (as in *FPHMA*), of sexuality (as in *FPHS*), of law (as in *FPHLAW*), of formal science (as in *FPHFS*), of aggression and pacificity (as in *FPHWP*), of normality and abnormality (as in *FPHP*). of morality and immorality (as in *BEPE*), and of the beginnings and ends of the universe (as in *BOCS*).

(b) In indirect applications, on the other hand, the theoretical insights can further be used to reveal other phenomena directly from them (viz., the theoretical insights) and therefore indirectly from the principles themselves.

A good illustration is of course the use of the theoretical insights on the freedom/unfreedom and equality/inequality dialectics for the understanding of the civilization/barbarity dialectics.

This distinction between direct and indirect applications may sound a bit academic, since even in indirect applications, the phenomena under study can still be directly related back to the principles themselves.

In the previous example, as an illustration, the civilization/barbarity dialectics can be directly related to the principles of existential dialectics without the intermediate role of the freedom/unfreedom and equality/inequality dialectics.

Multiple Levels of Application

There is another issue to be clarified, however. In other words, the theoretical insights can be applied to multiple levels of analysis—even though, in a given example, it may refer to one level or a few only.

For instance, in the example concerning the freedom/unfreedom dialectics, it can be used at the structural level (e.g., in relation to the theory of the cyclical progression of hegemony), but it can also be exploited for other levels (e.g., the theory of post-capitalism at the institutional level).

All these levels of application should not be misleadingly construed, as I stress this before, as a one-way street (that is, to use the ontological principles for theoretical application at multiple levels of specific analysis) but a two-way one, in which theoretical insights in praxis, when studied in more specific contexts, can in turn refine the nature of existential dialectics (for example, with the addition of new principles).

Sophisticated Methodological Holism

The summary of my original meta-theory on methodology is provided below, again, to be repeated *often verbatim* from my previous works—as this is something I regularly do in each new book to introduce my past works, which are in conversation with this current one.

This means that, for those readers who had read some of my previous books, the summary below is really not much new (*almost verbatim*, with only some updated revisions).

That clarified—this original contribution of mine to the study of methodology takes the form of an approach known as "the theory of methodological holism" or "methodological holism" in short, as already worked out in *FPHC* (2004).

I have made good use of this methodological approach of mine for all of my previous works.

The Confusion of Holism with Systems Approach

Yet, it is imperative to stress at the outset, as I often emphasized this in all my previous books, that my approach of "methodological holism" does not oppose or exclude "methodological individualism" (as some readers may be tempted to assume, as is conventionally understood) but actually includes it.

The confusion here has to do with equating holism with systems approach in general (be it about systems theory, chaos theory, complexity theory, or else), which I debunked, while learning something from it, as already elaborated in *The Future of Complexity* (2007)—and *The Future of Post-Human Formal Science* (2010) is another addition to this critique of the intellectual fad of systems approach.

In other words, my meta-theory of holism, or sophisticated methodological holism in short, makes a distinction between doing holistic work and doing systems work, for two reasons.

The first reason is that holism, in my theory, does not imply systems approach, since it rejects systems approach because of the myriad problems that the literature on systems approach has been unable to resolve, while learning something from it.

And the second reason is that my theory is related to the other theory of mine, that is, existential dialectics, which does not exclude methodological individualism (but includes it, because it simply goes beyond both reductionism and reverse-reductionism in a dialectic way).

For these reasons (and others too, as summarized in *Table 4.1*), my version of methodological holism is *sophisticated*—not *vulgar* as sometimes used by inapt scholars employing the same term.

The Ontological Constraints

Sophisticated methodological holism is subject to some ontological constraints, and five good examples include "the partiality-totality principle," "the predictability-unpredictability principle," "the explicability-inexplicability principle," "the absoluteness-relativeness principle," and "the formalness-informalness principle" in existential dialectics.

Of course, there can be more than five ontological principles in relation to method, but the five examples here are sufficient to illustrate the point in question.

With this caveat in mind—the first two were already analyzed in previous books of mine like *FC* and *FAE*, whereas the third one was addressed in *FPHU*, and the last two were accounted for in *FPHK* and *FPHML*, respectively.

For now, it suffices to show that sophisticted methodological holism, because of these ontological constraints on methodology, targets two major sins of methodology, namely, what I call, in the absence of better words, (a) *reductionism* and (b) *reverse-reductionism*—both of which come in all shapes and sizes, to be summarized below (and also shown in *Table 4.2*).

Against the Varieties of Reductionism

There are four versions of reductionism to be summarized here, which sophisticated methodological holism rejects, namely, (a1) conceptual, (a2) theoretical, (a3) methodological, and (a4) ontological forms of reductionism.

(a1) In conceptual reductionism, a good case in point concerns myriad dualities like mind vs. body, self vs. world, democracy vs non-democracy, and the like (as already addressed in *FHC*, *FPHC*, and *BDPD*, for instance).

(a2) In theoretical reductionism, an illuminating case study is best exemplified by what I originally called "the foundation fallacy" in *FPHST*, in any attempt to naively understand space-time from the physical perspective as the foundation and, consequently, to dangerously dismiss (or belittle) other perspectives.

In *FAE*, I elaborated further these versions of reductionism in the literature on aesthetics (e.g., form vs. content, representation vs. expression, critics vs. artists, and externalism vs. internalism).

In *FIA*, I revealed other forms of reductionism in the literature on information architecture (e.g., the constructivist argument and the representational argument).

In *FPHU*, I showed the persistent legacy of reductionism, this time, in the literature on anomalous experience (e.g., the obsession with physics, chemistry, and biology for explaining anomalous experience).

In *FPHE*, I examined another case of reductionism in action, in the context of engineering (e.g., technical constraints vs. normative constraints).

In *FPHMM*, I elaborated one more version of reductionism, in relation to the three domains of communication (e.g., the competing views on sending, connecting, and receiving).

In *FPHCT*, I explored another version of reductionism, in relation to invention and innovation (e.g., the bio-psychological argument vs. the socio-cultural arguments).

In *FPHG*, I identified another version of reductionism, in relation to infinity, symmetry, and dimensionality (e.g., the Euclidean argument vs. the non-Euclidean arguments).

In *FPHUP*, I analyzed another version of reductionism, in relation to density and void (e.g., the engineering argument and the ecology argument).

In *PFHL*, I examined another version of reductionism, in relation to structure and context (e.g., the structuralist argument and the contextualist argument).

In *PFHO*, I explored another version of reductionism, in relation to communcation, decision-making, and leadership (e.g., the rational-system argument vs. the natural-system argument vs. the open-system argument).

In *PFHMA*, I revealed another version of reductionism, in relation to the martial body and spirit (e.g., the spiritual argument vs. the materialist argument vs. the defensive argument).

In *PFHS*, I analyzed another version of reductionism, in relation to the sexual body and spirit (e.g., the naturalist argument vs. the constructivist argument).

In *PFHLAW*, I examined another version of reductionism, in relation to law (e.g., the necessity argument vs. the contengency argument).

In *FPHWP*, I showed another version of reductionism, in relation to war and peace (e.g., the aggressivist argument vs. the pacifist argument).

In *BEPE*, I analyzed another version of reductionism, in relation to morality and immorality (e.g., the objectivist argument vs. the non-objectivist argument vs. the skeptical argument).

And in *BCOS*, I examine another version of reductionism, in relation to the contested beginnings and speculative ends of the universe (e.g., the scientific argument vs. the religious argument vs. the esoteric argument vs. the metaphysical argument).

(a3) In methodological reductionism, a good illustration can be the debate between different versions of qualitative and quantitative methods (as already analyzed in *FC* and also *FHC*).

In *FPHML*, I examined similar reductionism, this time, in the literature on mathematical logic (e.g., the obsession with consistency, soundness, and completeness).

And in *FPHFS*, I explored the problems of reductionism in the context of formal science (e.g., the analytical argument, as opposed to the synthetic argument).

(a4) In ontological reductionism, an excellent instance involves another debate, this time between emergentism and reductionism in complexity theory (as addressed in *FC*) and also in psychology (as elaborated in *FPHC*, in the context of Being and Belonging).

But to target against the varieties of reductionism constitutes only a side of the same coin, and the other side concerns the varieties of reverse-reductionism.

Against the Varieties of Reverse-Reductionism

The other side of the same coin is the reverse version of reductionism, which is what I want to call, in the absence of better words, *reverse-reductionism.*

My sophisticated version of methodological holism targets against the varieties of reverse-reductionism (just as it also rejects the ones of reductionism).

Perhaps nothing expresses better the popularity of reverse-reductionim than the "anything-goes" mentality in postmodernism of our time, as shown in the following four versions.

(b1) In conceptual reverse-reductionism, any concept of "art" (e.g., fine art, applied art, outsider art, junk art) is welcome in postmodernism (as already analyzed in Ch.4 of *FHC*—and also in *FAE*).

(b2) In theoretical reverse-reductionism, a variety of art and literary theories co-exist. Take the case of literary studies, as there are now Literary Structuralism, Marxist Literary Criticism, New Criticism, Phenomenology, Hermeneutics, Language-Game Literary Criticism, Feminist Literary Criticism, Reception Theory, Reader Response Criticism, Poststructuralism, Semiotics, Pyschoanalytic Literary Criticism, just to cite some well-known ones, with no one being said to be better than any others (as detailedly analyzed in Ch.4 of *FHC*). (S. Raman 1997)

In *BNN*, I even introduced "the compromise fallacy" as another good illustration of theoretical reverse-reductionism, in misleadingly treating both genetic and environmental approaches as equally valid.

(b3) In methodological reverse-reductionism, multiple methodologies are deemed as acceptable in postmodernism (e.g., doing art without praxis, doing art with praxis, and doing art by sublation), as analyzed in Ch.4 of *FHC*.

And in *FPHFS*, I exposed the problems of reverse-reductionism in the context of formal science in relation to systems theory (e.g., the synthetic argument).

(b4) In ontological reverse-reductionism, no privileged ontology is allowed, and the door is open practically for anything in postmodernism (e.g., the equal status of the ontology of Being vs. that of Becoming, as already addressed in Ch.4 of *FHC*—and also in *FPHC*).

In *FAE*, I also introduced another version of reverse-reductionism, that is, "the pluralist fallacy," in the context of understanding aesthetic experience, for instance—although this fallacy has been committed not exclusively in relation to the ontological level (but also at the conceptual, theoretical, and methodological ones).

These dual dangers against reverse-reductionism (in this sub-section) and reductionism (in the previous sub-section) are something that sophisticated methodological holism rejects. And in *FPHFS*, I addressed these two problems further.

This point about the dual dangers is important enough, since many scholars often favor one against the other, but it is vital to target both.

The Holistic Organization of an Inquiry

With these dual dangers against reductionism and reverse-reductionism in mind—sophisticated methodological holism suggests that an inquiry of any given phenomenon is more complete, if treated in the context of a comprehensive analysis at all relevant levels, which challengingly encompass all the fields of human knowledge, ranging from the natural sciences and formal sciences through the social sciences to the humanities.

This section is something that I had already stressed in all my previous books and repeat (*often verbatim*) hereafter. But, for those readers who had read my previous books, this serves as a reminder.

With this reminder in mind—there are multiple ways to engage in a holistic inquiry with all relevent levels of analysis. Over the years, I have proposed different ways to fulfill this holistic methodological requirement, as repeated in my previous books.

Hereafter is a summary of four major ways, namely, (a) by discipline, (b) by domain, (c) by subject, and (d) by meta-analysis—all of which fulfill the holistic requirement of sophisticated methodological holism on a given issue.

(a) In a holistic organization by discipline, a good illustration is the multiple levels of analysis, namely, (a1) the micro-physical, (a2) the chemical, (a3) the biological, (a4) the psychological, (a5) the organizational, (a6) the institutional, (a7) the structural, (a8) the systemic, (a9) the cultural, (a10) the macro-physical (cosmological), and (a11) other relevant levels, which are either a combination of all these levels or the practical applications with a combination of them.

This holistic organization by discipline is used in *FCD* and *BCPC*, for instance.

(b) In a holistic organization by domain, some good candidates include the classification of different perspectives of analysis in relation to (b1) nature, (b2) the mind, (b3) society, and (b4) culture, as already worked out in *BCIV*, *FPHST*, *FC*, *FAE*, *FPHU*, *FPHK*, *FPHE*, and *FPHFS*.

Culture in (b4), in this re-classification, is the same as culture in (a9) and can be further divided into different sub-categories, with good examples like (a9i) epistemology (methodology), (a9ii) aesthetics, (a9iii) ethics, (a9iv) metaphysics (ontology), and (a9v) religion. For clarification, epistemology (methodology) in (a9i) and metaphysics (ontology) in (a9iv) overlap with the category of "meta-theory."

Society in (b3) includes the organizational in (a5), the institutional in (a6), the structural in (a7), and the systemic in (a8).

The mind (b2) has more to do with the chemical in (a2), the biological in (a3), and the psychological in (a4), although it can overlap with (a1), (b3) and (b4), for instance.

It should be stressed that the "mind" here also includes the categories like the "brain," "body," and the "spirit" (or the "soul")—whenever needed in a given context, so that one does not fall into the intellectual trap of the "mind-body dualism" in the history of modern (and especially, postmodern) philosophy, which I already went to great lengths to explain in *The Future of Human Civilization*. After all, in everyday's usage, the "mind" already implies the "brain" and the "body" (together with their interactions) and has different levels of consciousness which can be both "spiritual" and non-spiritual.

And nature in (b1) refers to the micro-physical in (a1) and the macro-physical (cosmological) in (a10), although it can also overlap with (a2), (a3) and (a4).

(c) In a holistic organization by subject, some excellent examples concern the analysis of consciousness in *FPHC* (i.e., on Having, Belonging, and Being); of civilization in *FHC* (i.e., the True, the Holy, the Everyday, the Technological, the Beautiful/Sublime, the Good, and the Just); of international politics in *BWT* (i.e., Hyper-Empires, Meso-Empires, and Micro-Empires); of nature and nurture in *BNN* (i.e., genes and memes); of information architecture in *FIA* (i.e., taxonomy and network); of mathematical logic in *FPHML* (i.e., consistency, soundness, and completeness); of communication in *FPHMM* (i.e., sending, connecting, and receiving); of creative thinking in *FPHCT* (i.e., invention and innovation); of geometry in *FPHG* (i.e., infinity, symmetry, and dimensionality); of urban planning in *FPHUP* (i.e., density and void), of language in *FPHL* (i.e., structure and context), and of organization in *FPHO* (i.e., communication, decision-making, and leadership), in *FPHMA* (i.e., the martial body and the martial spirit), in *FPHS* (i.e., the sexual body and the sexual spirit), in *FPHLAW* (i.e., necessity and contingency), in *FPHWP* (i.e., aggression and pacificity), in *FPHP* (i.e., normality and abnormality), in *BEPE* (i.e., morality and immorality), and in *BCOS* (i.e., the beginnings and ends of the universe)—by incorporating all the relevent levels of analysis as cited above in each.

(d) And in a holistic organization by meta-analysis, two good cases in point involve the works on post-capitalism in *BCPC* and on authoritarian liberal democracy in *ALD*, with the classification of analysis in theory and meta-theory.

The essential point to remember here, as I thus stress in all my previous books, is that the multiple levels of analysis can be reorganized in many different ways, insofar as none of the levels (if relevant to an inquiry in question) is ignored or dismissed, to avoid the danger of reductionism (and for that matter, the one of reverse-reductionism, in the opposite direction).

Three Clarifications

Three clarifications are needed here, however, to avoid any hasty misunderstanding.

Firstly, the entities in each classification are not mutually exclusive. For instance, in the holisitc organization by domain, nature in (b1) can also be linked to the chemical in (a2), the biological in (a3), and the systemic in (a8). And the mind (b3) can alternatively be related to the micro-physical in (a1) and the chemical in (a2), for example.

Secondly, there is also the important factor of luck (or uncertainty in conventional usage), but it is already implied or allowed in each of the organizations of an inquiry—especially in relation to the predictability-unpredictability principle and the order-chaos principle in existential dialectics. This topic about uncertainty was already addressed in *FHC*—and more extensively, in *FC*, in the context of order and chaos.

And lastly, the comparison in a classification is not absolute, but relative, as there are often some shades of gray, not exactly black or white (figuratively speaking).

Some further clarifications and qualifications of sophisticated methodological holism are shown in *Table 4.2*.

Some Distinctive Usefulness

In the end, sophisticated methodological holism—when applied, especially though not exclusively, as illustrated in my numerous works—can enrich the understanding of reality in some distinctive ways. Here are four examples (as revealed in each of my books).

Firstly, it provides a comprehensive analysis of a subject matter, from which much can be learned about reality, both in relation to the perspectives of the mind, nature, society, and culture.

Secondly, it offers a new classification of the subject matter.

Thirdly, it suggests some visions of the future in relation to the subject matter in question.

And fourthly, it proposes some insights on meta-theory (e.g., methodology and ontology) in general—with the clear understanding, however, of the dilemma of specific vs. general ontology (as shown in *Table 4.9* on the syntax of existential dialectics in the context of the dilemma of ontology).

For this very reason, all of these ways are important, without reducing one into the analysis of another. (*FPHCT*)

Chapter Outline

With this summary of my two original meta-theories (on methodology and ontology, *almost verbatim* from my previous works) in mind—this book is organized in four main parts, namely, (a) Part I on the introduction to this book, (b) Part II on the contested beginnings, (c) Part III on the speculative ends, and (d) Part IV on the conclusion of the book.

The book is thus organized in four chapters.

The introductory chapter here, that is, Chapter One titled "Introduction—The Awe of the Universe," begins by first challenging the euphoric framing of a worldview and then clarifying the distinction between cosmology and cosmogony, the different conceptions of the universe, and some deep questions in cosmology—together with the theoretical debate in the literature, my theory of post-cosmology, the relationship between theory and meta-theory, existential dialectics, sophisticated methodological holism, and some further clarifications.

Chapter Two titled "The Universe and Its Contested Beginnings" explores the contested beginnings of the universe and their aftermath from the four perspectives of the mind, nature, society, and culture, with a constructive critique of each.

Chapter Three titled "The Universe and Its Speculative Ends" examines the evolution of the universe and its speculative ends from the four perspectives of the mind, nature, society, and culture, with a deconstructive analysis of each.

The last chapter titled "Conclusion—The Future of Cosmology" summarizes the analysis in the book with fifteen major theses, namely, (a) the first thesis: the absoluteness-relativeness principle, (b) the second thesis: the predictability-unpredictability principle, (c) the third thesis: the explicability-inexplicability principle, (d) the fourth thesis: the preciseness-vagueness principle, (e) the fifth thesis: the simpleness-complicatedness principle, (f) the sixth thesis: the openness-hiddenness principle, (g) the seventh thesis: the denseness-emptiness principle, (h) the eighth thesis: the slowness-quickness principle, (i) the nineth thesis: the expansion-contraction principle, (j) the tenth thesis: the convention-novelty principle, (k) the eleventh thesis: the evolution-transformation principle, (l) the twelveth thesis: the symmetry-asymmetry principle, (m) the thirteenth thesis: the regression-progression principle, (n) the fourteenth thesis: the sameness-difference principle, and (o) the fifteenth thesis: the post-human odyssey—to be elaborated in the rest of the book, with a summary in the concluding chapter.

This seminal project is to fundamentally alter the way that we think about cosmology, from the combined perspectives of the mind, nature, society, and culture, with enormous implications for the human future and what I originally called its "post-human" fate.

Some Clarifications

But some additional clarifications are needed, before a formal analysis of the contested beginnings of the universe is to begin in Chapter Two.

These clarifications, lest any potential misunderstanding by the readers occurs, refer to (a) the conversation with my previous books, (b) the illustrative selection of case studies and examples, (c) detailed analysis versus overall analysis, (d) two distinctive features of using quotations, and (e) the use of neologisms.

These clarifications are something that I also regularly repeat in my previous books too, so these points are repeated hereafter, *often verbatim*—albeit in a different context this time.

In Conversation with My Previous Books

Firstly, my theory of post-cosmology is constructed on the theoretical foundation of my previous books (i.e., *FHC*, *FCD*, *FPHC*, *BDPD*, *BCPC*, *BCIV*, *FPHST*, *BNN*, *BWT*, *FC*, *FAE*, *ALD*, *FIA*, *FPHU*, *FPHK*, *FPHML*, *FPHE*, *FPHMM*, *FPHCT*, *FPHG*, *FPHUP*, *FPHL*, *FPHO*, *FPHMA*, *FPHS*, *FPHLAW*, *FPHFS*, *FPHWP*, *FPHP*, *BCOS*, and *BEPE*—as explicated in "The List of Abbreviations").

Just consider the following two illustrations, namely, (a) in relation to other related theoretical debates and (b) in relation to different visions as worked out in my previous books.

(a) The first illustration is that the theoretical debate on the nature of cosmology also involves a few other related theoretical issues.

These other related theoretical issues have to do with some deeply contested theoretical debates, which, however, I had already gone to great lengths to analyze in my numerous (previous) books, as summarized below. It should also be stressed, lest misunderstanding occurs, that each of the theoretical debates (as listed below) does not exhaust the issues covered in the books, as each book covers more issues than what is listed below (for illustration only).

- On Space and Time
  - –Ex: *The Future of Post-Human Space-Time*
  - –Ex: *The Future of Post-Human Geometry*
- On Infinity, Symmetry, and Dimensionality
  - –Ex: *The Future of Post-Human Geometry*
- On the Fate of the Universe
  - –Ex: *The Future of Post-Human Space-Time*
  - –Ex: *The Future of Post-Human Geometry*
  - –Ex: 2 volumes, *The Future of Human Civilization*
  - –Ex: *The Future of Capitalism and Democracy*
- On Uncertainty
  - –Ex: *The Future of Complexity*
  - –Ex: 2 volumes, *The Future of Human Civilization*
- On Systems Thinking
  - –Ex: *The Future of Post-Human Formal Science*
- On Mathematical Logic
  - –Ex: *The Future of Post-Human Mathematical Logic*
- On Epistemology and Ontology
  - –Ex: *The Future of Post-Human Knowledge*
  - –Ex: *The Future of Post-Human Consciousness*

–Ex: 2 volumes, *The Future of Human Civilization*
- On Consciousness and Unconsciousness
  –Ex: *The Future of Post-Human Consciousness*
  –Ex: *The Future of Post-Human Unonsciousness*
  –Ex: *The Future of Capitalism and Democracy*
- On Creativity
  –Ex: *The Future of Post-Human Creative Thinking*
  –Ex: *The Future of Aesthetic Experience*
- On Communication
  –Ex: *The Future of Post-Human Mass Media*
- On Cognition
  –Ex: *The Future of Post-Human Unconsciousness*
  –Ex: *The Future of Post-Human Consciousness*
  –Ex: *The Future of Capitalism and Democracy*
  –Ex: *The Future of Post-Human Creative Thinking*
- On Intuition
  –Ex: *The Future of Post-Human Unconsciousness*
  –Ex: *The Future of Post-Human Consciousness*
  –Ex: *Beyond Capitalism to Post-Capitalism*
  –Ex: *The Future of Capitalism and Democracy*
- On Being and Belonging
  –Ex: *The Future of Post-Human Consciousness*
  –Ex: 2 volumes, *The Future of Human Civilization*
- On Different Dimensions of Human Existence
  –Ex: 2 volumes, *The Future of Human Civilization*
- On Modernity, Post-Modernity, and After-Postmodernity
  –Ex: 2 volumes, *The Future of Human Civilization*
  –Ex: *Beyond Civilization to Post-Civilization*
- On Pre-Modernity
  –Ex: *Beyond Civilization to Post-Civilization*
- On No Freedom Without Unfreedom
  –Ex: 2 volumes, *The Future of Human Civilization*
  –Ex: *The Future of Capitalism and Democracy*
  –Ex: *Beyond Capitalism to Post-Capitalism*
  –Ex: *Beyond Democracy to Post-Democracy*
  –Ex: *Beyond Civilization to Post-Civilization*

Therefore, those who are interested in these other theoretical issues (and debates) which are related to the current context can consult these other books of mine for more details.

(b) And the second illustration is that one of the major theses of this book concerns the future, be it human and later post-human, which is closely linked with the different visions that I had worked out in my previous books, in relation to the mind, nature, society, and culture.

Of course, for the convenience of the readers, a summary of some of these visions will be provided whenever necessary, as shown in the two sections on meta-theory above (*often verbatim*) and in different tables at the end of this chapter and the concluding chapter, for instance.

Yet, I still expect the readers to read my previous books directly for more details—as this book is written in conversation with my previous ones.

Case Studies and Examples

Secondly, the case studies and examples in the chapters are not exhaustive but solely illustrative. Nor are they necessarily mutually exclusive, as they can be reclassified in a different way. And exceptions are allowed.

One is tempted to assume, however, that, since the case studies and their examples are not exhaustive, more case studies and examples are needed. But the problem here is that just adding more of them does not necessarily change the conclusions to be drawn and may even lead to redundancy.

After all, the case studies and examples used in this book are chosen with care and deemed sufficient for the purpose at hand, even though they are solely illustrative.

Detailed Analysis versus Overall Synthesis

Thirdly, there is a distinction between detailed analysis and overall synthesis, in that the former is, especially though not exclusively, more for the specialized viewpoint of a specific discipline, whereas the latter is, especially though not exclusively, more for the broad horizon of an interdisciplinary approach—relatively speaking, of course.

Or, by analogy, there is a distinction between depicting a "tree" (with tiny details) in a "forest" and viewing the entire "forest" itself (with the broader horizon).

This book, like all previous ones of mine, tries to strike a balance between the two forms of understanding, although this does not imply that the two forms should be exactly equally offered, in an exact 50-50 balance.

Instead, the book, more often than not, does not allow the tiny details to obscure the larger picture in a given case study, because the entire book is inter-disciplinary in nature and tries to look at the entire "forest," not being bogged down with the detailed analysis of an individual "tree" in the forest, by analogy.

In this sense, although detailed analysis can be provided whenever needed for a given case study, the important point to remember is the overall horizon with its broadness of scope.

Two Distinctive Features of Using Quotations

Fourthly, the way that quotes are used in this book (and others of mine) has two distinctive features, as explained below.

The first distinctive feature is that, at the end of each quotation, sometimes there are more than one reference—in which case the first reference is for the original citation, and the rest of the references are for the convenience of the reader for more information about the issue under quotation.

And the second feature is that quotations are used as often as possible, so as to let others speak for themselves, without the nuisance of putting my own words into their mouths. Even more importantly, this allows me to use the original text as an evidence for critical analysis at the end of each section.

The Use of Neologisms

And finally, I use different neologisms in my books, mostly to introduce my original concepts and theories, and good instances include those here or elsewhere in my previous books (e.g., "the theory of post-cosmology," "post-democracy," "hyper-spatial consciousness," "post-capitalism," and whatnot). Needless to say, they are used here solely for our current intellectual convenience, as they will be renamed differently in different ways in future history.

As I thus wrote in *FCD* (508-9), "all these terms 'post-capitalism,' 'post-democracy'...and other ones as introduced in...[the] project (e.g.,...'posthuman elitists,' and 'posthuman counter-elitists,' just to cite a few of them) are more for our current intellectual convenience than to the liking of future humans and post-humans, who will surely invent more tasteful neologisms to call their own eras, entities, and everything else, for that matter. But the didactic point here is to use the terms to foretell what the future might be like, not that its eras and entities must be called so exactly and permanently. After all, William Shakespeare (1995: Act II, Scene II, Line 47) well said long ago: 'What is in a name? That which we call a rose by any other name would smell as sweet.'"

As I had also stressed time and again before, each of the neologisms can be re-written as a different "X," only to be re-named differently by the powers that be in different eras of future history.

With these clarifications for thought in mind—I now proceed to Chapter Two on the contested beginngins of the universe.

Table 1.1. The Taxonomy of Eternal Cosmologies (Part I)

Types	*Thinkers*
"Eternal substance"	Thales, Anaximander, Xenophanes, Empedocles, Anaxagoras, Leucippus, Demokritus, Heraclitus, Parmenides, Plato, Stoa
"Eternity of the world, but emanation of the deity"	Plotinus, Proclus, Origen, Abu Nasr Al-Faràbi (Alfarabius), Abù'Alì al-Husayn Ibn Sinna (Avicenna)
"Timlessness of creation"	Johannes Scotus Eurigena, Meister Eckhart
"*Creatio continua*, continuous maintenance and re-creation of the world"	Anaximenes, Stoa, Cicero, Augustine of Hippo, Gregor the Great, Thomas Aquinas, Henry of Gent, Petrus Aureoli, René Descartes
"Absolute beginning, creation, *creatio ex nihilo*"	Johannes Philoponos, Augustine of Hippo, Yaaqub ibn Ishaq Al-Kindi (Alkindus); Hugo & Richard & Bernard of St. Viktor, Bernard & Thierry of Chartres, Wilhelm of Conches, Clarenbaldus of Arras, Abù Hamìd Muhammad Al-Ghazàli, Moses Maimonides, Albertus Magnus, Wilhelm of Auvergne, Robert Grosseteste, Roger Bacon, Bonaventura, Matthew of Aquasparta, Henry of Gent, John Pecham, Richard of Middleton, Peter Aureol, William of Alnwick, Henry Totting of Oyta, Marsilius of Inghen

(continued on next page)

Table 1.1. The Taxonomy of Eternal Cosmologies (Part II)

Types	*Thinkers*
"No beginning and no end (static. vs. evolutionary vs. revolutionary)"	• "static universe" by Albert Einstein (1917) • "empty expanding universe" by Willem de Sitter (1917) • "eternal expansion out of a static universe" by Arthur S. Eddington (1930) • "steady state" by Hermann Bondi, Thomas Gold & Fred Hoyle • "quasi-steady state" by Fred Hoyle, Geoffrey Burbidge & Jayant V. Narlikar • "chaotic inflation (global!)" by Andrei Linde (1983) • "Planckian cosmic egg (global!)" by Mark Israelit & Nathan Rosen (1989) • "big bounce" by Hans-Joachim Blome (1991) & Wolfgang Priester • "ekpyrotic and cyclic universe (global!)" by Paul Steinhardt & Neil Turok et al.

(continued on next page)

Table 1.1. The Taxonomy of Eternal Cosmologies (Part III)

Types	*Thinkers*
"Eternity: no beginning and no transience/destructibility"	Aristotle, Averroes, Thomas Aquinas, Siger of Brabant, Boetius of Dacia, Wilhelm Ockham, Giles of Rome, Godfrey of Fontaines, Henry of Harclay, Thomas of Wylton, Thomas of Strassburg; Pietro D'Abano, Johannes Jandunus, Pietro Pomponazzi, Giordano Bruno, Baruch de Spinoza
"no beginning, but an end"	• "collapse out of a static universe" by Arthur S. Eddington (1930)

Source: From Ch.2 of *FIA*. The table is reconstructed, based on the data from the work of Rudiger Vaas (2003).

Table 1.2. The Taxonomy of Initial Cosmologies (Part I)

Types	*Thinkers*
"Beginning and an end"	▪ "classical big bang/big crunch" by Alexander Friedmann, Stephen Hawking & Roger Penrose ▪ "quantum tunnel effect" by Alexander Vilenkin (1982) ▪ "no boundary instanton" by Stephen Hawking & James Hartle (1983)
"Beginning, but no end"	• "classical big bang/big whimper" by Alexander Friedmann, Georges Lemaître (1927), Stephen Hawking & Roger Penrose • "phoenix universe (global!)" by Georges Lemaître (1933) & Richard C. Tolman (1934) • "quantum tunnel effect and eternal inflation" by Alexander Vilenkin (1982) • "cosmic Darwinism" by Lee Smolin (1992) • "no boundary instanton" by Stephen Hawking & Neil Turok (1998) • "transcendentalism idealism" by Immanuel Kant
"time-loop with/without end "	• "self-creating universe" by John Richard Gott III & Li-Xin Li (1998)

(continued on next page)

Table 1.2. The Taxonomy of Initial Cosmologies (Part II)

Types	*Thinkers*
"pseudo-beginning with/without a local end "	*–background-dependent*: • "soft bang/emergent universe" by Eckard Rebhan (2000), George F. R. Ellis & Roy Maartens et al. • "quantum fluctuation, de Sitter instability, etc." by Edward Tryon, Robert Brout et al. (1978), Alexei A. Starobinsky, David Atkatz (1982) & Heinz R. Pagels, John Richard Gott III (1982), Mark Israelit (2002) • "pre-big bang" by Gabriele Veneziano & Maurizio Gasperini
	–background-independent: • "pregeometry" by John A. Wheeler, Peter W. Atkins (1981), & Stephen Wolfram (2002) • "loop quantum cosmology" by Abhay Ashtekar & Martin Bojowald et al.
"cycle (recurrence)"	• "oscillating universe (local!)" by Mark Israelit & Nathan Rosen (1989), & Redouane Fakir (1998) • "cyclic universe (local!)" by Paul Steinhardt & Neil Turok et al. (2002) • "circular time in a rotating universe" by Kurt Gödel (1949) • "big brunch/time-reversal" by Claus Kiefer & H. Dieter Zeh

Source: From Ch.2 of *FIA*. The table is reconstructed, based on the data from the work of Rudiger Vaas (2003).

Table 1.3. The Theoretical Debate on Space-Time (Part I)

- **Isaac Newton's Absolutist (Substantivist) Theory of Space-Time**
  —space and time are independent from each other. The structure of space-time is $E3 \times E1$ (with the structure of space, *P*, as the set of spatial locations in a three-dimensional Euclidean space, *E3*, and the structure of time as the set of temporal moments, *T*, in the one-dimensional real time, *E1*).
  —space and time are also independent from the effects of matter and events. The existence of space and time is possible even in a world absent of matter (and, for that matter, even in a world absent of events), as if they were material objects but with their total unchangingness thorough time.

- **Albert Einstein's Relativist Theory of Space-Time**
  —space and time are interchangeable (not absolute), just as matter and energy are equivalent (not independent) with the famous equation, $E = mc^2$ (as in the special theory of relativity in 1905).
  —space-time and matter-energy are also relative in a grand union (as in the general theory of relativity in 1915). Thus, each pair affects the other pair, as "matter 'tells' spacetime how to curve [in a non-Euclidean geometry] and curved spacetime 'tells' matter how to behave….Space contracts near mass and dilates away from it. Time dilates near mass and contracts away from it. Clocks positioned farther away from the mass of the earth run faster than clocks closer to the earth." (L. Shlain 1991: 328-330)

(continued on next page)

Table 1.3. The Theoretical Debate on Space-Time (Part II)

- **Peter Baofu's Perspectival Theory of Space-Time**
  —space and time can be understood from multiple perspectives, be they in relation to culture, society, nature, and the mind, with each perspective revealing something about the nature of space-time and simultaneously delimiting its view. This is subject to "the regression-progression principle" in existential dialectics.
  —each perspective of space and time exists in society and culture with good reasons, with some being more successful and hegemonic (dominant) than others. This is subject to "the symmetry-asymmetry principle" in existential dialectics.
  —space and time will not last, to be eventually superseded (altered) by post-humans in different forms (e.g.,stretching/shrinking space-time, engineering more dimensions of space-time, and manipulating multiverses), be they here in this universe or in multiverses. Thus, even the physical existence of space-time cannot last forever, with ever more transformations in the process. This is subject to "the change-constancy principle" in existential dialectics.
  —the conventional wisdom (especially by physicists) of treating the physical perspective of space and time as the foundation of all other perspectives (of space and time) and of regarding them as much less important is a form of reductionism,committing what I call *the foundation fallacy*, in misleadingly dismissing the multiple perspectives of space and time in relation to culture, society, nature, and the mind.

Notes: The examples in each category are solely illustrative (not exhaustive), and the comparison is relative (not absolute), nor are they necessarily mutually exclusive. Some can be easily re-classified elsewhere. As generalities, they allow exceptions.
Source: A summary of *Sec.1.1*, *Sec.1.2*, and *Sec.1.3* of *FPHST*—and, for that matter, the rest of *FPHST*

Table 1.4. Main Reasons for Altering Space-Time

- **The Need to Make New Energy-Matter**
  —Ex: manipulating molecular bonds for new materials
  —Ex: creating nanotechnologies on the atomic scale
  —Ex: engineering the atomic nucleus
  —Ex: restructuring most elementary particles
  —Ex: inventing new forms of matter and energy

- **The Need to Create New Space-Time**
  —Ex: creating "warp drive" (as in science fiction) for space travel
  —Ex: creating "pocket universes"

- **The Need to Conquer the Cosmos unto Multiverses**
  —Ex: spreading floating consciousness and hyper-spatial consciousness, besides other forms that humans have never known, in the cosmos and beyond unto multiverses for ultimate conquest

Notes: The examples in each category are solely illustrative (not exhaustive) nor necessarily mutually exclusive, and the comparison is relative (not absolute). As generalities, they allow exceptions. Also, it should be stressed that the three reasons are all related, in that they all contribute to the evolution of intelligent life in the cosmos unto multiverses in the most distant future beyond our current knowledge.
Sources: A summary of *Sec.6.2* of *FPHST*. See also *FHC*, *FCD*, *FPHC*, *FC*, *FPHU*, and *FPHG*, for example.

Table 1.5. The Technological Frontiers of the Micro-World

- **Type I-Minus**
  —Ex: building structures and mining

- **Type II-Minus**
  —Ex: playing with the genetic makeups of living things

- **Type III-Minus**
  —Ex: manipulating molecular bonds for new materials

- **Type IV-Minus**
  —Ex: creating nanotechnologies on the atomic scale

- **Type V-Minus**
  —Ex: engineering the atomic nucleus

- **Type VI-Minus**
  —Ex: restructuring most elementary particles

- **Type Ω-Minus**
  —Ex: altering the structure of space-time

Notes: As already indicated in *Sec.4.4.2.2* of *FPHC*, the problem with this micro-classification (from Barrow's work) is that the civilization types (with the exception of Type Ω-Minus, for example) are not quite distinct, since many of them can be achieved more or less in a civilization, to the extent that Type II-minus and Type III-minus, just to cite two plausible types, can be historically contemporaneous, relatively speaking, unlike the vast historical distance between, say, Type O and Type I (or Type I and Type II) civilizations. In other words, the micro-classification here is not very useful to understand civilization types but is revealing to see the technological frontiers of the micro-world.
Sources: A reconstruction from J.Barrow (1998:133), as originally shown in *Table 4.7* of *FPHC*. See *FPHC* for more info.

Table 1.6. Finity, Transfinity, and Infinity (Part I)

- **Infinity**
  —The term "infinity," which is "symbolically represented with ∞," derives from the Latin *infinitas* (or "unboundedness"), to refer to "several distinct concepts—usually linked to the idea of 'without end'—which arise in philosophy, mathematics, and theology." (WK 2008c)
  —In the context of mathematics, "infinity" is often "treated as if it were a number (i.e., it counts or measures things: 'an infinite number of terms') but it is a different type of 'number' from the real numbers. Infinity is related to limits,...large cardinals,...projective geometry, extended real numbers and the absolute Infinite," for instance. (WK 2008c)

- **Transfinity**
  —To avoid confusion, my usage of the word "transfinite" (as used here in the ontological principle) is not the same as the one which was originally coined by Georg Cantor to refer to "cardinal numbers or ordinal numbers that are larger than all finite numbers, yet not necessarily absolutely infinite" (in the mind of God), as opposed to "relative" infinity (in the mind of Man). (WK 2008d) In other words, for Cantor, there are two kinds of infinity, "absolute" and "relative"—with "relative" infinity or "transfinity" (in the mind of Man) being dependent on "absolute" infinity (in the mind of God) for its existence.
  —By contrast, my usage of the word "transfinite" here differs radically from the one used by Cantor (and other mathematicians) and is more limited, in light of the problems confronting any attempt to understand the idea of infinity, be it by intuition, imagination, and conception (as already shown in the prevous three sub-sections).
  —With this clarification in mind, I allow numbers which can be many times larger—or smaller, for that matter—than the finite things that we encounter in daily life, but they do not have to be related to the idea of infinity at all (which may not exist).

(continued on next page)

Table 1.6. Finity, Transfinity, and Infinity (Part II)

• **Transfinity** ***(cont'd)***

—Of course, there may be some *borderline* cases, in which it is not clear whether the number in question is transfinite (in my usage) or simply a mathematical convenience. A good example of a borderline case is the Planck unit of length for "the smallest space possibly measured in nature," which is "less than billionths of trillionths of trillionths of an inch" (or something like 1.6×10^{-35} meters). (P. Baofu 2006a; N. McAleer 1987: 219; WKV 2008; D. Corbett 2008)

—The ideas concerning symmetry and dimensions in the examples above were addressed in Chapter Three and Chapter Four of *FPHG*. For now, it suffices to further add that it is mind-boggling to even conceive or imagine extreme numbers like the Planck unit of length (or others). Does it really exist at all? Or is it merely a mathematical convenience to describe a physical situation that the human mind does not understand properly?

—In any event, "transfinity" (in my usage) can be used as a further extension (of number) to the finite numbers in daily life—albeit with some borderline cases in mind.

• **Finity**

—It refers to numbers which are bounded (that is, with an end), especially (though not exclusively) in relation to things in everyday life.

Source: From *Sec. 1.4* and *Sec. 2.2.3* of *FPHG*

Table 1.7. Theoretical Speculations of Multiverses

• **"Baby Universes" (Ex: Andre Linde and others)**
—Ex: In a flat universe theory, "even if our part of it eventually collapses,...some spots in the cosmos would suddenly start inflating on their own, creating brand-new 'baby universes.'" (P. Baofu 2000: 623)

• **"Parallel Universes" (Ex: Stephen Hawking and others)**
—Ex: In quantum cosmology, there allows the existence of infinite numbers of parallel universes, with tunneling among them. (M. Kaku 1994: 256) Hawking later revised his views on this.

• **"Pocket Universes" (Ex: Alan Guth)**
—Ex: "As the pocket universes live out their lives and recollapse or dwindle away, new universes are generated to take their place....While life in our pocket universe will presumably die out, life in the universe as a whole will thrive for eternity." (A. Guth 1997: 248; P. Baofu 2002: 482)

• **"Brane Worlds" (Ex: Warren Siegel, Lisa Randall, and others)**
—Ex: Our universe is stuck on a membrane of space-time embedded in a larger cosmos, with different brane worlds connecting and/or colliding with each other.

Notes: These examples are solely illustrative (not exhaustive), and some of the items can be reclassified somewhere else. Nor are they always mutually exclusive. Since they are generalities, exceptions are expected.
Source: From *Table 4.8* of *FPHC*

Table 1.8. The Confusion between "Many Worlds" and "Multiverse" (Part I)

• **First Problem**

—The first problem in this second confusion concerns a debate between the "many-worlds" interpretation of quantum mechanics by Hugh Everett (1957) and others, and the Copenhagen interpretation by Niels Bohr and Werner Heisenberg on the effects of quantum states after interacting with an external environment (e.g., by a measuring observer).

—On the one hand, the Copenhagen interpretation of quantum mechanics is more indeterministic in its concern with the probabilities of the outcomes of a wave function after interacting with an external environment (e.g., by a measuring observer using light to detect, for example, "an electron in a particular region around the nucleus at a particular time." (WK 2008e) Each wave function for the state of a particle refers to "a mathematical representation used to calculate the probability for it to be found in a location, or state of motion." (WK 2008f) And the "wavefunction collapse" means that "the act of measurement causes the calculated set of probabilities to 'collapse' to the value defined by the measurement." (WK 2008f)

—On the other hand, the Everett's interrpetation is more deterministic, in rejecting "the objective reality of wavefunction collapse" and "instead explaining the subjective appearance of wavefunction collapse with the mechanism of quantum decoherence." (WK 2008g) The words "quantum decoherence" here refers to "the mechanism by which quantum systems interact with their environments to exhibit probabilistically additive behavior," in such a way that "the quantum nature of the system is simply 'leaked' into the environment." (WK 2008g) For example, "[a]s a result of an interaction, the wave functions of the system and the measuring device become entangled with each other. Decoherence happens when different portions of the system's wavefunction become entangled *in different ways* [italic added] with the measuring device...." (WK 2008g) Thus, for Everett, the appearance of "wavefunction collapse" can be explained by "quantum decoherence" for quantum states to branch out in different independent worlds after interacting with an external environment.

(continued on next page)

Table 1.8. The Confusion between "Many Worlds" and "Multiverse" (Part II)

- **First Problem *(cont'd)***
  —Yet, a major criticism against Everett's interpretation is that these different independent worlds in Everett's sense "will never be accessible to us," because of the interference by the measuring observer in an external environment. (WK 2008g)
  —For illustration, "once a measurement is done, the measured system becomes entangled with both the physicist who measured it and a huge number of other particles, some of which are photons flying away towards the other end of the universe; in order to prove that the wave function did not collapse one would have to bring all these particles back and measure them again, together with the system that was measured originally. This is completely impractical, but even if one can theoretically do this, it would destroy any evidence that the original measurement took place (including the physicist's memory)." (WK 2008g)
  —But then, the critics ask, If these independent worlds in Everett's sense are inaccessible to us, how do we know much of anything really about them?

- **Second Problem**
  —What is relevant at the quantum mechanical level may not necessarily apply to the world at the physicsl level.
  —In fact, the persistent inability among contemporoary physicists to unifiy the theory of relativity for large objects at the physical level and quantum mechanics for tiny objects at the quantum mechanical level should warn us about the danger of reductionism, by reducing the understanding of different universes ("parallel universes") at the physical level from a generalization of "many worlds" at the quantum mechanical level.

- **Third Problem**
  —There is no empirical evidence to suggest that there are infinite (or extremely many) possible worlds in the cosmos as the many-worlds model by Everett would have us to believe.

Source: From *Sec.4.4.1* of *FPHG*

Table 1.9. Hyperspace and Its Challenge (Part I)

- **First Problem**
  —Hyperspace for intra- and inter-universal travel is assumed to be possible, on the basis of four main assumptions; although there may be others, these are the often cited ones.
  —The first assumption concerns the existence of "entry" and "exit" points in hyperspace, which is easier said than done to enter, go through, and then exit hyperspace.
  —For instance, using hyperspace "requires complex calculations…of single or multiple hyperspace jumps and the control of the jump….Peturbations such as those experienced by ship in space from the gravitional field around an object such as a planet or even a star are exacerbated in hyperspatial travel, since mass in real space distorts hyperspace in an equal measure. 'Jumping' near to a gravitational mass is likely to make resulting exit from hyperspace to be highly uncertain, with the level of improbability *i* increasing with the square of the distance to the nearest gravitional 'well.'" (WK 2008h)
  —Besides, "[o]nly spaceships equipped with a special force field can enter hyperspace, because exposure to hyperspace even for short period of time is hazardous to unprotected humans." (WK 2008h)

- **Second Problem**
  —The second assumption concerns the faster-than-light speed. It is well known that Einstein argued for the speed of light as the ultimate limit, but nowadays phyiscists do allow the violation of this limit under certain conditions (like the hypothetical existence of tachyons, as will be explained further in a later sub-section on time travel).
  —Even then, there is a challenge to explain in science fiction "why ships can travel faster than light in hyperspace," and two good arguments are that "hyperspace may be smaller than real space and therefore a star ship's propulsion seems to be greatly multiplied, or else the speed of light in hyperspace is not a barrier as it is in real space. Whatever the reasoning, the general effect is that ships traveling in hyperspace seem to have broken the speed of light, appearing at their destinations much more quickly and without the shift in time that the Theory of Relativity would suggest." (WK 2008h)

(continued on next page)

Table 1.9. Hyperspace and Its Challenge (Part II)

• **Third Problem**

—The third assumption concerns the speed of time (to be measured, say, by a clock), but this is elaborated in a different section on the related issue of "time travel" (as shown in a different table on time travel).

• **Fourth Problem**

—The fourth assumption concerns spatial shortcut, as it is possible, because of the curvature of space, in that "the manifold of ordinary three-dimensional space is curved in four or more 'higher' spacial dimensions (a 'hyperspace' in the geometric sense…). This curvature causes certain widely separated points in three-dimensional space to nonetheless be 'adjacent' to each other four-dimensionally. Creating an aperture in 4D space (a wormhole) between these locations can allow instantaneous transit between the two locations; a common comparison is that of a folded piece of paper, where a hole punched through two folded sections is more direct than a line drawn between them on the sheet. This idea probably arose out of certain popular descriptions of General Relativity and/or Riemannian manifolds, and may be the original form from which later concepts of hyperspace arose." (WK 2008h)

—But this does not mean that hyperpsace is always safe. In fact, "in some science fiction, the danger of hyperspace travel is due to the chance that the route through hyperspace may take a ship too close to a celestial body with a large gravitational field, such as a star. In such scenarios, if a starship passes too close to a large gravitational field while in hyperspace, the ship is forcibly pulled out of hyperspace and reverts to normal space. Therefore, certain hyperspace 'routes' may be mapped out that are safe, not passing too close to stars or other dangers." (WJ 2008h)

Source: From *Sec.4.4.1* of *FPHG*

Table 1.10. The Problems of Time Travel into the Future (Part I)

- **First Problem**

—The first problem concerns "time dilation under the Theory of Special Relativity" (which separates space-time from mass-energy as special cases). Suppose you are on a moving train and an observer outside the train is watching you along the tracks. As your train moves along at the speed of light (or something close to it), "time, as measured by your watch, ticked along at a slower pace than time measured by the observer. Not only that, distance changed, too. For the observer, a one-foot ruler whizzing by on the train would have measured less than a foot." (NO 2000)

—In fact, you may not even notice that this slower clicking of the clock; in fact, everything looks normal to you inside the train as it was before: "The weird thing is that, for you on the train, time wouldn't seem to be moving slower and your ruler wouldn't be shorter—all would appear normal." (NO 2008) By contrast, you instead think that "time on the rest of the Earth would appear to be ticking along slower and its rulers would be shorter." (NO 2000) So, both of you insist that the other clock (not his) is clicking more slowly—assuming, of course, an inertial frame of reference.

—Now, change the story a bit, in that, this time, instead of an "inertial reference frame," acceleration is allowed, and you are on your way to board a spaceship. As an illustration, suppose "[y]ou board a spaceship and take off for deep space. The ship approaches the speed of light. Time for you seems to pass as it always has. It takes you about five seconds to tie your shoe. But to an observer on Earth (assuming he or she could watch you), you are moving at a snail's pace. It takes hours to tie your shoe." (NO 2000) In any event, "you continue on your journey. You slow down, stop, and accelerate back to Earth. You arrive home. You have aged two years during your flight. Two hundred years have passed on Earth. You have successfully travelled forward through time." (NO 2000) In this case, both the observer (if he is still alive) and you agree that you have aged less.

(continued on next page)

Table 1.10. The Problems of Time Travel into the Future (Part II)

- **First Problem** ***(cont'd)***
  —In other words, when you are back to Earth, a lot of things have changed since, and you may find rather strange to be in a new environment which you hardly recognize, when compared with what you used to think as your city, or the Earth (two years ago, well, in accordance to your time of measurement). This means that you are in the future (two hundred years later), even though you are still relatively young (only two years older). But, can you go back? Well, no. The reason is that "[a]ccording to relativity, you can only move through time in one direction." (NO 2000)

- **Second Problem**
  —The second problem concerns "time dilation under the Theory of General Relativity" (which unifies space-time with mass-energy in a general theory), the same result of time dilation occurs.
  —For instance, when "one clock is deeper in a gravity well than the other," a result is that "the clock deeper in the well" will "tick…more slowly; this effect must be taken into account when calibrating the clocks on the satellites of the Global Positioning System, and it could lead to significant differences in rates of aging for observers at different distances from a black hole." (WK 2008i)
  —More precisely, by way of another example, "it has been calculated that, under general relativity, a person could travel forward in time at a rate four times that of distant observers by residing inside a spherical shell with a diameter of 5 meters and the mass of Jupiter. For such a person, every one second of their 'personal' time would correspond to four seconds for distant observers. Of course, squeezing the mass of a large planet into such a structure is not expected to be within our technological capabilities in the near future." (WK 2008i; R. Gott 2002)
  —The point here, however, is that time travel into the future, in this sense, is allowed, theoretically speaking, but in a very limited way, from the theory of relativity. But one can also ask, What if the theory of relativity were wrong? Well, in that case, there would be a need to find a better theory to explain time travel into the future.

Source: From *Sec.4.4.2* of *FPHG*

Table 1.11. The Problems of Time Travel into the Past (Part I)

- **First Problem**
  - —The first problem concerns the faster-than-light-speed, since, in accordance to the theory of special relativity, when an object goes faster than the speed of light, something weird can happen, in that, instead of having the clock clicking more slowly, it is moving backward, in relation to some inertial frame of reference—which then raisies the issue of whether time travel into the past is possible at all.
  - —But the problem here is about the violation of the law of casuality. In everyday language, an excellent illustration of causality violation is that, suppose you travel back into the past and then kill your parents, this then means that you could not have been born. But this is not true, since you are alive.
  - —Yet, "in the case of a hypothetical signal moving faster than light, there would always be some frames in which the signal was received before it was sent, so that the signal could be said to have moved backwards in time." The violation of casuality exists in this case.
  - —The first part of the problem is that, in accordance to the theory of special relativity, "it would take an infinite amount of energy to accelerate a slower-than-light object to the speed of light"—let alone the energy needed to propel an object to go faster than the speed of light. (WK 2008i) Although some suggest the possibility of "negative energy," it remains to be seen to what extent this is true.
  - —And the second part of the problem is that, "although relativity does not forbid the theoretical possibility of tachyons which move faster than light at all times, when analyzed using quantum field theory it seems that it would not actually be possible to use them to transmit information faster than light and there is no evidence for their existence." (WK 2008i; S. Chase 1993)

(continued on next page)

Table 1.11. The Problems of Time Travel into the Past (Part II)

- **Second Problem**
  —The second problem concerns cosmic string and black holes, since it is the theory of general relativity which "extends the special theory to cover gravity, illustrating it in terms of curvature in spacetime caused by mass-energy and the flow of momentum." (WK 2008i)
  —On the other hand, "[g]eneral relativity describes the universe under a system of field equations…and there exist solutions to these equations that permit what are called 'closed time-like curves,' and hence time travel into the past….The first of these was proposed by Kurt Gödel, a solution known as the Gödel metric, but his (and many others') example requires the universe to have physical characteristics that it does not appear to have." (WK 2008i; K. Thorne 1994)
  —But then, the problem now becomes: "Whether general relativity forbids closed time-like curves for all realistic conditions is unknown." (WK 2008i)

- **Third Problem**
  —The third problem concerns wormholes and related ones, and "[w]ormholes are a hypothetical warped spacetime which are also permitted by the Einstein field equations of general relativity, although it would be impossible to travel through a wormhole unless it was what is known as a traversable wormhole." (WK 2008i; M. Visser 1995) In other words, "a wormhole is a hypothetical topological feature of spacetime that is fundamentally a 'shortcut' through space and time. Spacetime can be viewed as a 2D surface, and when 'folded' over, a wormhole bridge can be formed. A wormhole has at least two mouths which are connected to a single throat or tube. If the wormhole is traversable, matter can 'travel' from one mouth to the other by passing through the throat." (WK 2008j)
  —But the problem here is three-fold, as shown in what follows.

(continued on next page)

Table 1.11. The Problems of Time Travel into the Past (Part III)

- **Third Problem** ***(cont'd)***
  —The first part of the problem is that "there is no observational evidence for wormholes." (WK 2008j)
  —The second part of the problem is that it is not clear how exactly wormholes (and black holes, for that matter) could allow time travel into the past at all.
  —And the third part of the problem is that if using wormholes (and black holes, for that matter) to travel into the past requires something to go faster than light too, then this raises another issue concerning the two problems as previously indicated in the sub-section on the faster-than-light-speed.

- **Fourth Problem**
  —The fourth problem is that some physicists have argued that, even should time travel into the past be possible, by way of these three methods, the person still would not return to the same past history that he exactly experienced before, but in a different one. (WK 2008i)
  —In other words, "strictly speaking," time travel into the past is not "really" possible, in light of these criticisms, so there is no need to worry that one could travel into the past and kill one's parent, as a way to violate the law of casuality.

Source: From *Sec.4.4.2* of *FPHG*

Table 1.12. The Conceptual Dimensions of Consciousness (and Other Mental States)

- **On Heredity and Time**
  - —Heredity and the Environment
  - —The Past and the Present
- **On Layers of Mental States and Abnormality**
  - —Consciousness, Unconsciousness, and Preconsciousness
  - —Normality and Paranormality
- **On Organicity and Motivation**
  - —Mechanicity and Organicity
  - —Primary Motivations and Multiple Motivations
- **On Other (Mostly Epistemic) Considerations**
  - —Synthesis and Analysis
  - —Situation and the Subject
  - —Process and Outcome
  - —Reasoning and Other Modes of Thinking
  - —Meta-Conceptual Nominalism and Realism

Notes:: These examples are solely illustrative (not exhaustive), and some of the items can be reclassified somewhere else. Nor are they always mutually exclusive. Since they are generalities, exceptions are expected.
Sources: From *FPHC*. A re-construction, but with my own contribution, originally from G.Lindzey & C.Hall, *Introduction to Theories of Personality* (NY: John Wiley & Sons, 1985).

Table 1.13. The Theoretical Levels of Consciousness (and Other Mental States) (Part I)

- **At the Micro-Physical Theoretical Level**
  - —Quantum-Mechanics
  - —Electromagnetism

- **At the Chemical Theoretical Level**
  - —Biochemistry

- **At the Biological Theoretical Level**
  - —Evolutionary Biology
  - —Neuroscience
  - —Artificial Intelligence

- **At the Psychological Theoretical Level**
  - —Psychodynamic Psychology
    - Psychoanalytical Psychology
    - Analytical Psychology
    - Socially Oriented Psychology
  - —Experimental (Behavioral) Psychology
    - Operant Reinforcement Theory
    - Stimulus-Response Theory
    - Social Learning Theory
  - —Cognitive (Gestalt) Psychology
    - Humanist Psychology
    - Existential Psychology
    - Field Theory
  - —Psychometric Psychology
    - Idiographics
    - Constitutional Psychology
  - —Social Psychology
    - Symbolic Interactive Theory
    - Social Exchange Theory

(continued on next page)

Table 1.13. The Theoretical Levels of Consciousness (and Other Mental States) (Part II)

- **At the Organizational Theoretical Level**
  - —Managerial-Bureaucratic Theory
  - —Oligarchic Theory
  - —Network Theory

- **At the Institutional Theoretical Level**
  - —Functionalist Theory
  - —Anomic Theory

- **At the Structural Theoretical Level**
  - —Conflict Theory
    - Marxian Theory
    - Critical Theory
    - Weberian Theory
  - —Games Theory (in Formal Theory)
  - —Feminist Theory
    - Feminist Concerns
    - Feminist Hopes

- **At the Systemic Theoretical Level**
  - —Equilibrium Theory
  - —System Theory
  - —Chaos Theory

- **At the Cosmological Theoretical Level**
  - —Superluminal Model
  - —The Theory of Floating Consciousness

(continued on next page)

Table 1.13. The Theoretical Levels of Consciousness (and Other Mental States) (Part III)

- **At the Cultural Theoretical Level**
  - —Substantive Theories
    - • Structuralist Theory
    - • Post-Structuralist Theory (in Postmodernism)
  - —Meta-Theories
    - • Epistemic Objectivism vs. Epistemic Historicism
    - • Epistemic Subjectivism vs. Epistemic Non-Subjectivism
      - –Phenomenology
      - –Ethnomethodology
      - –Hermeneutics
    - • Epistemic Relativism vs. Epistemic Absolutism
    - • Epistemic Reductionism vs. Epistemic Emergencism

- **At Other Levels**
  - —Structuration Theory
  - —Reflexive Socioanalysis

Notes: These examples are solely illustrative (not exhaustive), and some of the items can be reclassified somewhere else. Nor are they always mutually exclusive. Since they are generalities, exceptions are expected.
Source: From many different sources as indicated in *FPHC*

Table 1.14. The Thematic Issues of Consciousness (and Other Mental States)

- **The Factor of History**
  - —Pre-Modernity
  - —Modernity
  - —Postmodernity
  - —After-Postmodernity

- **The Importance of Needs**
  - —Having (e.g., the everyday, the technological)
  - —Belonging (e.g., the just and the good)
  - —Being (e.g., the true, the holy, the sublime/beautiful, and the good)

Notes: These examples are solely illustrative (not exhaustive). Nor are they always mutually exclusive. Since they are generalities, exceptions are expected. Also, some categories overlap with others; for instance, the good is also related with being-ness, just as the everyday can be connected with belong-ness.

Sources: From *FPHC*, based on *FHC* and *FCD*

Table 1.15. Having, Belonging, and Being in Consciousness

- **Having (e.g., About the Everyday and Technological)**
  —Ex: Physiological needs (e.g., thirst, hunger, sex)
  —Ex: Safety (e.g., stability, freedom from fear and anxiety)

- **Belonging (e.g., About the Good and Just)**
  —Ex: Social interaction
  —Ex: Friendship, acquaintance
  —Ex: Love, family
  —Ex: Self-respect, respect from others

- **Being (e.g., About the True, Holy, Beautiful/Sublime, and Good)**
  —Ex: Understanding of reality about self and world
  —Ex: Spiritual quest for holiness, beauty, sublimity, and goodness

Notes: These examples are solely illustrative (not exhaustive), and some of the items can be reclassified somewhere else. Nor are they always mutually exclusive. Since they are generalities, exceptions are expected.
Sources: From *FPHC*, based on *FHC* and *FCD*

Table 1.16. The Having-Ness of Consciousness (and Other Mental States)

- **At the Micro-Physical Level**
  —Ex: The finer physics of nutrition
- **At the Chemical Level**
  —Ex: Food guide pyramid
- **At the Bio-Psychological Level**
  —Ex: The hierarchy of needs
- **At the Institutional Level**
  —Ex: The production of commodity fetishism
- **At the Organizational Level**
  —Ex: Mass standardization (and customization); bureaucratic rationality
- **At the Structural Level**
  —Ex: The divide between the haves and the have-nots
- **At the Cultural Level**
  —Ex: The bondage of greater expectations
- **At the Systemic Level**
  —Ex: The technological transformation of poverty with transvaluations
- **At the Cosmological Level**
  —Ex: Floating consciousness in deep space without the human physiological needs

Notes: These examples are solely illustrative (not exhaustive), and some of the items can be reclassified somewhere else. Nor are they always mutually exclusive. Since they are generalities, exceptions are expected.
Sources: From *FPHC*. See also *FCD* and *FHC*.

Table 1.17. The Belonging-Ness of Consciousness (and Other Mental States)

- **At the Micro-Physical Level**
  —Ex: Statistical physics and social networks

- **At the Chemical Level**
  —Ex: Olfaction and social relations

- **At the Bio-Psychological Level**
  —Ex: Dual human nature on relationships

- **At the Institutional Level**
  —Ex: Private property and social cleavage

- **At the Organizational Level**
  —Ex: Legal formalism and iron cage

- **At the Structural Level**
  —Ex: Class conflicts and social alienation

- **At the Cultural Level**
  —Ex: Ruthless society and the culture of law

- **At the Systemic Level**
  —Ex: Human extinction and post-human relations

- **At the Cosmological Level**
  —Ex: The transcendence of equality in the cosmos and beyond

Notes: These examples are solely illustrative (not exhaustive), and some of the items can be reclassified somewhere else. Nor are they always mutually exclusive. Since they are generalities, exceptions are expected.
Sources: From *FPHC*. See also *FCD* and *FHC*.

Table 1.18. The Being-Ness of Consciousness (and Other Mental States) (Part I)

- **At the Micro-Physical Level**
  —Ex: Space-time and matter-energy in classical mechanics, quantum mechanics, and the theory of relativity

- **At the Chemical Level**
  —Ex: Drugs and well being

- **At the Biological Level**
  —Ex: Neuroscience and mystical experiences

- **At the Psychological Level**
  —Ex: Reductionism and emergencism

- **At the Institutional Level**
  —Ex: The proliferation of cults and sects, and the corporate-art industry

- **At the Organizational Level**
  —Ex: The organization of ideas in the scheme of things

- **At the Structural Level**
  —Ex: Disciplinary power and control in the order of things

- **At the Systemic Level**
  —Ex: Mutualism and ecological psychology

- **At the Cultural Level**
  —Ex: The recurrence of competing spirits, and floating consciousness as a climax of evolution

(continued on next page)

Table 1.18. The Being-Ness of Consciousness (and Other Mental States) (Part II)

- **At the Cosmological Level**
  —Ex: The search for elsewherewhen, and the evolution of other consciousnesses (e.g., the hyper-spatial consciousness)

Notes: These examples are solely illustrative (not exhaustive), and some of the items can be reclassified somewhere else. Nor are they always mutually exclusive. Since they are generalities, exceptions are expected.
Sources: From *FPHC*. See also *FCD* and *FHC*.

Table 1.19. Cognitive Partiality in Different Mental States

- **The Conscious**
  - —*Biased*
    - Ex: Self-overratingness
    - Ex: Self-modesty
    - Ex: Self-fulfilling prophesy
  - —*Shallow*
    - Ex: Attribution error
    - Ex: Heuristics
    - Ex: Illusory thinking
  - —*Short-term*
    - Ex: Deindividuation
  - —*Materialistic*
    - Ex: Hierarchy of needs

- **The Subconscious**
    - Ex: mental reconstruction
    - Ex: stereotyping

- **The Unconscious**
    - Ex: automatic processing
    - Ex: cognitive intuition

Notes: These examples are solely illustrative (not exhaustive), and some of the items can be reclassified somewhere else. Nor are they always mutually exclusive. Since they are generalities, exceptions are expected. And the comparison is relative, not absolute.
Source: From *BCPC*

Table 1.20. Emotional Non-Neutrality and Behavioral Alteration in Different Mental States

- **The Emotional**
  - —*Conscious*
    - Ex: Feelings
  - —*Subconscious*
    - Ex: Moods
  - —*Unconscious*
    - Ex: Empathy
    - Ex: Instincts

- **The Behavioral**
  - —*Conscious*
    - Ex: Foot-in-the-door
    - Ex: Role playing
    - Ex: Saying-is-believing
  - —*Subconscious*
    - Ex: Discrimination
  - —*Unconscious*
    - Ex: Non-Verbal Communication

Notes: These examples are solely illustrative (not exhaustive), and some of the items can be reclassified somewhere else. Nor are they always mutually exclusive. Since they are generalities, exceptions are expected. And the comparison is relative, not absolute.
Source: From *BCPC*

Table 1.21. The Limits of Intuition in Unconsciousness

- **Cognitive Intuition (or Intuitive Thinking)**
  - —*Features*
    - Ex: mental schemas
    - Ex: expertise
    - Ex: blindsight
  - —*Problems*
    - Ex: inaccurate
    - Ex: non-explanatory

- **Emotional Intuition**
  - —*Features*
    - Ex: empathy
    - Ex: instincts
  - —*Problems*
    - Ex: over-generalized
    - Ex: non-explanatory

- **Behavioral Intuition**
  - —*Features*
    - Ex: prosemics
    - Ex: kinesics
  - —*Problems*
    - Ex: unreliable
    - Ex: non-explanatory

Notes: These examples are solely illustrative (not exhaustive), and some of the items can be reclassified somewhere else. Nor are they always mutually exclusive. Since they are generalities, exceptions are expected. And the comparison is relative, not absolute.
Source: From *BCPC*

Table 1.22. The Wealth/Poverty Dialectics in Different Mental States: The Case of Cognition

- **The Conscious**
  - —*Biased*
    - Ex: If richer in self-serving bias, then poorer in understanding others
  - —*Shallow*
    - Ex: If richer in heuristic thinking, then poorer in comprehending reality
  - —*Short-term*
    - Ex: If richer in short term tendency, then poorer in long-term planning
  - —*Materialistic*
    - Ex: If richer in physiological obsession, then poorer in spiritual enlightenment

- **The Subconscious**
  - Ex: If richer in mental reconstruction, then poorer in accuracy of recall
  - Ex: If richer in stereotyping, then poorer in understanding groups

- **The Unconscious**
  - Ex: If richer in automatic processing, then poorer in depth of analysis
  - Ex: If richer in cognitive intuition, then poorer in reliability of judgments

Notes: These examples are solely illustrative (not exhaustive), and some of the items can be reclassified somewhere else. Nor are they always mutually exclusive. Since they are generalities, exceptions are expected. And the comparison is relative, not absolute.
Source: From *BCPC*

Table 1.23. The Wealth/Poverty Dialectics in Different Mental States: The Case of Emotion and Behavior

- **Emotion**
  - —*Conscious*
    - Ex: If richer in relying on familiarity and similarity for social attraction, then poorer in having diverse social relations
  - —*Subconscious*
    - Ex: If richer in moodiness, then poorer in sober thinking
  - —*Unconscious*
    - Ex: If richer in empathy, then poorer in task-oriented efficiency
    - Ex: If richer in instincts, then poorer in critical thinking

- **Behavior**
  - —*Conscious*
    - Ex: If richer in foot-in-the-door technique, then poorer in instant Gratification
    - Ex: If richer in role playing, then poorer in identity stability
  - —*Subconscious*
    - Ex: If richer in discriminating, then poorer in social harmony
  - —*Unconscious*
    - Ex: If richer in kinesics and proxemics, then poorer in verbal skills, relatively speaking.

Notes: These examples are solely illustrative (not exhaustive), and some of the items can be reclassified somewhere else. Nor are they always mutually exclusive. Since they are generalities, exceptions are expected. And the comparison is relative, not absolute.

Source: From *BCPC*

Table 1.24. The Theoretical Debate on Nature vs. Nurture (Part I)

- **The Environmental Approach**
  —*Thesis*: It focuses, relatively speaking, more on the environment (culture and society) in explaining the achievement gap among individuals and for that matter, countries or regions, when contrasted with the natural factors.
  —*Discourse*: Examples include Jose Ortega y Gasset ("Man has no nature; what he has is history"), Ashley Montagu ("Man is man because he has no instincts, because everything he is and has become he has learned from his culture, from the man-made part of the environment, from other human beings"), Stephen Jay Gould ("[The] brain [is] capable of a full range of behaviors and predisposed to none"), and Jesse Jackson (who blames white racism for the failure of blacks to close the achievement gap between whites and blacks in America). The works on dependency theory in international political economy (with a Marxian influence) and on the Protestant work ethic (by Max Weber) also point to this environmental direction.

- **The Genetic Approach**
  —*Thesis*: It focuses instead, relatively speaking again, on hereditory factors (e.g., genes and evolution) in explaining the achievement gap among individuals and for that matter, countries or regions, when contrasted with the envrionmental factors.
  —*Discourse*: Examples are Hans Eysenck and William Sheldon (in constitutional psychology), Konrad Lorenz (in his work on innate aggressive human nature), Gary Marcus (on the complexities of human thought by a tiny number of genes) and Robert Plomin, together with Michael Owen and Peter McGuffin (on the genetic basis of complex human behaviors).

(continued on next page)

Table 1.24. The Theoretical Debate on Nature vs. Nurture (Part II)

- **The Compromise Approach**
  —*Thesis*: It seeks the middle-of-the-road argument in regard to nature and nurture and regards all differences among individuals and groups as the result of the mixture of both nature and nurture, more or less equally.
  —*Discourse*: Examples include C. Murray and R. Herrnstein ("It seems highly likely to us that both genes and environment have something to do with this issue") and Dan Dennett ("Surely 'everyone knows' that the nature-nurture debate was resolved long ago, and neither side wins since everything – is – a – mixture – of – both – and - it's – all – very -complicated…."

- **The Transcendent Approach**
  —*Thesis*: It goes beyond both nature and nurture (without, however, committing the compromise fallacy) in showing their closely intertwined interactions in producing the behavioral differences as often seen in individual human endeavors on the micro scale, and for that matter, in country (or regional) endeavors on the macro one—in the context of my five theses, namely, (a) the compromise fallacy, (b) no oppression without self-oppression, (c) no success without failure, (d) the factor of randomness, and (e) the post-human vision, to be elaborated in Chapter Six.
  —*Discourse*: Peter Baofu proposed this approach on the basis of his "theory of contrastive advantages" (as an original theoretical contribution to the debate, which was first proposed and analyzed in *The Future of Capitalism and Democracy*). In the end, the human genes will not last, to be eventually superseded by post-human life forms, so the debate between genes and memes has obscured something profoundly important about the future that the world has never known. And the debate is also misleading and faulty in its dichotomy.

Source: From *BNN*. See the book for citations and details.

Table 1.25. Physical Challenges to Hyper-Spatial Consciousness

---

- **The Understanding of a Higher-Dimensional World of Space-Time**
  —Ex: 4 for traditional aspects of space-time (e.g., length, width, breadth and time) plus 6 more new dimensions in theory of hyper-space, with profound implications for practical applications to new forms of consciousness.

- **The Mastering of Dark Matter and Dark Energy**
  —Ex: "ordinary matter" (e.g., atoms, molecules) as a mere 4.4% of the universe, with 23% made of "cold dark matter" and the rest (about 73%) of mysterious "dark energy," with fundamental significance to questions about the limit of the speed of energy (or info), the availability of energy for use, and the nature of space-time, just to cite some examples.

- **The Exploration of Multiverses**
  —Ex: theoretical speculation of other universes (e.g., "baby universes," "gateways" in black holes, "wave function of the universe," "many worlds," "brane worlds"), with potentially seminary discoveries of different physical laws in relation to matter-energy and space-time, and vital differences to the future of post-human conquest of other universes (for the emergence of new forms of consciousness).

Notes: These examples are solely illustrative (not exhaustive), and some of the items can be reclassified somewhere else. Nor are they always mutually exclusive. Since they are generalities, exceptions are expected. The point here is to give a rough picture of the evolution of consciousness to the hyper-spatial consciousness and others totally unknown to current earthlings. As a note of clarification, it makes no difference to my argument as to whether or not the hyper-spatial consciousness may emerge before, during, and after floating consciousness.
Source: From *Table 4.5* of *FPHC*

Table 1.26. The Theory of Floating Consciousness (Part I)

- **At the Micro-Physical Level**
  —Ex: intelligent life without the human physical-chemical system

- **At the Chemical Level**
  —Ex: space radiation and toxins

- **At the Bio-Psychological Level**
  —Ex: exo-biological evolution in deep space
  —Ex: genetic engineering of new beings

- **At the Institutional Level**
  —Ex: post-capitalism
  —Ex: post-democracy

- **At the Organizational Level**
  —Ex: less legal-formalistic routines

- **At the Structural Level**
  —Ex: alien forms of violence

- **At the Cultural Level**
  —Ex: transcending freedom
  —Ex: transcending equality

(continued on next page)

Table 1.26. The Theory of Floating Consciousness (Part II)

- **At the Cosmological Level**
  —Ex: parallel universes
  —Ex: pocket universes

- **At the Systemic Level**
  —Ex: space habitats (in zero-gravity)

Notes: Each example draws from the works of different scholars in the field. For instance, at the cosmological level, the idea of parallel universes is from the theoretical speculation in quantum cosmology by Stephen Hawking and others, while the one of pocket universes comes from the theoretical work of Allan Guth at MIT. And at the institutional level, I proposed post-capitalism and post-democracy in *FCD* (and later, from *BDPD* and *BCPC*). In addition, the examples are solely illustrative (not exhaustive), and some of the items can be reclassified somewhere else. Nor are they always mutually exclusive. Since they are generalities, exceptions are expected.
Sources: From *FPHC*—and, originally, from *FCD*

Table 1.27. The Potential of Unfolding Unconsciousness

- **Superior Senses**
  - —*Through space*
    - Ex: clair-sentience (in feeling)
    - Ex: clair-voyance (in seeing)
    - Ex: clair-audience (in hearing)
    - Ex: clair-austance (in tasting)
    - Ex: clair-alience (in smelling)
    - Ex: clair-cognizance (in knowing)
  - —*Through time*
    - Ex: pre-cognition (in knowing), pre-sentiment (in feeling)
    - Ex: retro-cognition (in knowing), retro-sentiment (in feeling)

- **Superior Contacts**
  - —*With living minds*
    - Ex: telepathy
  - —*With the dead*
    - Ex: mediumship
  - —*With other-living entities*
    - Ex: telekinesis

Notes: The categories and examples are solely illustrative (not exhaustive), and the comparison is relative (not absolute), nor are they necessarily mutually exclusive. And some can be easily re-classified elsewhere. As generalities, they allow exceptions.
Source: A summary of Ch.5 of *FPHU*

Table 1.28. The Future Exploration of Unfolding Unconsciousness

- **In the Context of Nature**
  - —Ex: the higher-dimensional space-time in hyperspace theory
  - —Ex: the particle-wave interactions in quantum mechanics
  - —Ex: the study of brain waves in electromagnetic theory
- **In the Context of the Mind**
  - —Ex: the unconscious fantasies in psychoanalysis
  - —Ex: the mind-altering drugs in biochemistry
  - —Ex: the manipulation of neural activity in neurobiology
- **In the Context of Culture**
  - —Ex: the correlation between anomalous experience and moral/religious/aesthetic interests in culture studies
  - —Ex: the role of myths and language in epistemology
- **In the Context of Society**
  - —Ex: different organizational agendas in anomalous research
  - —Ex: the scholarly divide on anomalous study
  - —Ex: interests and propaganda in institutioanl research
  - —Ex: the transformative power of technology in research on social systems
  - —Ex: the natural and non-natural selection of other worlds in exo-biology
  - —Ex: the role of the post-human conscious mind in qualitative demography

Notes: The examples in each category are solely illustrative (not exhaustive), and the comparison is relative (not absolute), nor are they necessarily mutually exclusive. And some can be easily re-classified elsewhere. As generalities, they allow exceptions.
Source: A summary of Chs.2-5 of *FPHU*, in conjunction with the rest of the book

Table 1.29. Creative Techniques and Traits

- **Creative Techniques**
  - —*Comprehensiveness*
    - Ex: Broadness of Scope
    - Ex: In-Depth Analysis
    - Ex: Multifaceted Taxonomy
    - Ex: Numerous Theories
  - —*Arrangement*
    - Ex: Separation (or Subtraction)
    - Ex: Combination (or Addition)
    - Ex: Replacement
    - Ex: Reversal
    - Ex: Accommodation
    - Ex: Stretching
  - —*Serendipity*
    - Ex: Play
    - Ex: Chance

- **Creative Traits**
  - —*Openness / Risk*
    - Ex: Ability to welcome new ideas or to do new things, with risk
  - —*Discipline*
    - Ex: Ability to sacrifice whatever it takes to succeed
  - —*Resilience / Confidence*
    - Ex: Ability to take failures to mind, not to heart
  - —*Trust*
    - Ex: Ability to cooperate with others for team work, if necessary

Notes: The examples in the categories are solely illustrative (not exhaustive), and the comparison is relative (not absolute), nor are they necessarily mutually exclusive. And some can be easily re-classified elsewhere. As generalities, they allow exceptions.
Source: From *Sec.4.2* of *FPHCT*. For more details, also read the rest of the book.

Table 1.30. The Desirability of Creativity, and Its Dark Sides (Part I)

- **In Relation to Instrumental Rationality**
  —*Desirability*
  - Ex: For the sake of efficiency and effectiveness in this capitalist era of our time.

  —*Undesirability*
  - Ex: What Max Weber (1930) called the "iron cage" of capitalism, where live the "sensualists without heart, specialists without spirit"—or something which the Frankfurt School has forcefully asked us to be freed from.

- **In Relation to Substantive Rationality**
  —*Desirability*
  - Ex: For the sake of God, the King, Motherland, or other comparable ideals.

  —*Undesirability*
  - Ex: Domination and oppression of various forms—something which the moderns had struggled to free themselves from in the first place, since the dawn of modernity.

- **In Relation to Autonomous Rationality**
  —*Desirability*
  - Ex: For the sake of itself, the autonomy of creative endeavor.

  —*Undesirability*
  - Ex: The decadence of its own degeneration (e.g., the donwsides of postmodernism in our postmodern times, or what Nietzsche once called the unbearable "unreality" and "falseness" of the autonomous artist's "innermost existence"—and, in other cases, the painful suffering from different mental illnesses for those highly creative individuals).

(continued on next page)

Table 1.30. The Desirability of Creativity, and Its Dark Sides (Part II)

Notes: The examples in the categories are solely illustrative (not exhaustive), and the comparison is relative (not absolute), nor are they necessarily mutually exclusive. And some can be easily re-classified elsewhere. As generalities, they allow exceptions.
Also, I already went to great lenghts to explain the desirability and its dark sides of all these three rationalities (although there may be others too) in the two volumes of *FHC*—and, for that matter, *FCD*.
Source: From *Sec.4.9* of *FPHCT*. For more details, also read the rest of the book.

Table 1.31. Posthuman-Ism, Post-Humanism, and Trans-Humanism

- **Post-Humanism**
  —The neologism "post-human" used in my books should *not* be confused with another term which looks similar but has a totally different meaning in the literature of postmodernism, namely, "post-humanism"—which constitutes a critique of "humanism" as tradionally understood (especially, though not exclusively, in relation to the idea of progress in science and reason in the Enlightenment project). (WK 2008)
  —My works reject the project of "postmodernism" and propose the future world of what I orignally called "after-postmodernity" in *FHC* and *FCD*, for instance.

- **Trans-Humanism**
  —Also, the neologism "post-human" used in my books should *not* be confused with a similar term which is used to champion the ideology of technology for the fturue co-existence between humans and cyborgs in "trans-humanism." (WK 2008a)
  —Instead, my term "post-human" in relation to "posthuman-ism" also rejects "trans-humanism" (especially, though not exclusively, in relation to the promises of technology) and refers to something else altogether, that is, the future extinction of humans and its post-human successors in deep space and beyond unto multiverses.
  —Chapter Two of my book on ethics (*BEPE*) provides a more comprehensive critique of trans-humanism.

- **Posthuman-Ism**
  —With these two clarifications in mind, the "post-humans" (as envisioned in my books) can take different forms, and I proposed, in my different books, some of them, such as "unfolding unconsciousness," "floating consciousness," "hyper-spatial consciousness," "thinking machines," "thinking robots," "genetically altered superior beings," "cyborgs," and others (as already analyzed in my previous works).
  —They will evolve in conjunction with other visions of mine in relation to nature, society, and culture.

Sources: From *Sec.1.6* of *FPHE* (and also *BEPE*). For more details, also read the rest of the books (and other books of mine).

Table 1.32. Three Great Future Transformations of the Martial Body (Part I)

- **Virtual Battle**
  —Ex: virtual fight
  —Ex: virtual memory

- **The Hyper-Martial Body**
  —*In relation to the physical dimension*
  - Ex: striking (e.g., punching, kicking, trapping, acupressure-striking, and so on in an amazing way)
  - Ex: grapping (e.g., throwing, pinning, joint-locking, and so on in an amazing way)
  - Ex: running and walking (e.g., on different surfaces)
  - Ex: jumping (e.g., through walls, trees, etc.)
  - Ex: flying (e.g., through space)
  - Ex: enduring pain (e.g., absorbing punches, kicks, and so on in an amazing way)
  - Ex: using extremely advanced hi-tech weapons
  - Ex: healing (e.g., with special energy, etc.)
  - Ex: entering (e.g., into different dimensions of space)
  - Ex: sensing (e.g., in extraordinary ways through space and time, etc.)
  - Ex: wearing hi-tech suits for defensive and offensive purposes (e.g., like a hi-tech armor)

  —*In relation to the mental dimension*
  - Ex: the ability to relax the martial body in a way that current humans cannot.
  - Ex: the ability to concentrate the martial mind in a way that current humans are not accustomed to.
  - Ex: the ability to breathe and be timely in preparing the martial body for fighting in a way that current humans are not good at.

- **The Post-Martial Floating Mind**
  —Ex: the mind without the human body
  —Ex: the quest for spirituality beyond martial arts

(continued on next page)

Table 1.32. Three Great Future Transformations of the Martial Body (Part II)

Notes: The examples in the categories are solely illustrative (not exhaustive), and the comparison is relative (not absolute), nor are they necessarily mutually exclusive. And some can be easily re-classified elsewhere. As generalities, they allow exceptions.
Source: From *Sec. 2.5.1* of *FPHMA*. See text for citations.

Table 1.33. Three Great Future Transformations of the Sexual Body

- **Virtual Pleasure**
  —Ex: virtual Sex
  —Ex: virtual Recall

- **The Hyper-Sexual Body**
  —*In relation to the physical dimension*
  - Ex: the ability to perform differenet sexual positions without the various physical limitations that humans currently have.
  - Ex: the ability to produce unusual amount of sexual energy to endure in sexual acts in a way that humans currently cannot.
  - Ex: the ability to engage in multiple sexual experiences comparable to (but different from) non-human sexuality (e.g., hermaphrodite, monoecious, sex-switching, and many other types) that current humans cannot.

  —*In relation to the mental dimension*
  - Ex: the ability to relax the sexual body in a way that current humans are not accustomed to.
  - Ex: the ability to concentrate the sexual mind in a way that current humans are not good at.
  - Ex: the ability to breathe and be patient in preparing the sexual body for sexuality in a way that current humans are not able to.

- **The Post-Sexual Floating Mind**
  —Ex: the mind without the human body
  —Ex: the quest for spirituality beyond sexuality

Notes: The examples in the categories are solely illustrative (not exhaustive), and the comparison is relative (not absolute), nor are they necessarily mutually exclusive. And some can be easily re-classified elsewhere. As generalities, they allow exceptions.
Source: From *Sec. 3.4.2* of *FPHS*. See text for more info.

Table 1.34. The Trinity of Pre-Modernity

- **Pre-Free-Spirited Pre-Modernity (Pre-Modernism) and Its Internal Split**
  —Competing worldviews and values both within and between linear centric (e.g., Islamic, Christian, Judaic, Imperial Roman) and cyclical-centric (e.g., Confucian, Taoist, Hindu, and Buddhist) orientations
  —Compare modernism with pre-modernism here in relation to the seven dimensions of human existence like the true and the holy (e.g., different versions of epistemic dogmas and religious superstitions), the everyday and the technological (e.g.,different versions of non-technophilism and non-consumerism), the beautiful/sublime (e.g., different versions of aesthetic non-autonomy), and the good and the just (e.g., different versions of moral particularism).

- **Pre-Capitalist Pre-Modernity (Pre-Modernization) and Its Own Discontents**
  —Competing versions of societal arrangements (e.g., feudalism, monarchism, and the holy order)

- **Hegemonic Pre-Modernity and Its Countervailing Forces**
  —Different power centers and their enemies (e.g., the Roman Empire and the "barbarian hordes," the "Holy Crusades" and the Muslims, the Middle Kingdom and the invading tribes, different social castes in India, and warring Greek city-states)

Notes: The examples in each category are solely illustrative (not exhaustive) nor necessarily mutually exclusive, and the comparison is relative (not absolute). As generalities, they allow exceptions. Also, it does not matter what the "base" era is in the analysis of any trinity. And in the present context, the "base" era is modernity (for instance, with its "free-spirited modernity" and the other two parts). So, for pre-modernity, the trinity takes the form of, say, "pre-free-spirited pre-modernity," together with the other two parts.
Sources: From Ch.2 of *BCIV* and also the 2 volumes of *FHC*

Table 1.35. The Trinity of Modernity (Part I)

- **Free-Spirited Modernity (Modernism) and Its Internal Split**
  —*On the True and the Holy*
    - The freedom from the dogmas of the past to the better understanding of, and union with, the world and self (Ch.3 of *FHC*)
    - Alternative discourses: about the true (e.g., anti-science discourses) and the holy (non-mainstream theologies) (Ch.3 of *FHC*)

  —*On the Technological and the Everyday*
    - The freedom from life harshness to the higher technophilic, consumeristic lifeform (Ch.2 of *FHC*)
    - Alternative discourses: about the everyday (e.g., transcendental mindsets) and the technological (e.g., Arcadianism) (Ch.2 of *FHC*)

  —*On the Good and the Just*
    - The freedom from the theo-aristocratic tyranny to the moral universality for a just society (Ch.5 of *FHC*)
    - Alternative discourses: about the just (e.g., Communism, Anarchism) and the good (e.g., Nazism/Fascism, and Zarathustrianism) (Chs.5-6 of *FHC*)

  —*On the Beautiful and the Sublime*
    - The freedom from the external distortion of aesthetic pleasure to the boundless infinity of totality in artistic autonomy (Ch.4 of *FHC*)
    - Alternative discourses: about the beautiful/sublime (e.g., kitsch and historical avant-gardism) (Ch.4 of *FHC*)

- **Capitalist Modernity (Modernization) and Its Own Discontents**
  —*During the Industrial Revolution*
    - Ex: Marx on the institution of inequality (Ch.1 of *FHC*)

  —*During the Modern Rational-Instrumental Epoch*
    - Ex: Weber on the politics of soft liberal institutions (Ch.5 of *FHC*)

  —*During the Great Depression*
    - Ex: Keynes on the myth of the free market (Chs.1-3 of *FHC*)

  —*During the Cold War*
    - Ex: Lasch on the narcissistic culture industry (Chs.2-3 of *FHC*)

(continued on next page)

Table 1.35. The Trinity of Modernity (Part II)

- **Hegemonic Modernity and Its Countervailing Forces**
  —*The Legacies of Colonialism and Imperialism*
    - Ex: European colonization of most of the modern world (Ch.1 of *FHC*)

  —*The Struggle for Decolonialization*
    - Ex: The countervailing forces of resentment, rechantment, and regionalism (Chs.1 & 6 of *FHC*)

Notes: The examples in each category are solely illustrative (not exhaustive) nor necessarily mutually exclusive, and the comparison is relative (not absolute). As generalities, they allow exceptions.
Sources: From the 2 volumes of *FHC*—and also from *FCD*

Table 1.36. The Trinity of Post-Modernity

- **Free-Spirited Postmodernity (Postmodernism) and Its Internal Split**
  —*On the True and the Holy*
    - Postmodern performative turn for knowing and its enemies (Ch.3 of *FHC*)
    - Postmodern comparative theology and its opponents (Ch.3 of *FHC*)

  —*On the Technological and the Everyday*
    - Postmodern corporate technological mindset and its adversaries (Ch.2 of *FHC*)
    - Postmodern postmaterialism and its critics (Ch.2 of *FHC*)

  —*On the Good and the Just*
    - Postmodern politics of difference and its foes (Ch.5 of *FHC*)

  —*On the Beautiful and the Sublime*
    - Postmodern deconstruction and its dissenters (Ch.4 of *FHC*)

- **Capitalist Postmodernity (Postmodernization) and Its Own Discontents**
  —*During the Post-Cold War and Beyond*
    - Ex: post-Fordism and its shortcomings (Ch.6 of *FHC*; Chs.6-7 of *FCD*)

- **Hegemonic Postmodernity and Its Countervailing Forces**
  —*The Debate on the Global Village*
    - Ex: uni-civilizationalism vs. multi-civilizationalism (Ch.6 of *FHC*)

  —*The Resistance Movement*
    - Ex: rechantment and the politics of civilizational claims (e.g., Islamic, Confucian and other ethos in relation to the Same) (Ch.6 of *FHC*; Ch.10 of *FCD*)
    - Ex: resentment and the politics of resurgence (e.g., the rising Chinese superpower, the growing EU, and other players in relation to the U.S. and her allies) (Ch.6 of *FHC*; Ch.8 of *FCD*)
    - Ex: regionalism and the politics of inequality (e.g., trans- or inter-national blocs, the North-South divide, NGO's) (Ch.6 of *FHC*; Ch.5 of *FCD*)

Notes: The examples in each category are solely illustrative (not exhaustive) nor necessarily mutually exclusive, and the comparison is relative (not absolute). As generalities, they allow exceptions.
Sources: From *FCD* and the 2 volumes of *FHC*

Table 1.37. The Trinity of After-Postmodernity

- **Free-Spirited After-Postmodernity (After-Postmodernism) and Its Internal Split**
  —The discourse of naked contingency (Ch.10 of *FCD*; Ch.4 of *FPHC*)

- **Post-Capitalist After-Postmodernity (After-Postmodernization) and Its Own Discontents**
  —Different versions of post-capitalism and post-democracy, and their enemies (Ch.10 of *FCD*; Chs.3-4 of *FPHC*)

- **Hegemonic After-Postmodernity and Its Countervailing Forces**
  —The Cyclical Progression of Hegemony in Multiverses (Chs.9-10 of *FCD*; Ch.4 of *FPHC*)

Notes: The examples in each category are solely illustrative (not exhaustive) nor necessarily mutually exclusive, and the comparison is relative (not absolute). As generalities, they allow exceptions.
Sources: From *FCD* and also *FHC*

• PART TWO •

The Contested Beginnings

CHAPTER 2
THE UNIVERSE AND ITS CONTESTED BEGINNINGS

> An infinite temporal regress of events cannot exist.
> —The Doctrine of Creation
> (WK 2010b)

The Persuasion of Its Contested Beginnings

Whether or not, or to what extent, the universe has a beginning has long been one of the most profound but hotly contested questions in the history of cosmology. After all, different thinkers over the ages had given different answers to this very question (as summarized in *Table 1.1* and *Table 1.2*).

With this contested history in mind—this chapter thus takes the challenging task to examine the opposinog sides of this hotly contested debate about the possible beginnings of the universe.

A good point of departure to examine the opposing sides of this hotly contested debate is by way of the evaluation of the extent to which the contested beginnings of the universe are in fact both possible and desirabe.

With this task in mind, this chapter provides a comprehensive analysis of the contested beginnings of the universe (and their aftermath) from the four perspectives of the mind, nature, society, and culture (in accordance to my sophisticated methodological holism, as explained in *Sec. 1.9*).

In other words, this chapter is organized in four main sections, in relation to (2.2) its contested beginnings—and the mind, (2.3) its contested beginnings—and nature, (2.4) its contested beginnings—and society, and (2.5) its contested beginnings—and culture, to be addressed in what follows, in that order (and summarized in *Table 2.1*).

Its Contested Beginnings—and the Mind

The contested beginnings of the universe can be analyzed from the perspective of the mind, as it can show us something illuminating about the universe in a distinctive way.

Two case studies are provided hereafter, namely, (2.2.1) acosmism, and the illusion in the creation of the universe, and (2.2.2) idealism, the mind, and the problem of universals—to be analyzed in what follows, in that order.

Acosmism, and the Illusion in the Creation of the Universe

A good way to begin the evaluation of the possibility and desirability of the contested beginnings of the universe from the perspective of the mind is to introduce the idea of "acosmism."

The term "acosmism" refers to the denial of "the reality of the universe, seeing it as ultimately illusory" (with "the prefix '*a*-' in Greek meaning negation; like '*un*-' in English") and accepts "only the infinite unmanifest Absolute as real." (WK 2010v)

In acosmism, although the universe is created, its reality is not quite true, because it does not contitute "the infinite unmanifest Absolute," which is the only ultimately "real" entity. (WK 2010v)

Thus, acosmism "begins with the premise that there is only one Reality that is infinite, non-dual, and blissful. But the phenomenal reality of which we are normally aware is just the opposite: finite, dualistic, and full of pain and suffering. And since the Absolute is the only reality, that means that everything that is not-Absolute cannot be real. Thus, according to this viewpoint, the phenomenal dualistic world is ultimately an illusion (maya to use the technical Indian term), irrespective of the apparent reality it possesses at the mundane or empirical level." (WK 2010v)

A good illustration of acosmism in more detail concerns a school of thought known as "Advaita Vedanta" (or "अद्वैत वेदान्त" in Sanskrit), which is "a sub-school of the Vedānta...school of Hindu philosophy." (WK 2010u; E. Deutsch 1980)

The first word "Advaita" means "non-duality" in the context of "a monistic system of thought," which advocates "the identity of the Self (Atman) and the Whole (Brahman)." (WK 2010v) And the second word 'Vedanta" literally means "the end or the goal of the Vedas" in Sanskrit. Therefore, "Advaita Vedanta" refers to the end of the goal of the Vedas for "non-duality" in "the identity of the Self (Atman) and the Whole (Brahman)."

Advaita is one of the three major Vedānta school of Hindu philosophy; the other two sub-schools of the Vedanta are "Dvaita" and "Viśishṭādvaita."

Yet, all three sub-schools of Vedānta share the key source texts known as "the Prasthanatrayi—the canonical texts consisting of [a] the Upanishads, [b] the Bhagavad Gita and [c] the Brahma Sutras. The first person to explicitly consolidate the principles of Advaita Vedanta was Adi Shankara." (WK 2010u; D. Collinson 1994)

The Advaita Vedanta Theory of Creation

For Adi Shankara, who is the first person "to explicitly consolidate the principles of Advaita Vedanta"—the universe was created in a subtle way, in that the effect (i.e., the universe) pre-existed in the cause (i.e., "the infinite unmanifest Absolute"). (WK 2010u)

This Hindu view of creation is also related to the idea of "emanation" in esoteric cosmology (which, however, will be postponed for analysis until *Sec. 3.2.1* on involution and the cosmic return to the Absolute).

With this clarification in mind—the Advaita Vedanta theory of creation thus treats the universe as an effect (of creation), which "is merely an apparent transformation of its cause—like illusion"; for instance, "[i]n darkness, a man often confuses a rope to be a snake. But this does not mean that the rope has actually transformed into a snake." (WK 2010u)

In Advaita, the cause and the effect are similar with each other and yet different from each other.

On the one hand, the cause and the effect are similar in relation to each other: "If the cause is destroyed, the effect will no longer exist. For example, if from the effect, cotton cloth, the cause, threads, are removed, there will be no cloth....Similarly if in the effect, thread, the cause, cotton, is removed, there will be no thread." (WK 2010u; S. Sivananda 2010)

On the other hand, the cause and effect are different from each other: "Despite the non-difference of cause and effect, the effect has its self in the cause but not the cause in the effect. The effect is of the nature of the cause and not the cause the nature of the effect. Therefore the qualities of the effect cannot touch the cause." (WK 2010u; S. Sivananda 2010)

In the context of Advaita, this difference between the cause and the effect means that the universe (the effect) "is merely an apparent transformation of its cause" (i.e., "the infinite unmanifest Absolute") and thus is not real (true) in the ultimate sense.

Three Levels of Truth (or Reality)

To understand this distinction between reality and illusion requires an explanation of three levels of truth (or reality) in Advaita Vedanta, as shown below (and summarized in *Table 2.2*).

- "The Transcendental or the Pāramārthika Level"—this is the level of truth (or reality) where "Brahman is the only reality and nothing else." (WK 2010u)
- "The Pragmatic or the Vyāvahārika Level"—this is the level of truth (or reality) where "both Jiva (living creatures or individual souls) and Iswara are true; here, the material world is also true." (WK 2010u)
- "The Apparent or the Prāthibhāsika Level"—this is the level of truth (or reality) where "material world reality is actually false, like illusion of a snake over a rope or a dream." (WK 2010u)

These three levels are to be understood in relation to each other.

(a) At the Transcendental Level

At the transcendental level—"God, the Supreme Cosmic Spirit, or Brahman…is the One, the whole and the only reality. Other than Brahman, everything else, including the universe, material objects and individuals, are false. Brahman is at best described as that infinite, omnipresent, omnipotent, incorporeal, impersonal, transcendent reality that is the divine ground of all Being." (WK 2010u)

However, "Brahman cannot be correctly described as this or that. 'It' (grammatically neutral, but exceptionally treated as masculine) is the origin of this and that, the origin of forces, substances, all of existence, the undefined, the basis of all, unborn, the essential truth, unchanging, eternal, the absolute. How can it be properly described as something in the material world when itself is the basis of reality? Brahman is also beyond the senses, it would be akin a blind man trying to correctly describe color. It, though not necessarily a form of physical matter, is the substrate of the material world, which in turn is its illusory transformation. Brahman is not the effect of the world. Brahman is said to be the purest knowledge itself, and is illuminant like a source of infinite light." (WK 2010u)

(b) At the Pragmatic Level

At the pragmatic level—"when man tries to know the attributeless Brahman with his mind, under the influence of Maya, Brahman becomes the Lord [or Ishvara]. Ishvara is Brahman with Maya—the manifested form of Brahman. Adi Shankara uses a metaphor that when the 'reflection' of the Cosmic Spirit falls upon the mirror of Maya, it appears as the

Ishvara or Supreme Lord. The Ishvara is true only in the pragmatic level. God's actual form in the transcendental level is the Cosmic Spirit." (WK 2010u)

But did Ishvara create the world in the first place, as "the manifested form of Brahman"? Here is the answer in Advaita Vedanta: "If one assumes that Ishvara creates the world for any incentive, this slanders the wholeness and perfection of Ishvara. For example, if one assumes that Ishvara creates the world for gaining something, it would be against His perfection. If we assume that He creates for compassion, it would be illogical, because the emotion of compassion cannot arise in a blank and void world in the beginning (when only Ishvara existed). So Adi Shankara assumes that Creation is recreation or play of Ishvara. It is His nature, just as it is man's nature to breathe." (WK 2010u)

(c) At the Apparent Level

At the apparent level—"Māyā is the complex illusionary power of Brahman which causes the Brahman to be seen as the material world of separate forms. Maya has two main functions—one is to 'hide' Brahman from ordinary human perception, and the other is to present the material world in its (Brahmam) place." (WK 2010u)

In addition, "Māyā is…said to be indescribable, though it may be said that all sense data entering ones awareness via the five senses are Māyā, since the fundamental reality underlying sensory perception is completely hidden….Its shelter is Brahman, but Brahman itself is untouched by the illusion of Māyā, just like a magician is not tricked by his own magic. Māyā is temporary and is transcended with 'true knowledge,' or perception of the more fundamental reality which permeates Māyā." (WK 2010u)

In Advaita Vedanta, "suffering is due to Maya, and only knowledge (called Jnana) of Brahman can destroy Maya. When Maya is removed, there exists ultimately no difference between the Jiva-Atman and the Brahman. Such a state of bliss when achieved while living is called Jivan mukti." (WK 2010u)

The Goal of Salvation

This state of bliss is one of "non-duality," in which "there exists ultimately no difference between the Jiva-Atman and the Brahman." (WK 2010u)

But how can this non-duality between Atman and the Brahman be achieved? After all, "how [can] the individual soul, which is limited and one in each body,…be the same as Brahman? Adi Shankara explains that the Self is not an individual concept. Atman is only one and unique.

Indeed Atman alone is....It is a false concept that there are several Atmans." (WK 2010u)

As an analogy, "just as the same moon appears as several moons on its reflections on the surface of water covered with bubbles, the one Atman appears as multiple atmans in our bodies because of Maya. Atman is self-proven, however, some proofs are discussed—eg., a person says 'I am blind,' 'I am happy,' 'I am fat,' etc. The common and constant factor, which permeates all these statements is the 'I' which is but the Immutable Consciousness. When the blindness, happiness, fatness are inquired and negated, 'I' the common factor which, indeed, alone exists in all three states of consciousness and in all three periods of time, shines forth. This proves the existence of Atman, and that Consciousness, Reality and Bliss are its characteristics. Atman, being the silent witness of all the modifications, is free and beyond sin and merit. It does not experience happiness or pain because it is beyond the triad of Experiencer, Experienced and Experiencing." (WK 2010u)

In the end, at "the transcendental level, only the one Atman, equal to Brahman, is true." (WK 2010u) And here is where salvation is achieved, in "a state of...universal, transcendent, transpersonal existence. In that state, according to Adi Shankara, there is no difference between Self and God; there is only the Absolute (Brahman)"—and such is the state of bliss in salvation. (WK 2010v)

A Sober Reflection

But the promise of the state of bliss in Advaita Vedanta has its own critics. Consider a few criticisms below, for illustration.

Firstly, a major criticism of Advaita Vedanta is that the very idea of "God, the Supreme Cosmic Spirit, or Brahman...cannot be correctly described as this or that." (WK 2010u) But if it cannot be correctly described, then how can one be sure that it is what the advocates of Advaita Vedanta would like us to believe?

Secondly, another serious criticism is that, while it makes sense to suggest that the cause and the effect are not the same, it is a leap of logic to conclude that the cause must be "the infinite unmanifest Absolute," because the effect is the finite material world. In fact, if the effect is the finite material world, the cause may well be finite and manifest too—just as the old saying goes, "Like father, like son."

And thirdly, still another major criticism questions the desirability of this so-called state of "bliss" in the "non-duality" between the "Self and God" (or between "Atman and the Brahman"). The critique of Christian asceticism by Friedrich Nietzsche is worth mentioning here, in that the search for spiritual salvation by way of the suppression of the "self" produces the opposite outcome—"life-denying," instead of "life-affirming."

However, all these criticisms do not mean to dismiss the Advaita Vedanta theory of creation but to let us hear the voices of both sides in the debate on the contested beginnings of the universe and the aftermath from the perspective of the mind.

In the process, this analysis can also teach us something valuable about some ontological principles in existential dialectics, and good examples include the absoluteness-relativeness principle, the partiality-totality principle, the predictability-unpredictability principle, the explicability-inexplicability principle, the finiteness-transfiniteness principle, the preciseness-vagueness principle, the simpleness-complicatedness principle, the openness-hiddenness principle, the denseness-emptiness principle, the slowness-quickness principle, the expansion-contraction principle, the theory-praxis principle, the convention-novelty principle, the evolution-transformation principle, the symmetry-asymmetry principle, the softness-hardness principle, the regression-progression principle, and the same-difference principle.

For instance, in relation to the absoluteness-relativeness principle, if there is absolute viewpoint (e.g., the absolute view in Advaita Vedanta that there is "the infinite unmanifest Absolute" at the transcendental level of truth), there is relative viewpoint (e.g., what is true for Advaita Vedanta in regard to "the infinite unmanifest Absolute" at the transcendental level is not necessarily so for astrophysics in regard to the beginning of the universe as the big bang, without the appeal to any transcendental level of reality). And the reverse direction also holds true.

In relation to the partiality-totality principle, if there is partiality (e.g., the partial view of each individual self, in Advaita Vedanta), there is totality (e.g., the holistic view of the Brahman, in Advaita Vedanta, such that the whole is not simply the sum of its parts). And the reverse direction also holds true.

In relation to the predictability-unpredictability principle, if there is predictability (e.g, the predictable tendency of a disciple of Advaita Vedanta to learn the Vedic scriptures, with the guidance of a Guru, in the quest for "moksha" as a "liberation from the cycle of births and deaths," with its recurrence of pain and suffering), there is unpredictability (e.g, the more difficult task to predict which particular disciple of Advaita Vedanta will suceed in achieving salvation, since it requires some abilities which not everyone equally has, like "the reunuciation of enjoyments of objects" in the material world, concentration on "meditation," the discrimination between the "eternal" and the "transitory" in illusion, etc.). (WK 2010u) And the reverse direction also holds true.

In relation to the explicability-inexplicability principle, if there is explicability (e.g., the explanation, in Advaita Vedanta, of the fact that "the effect has its self in the cause but not the cause in the effect," on the basis of the temporal sequence of cause and effect), there is inexplicability (e.g., the lack of convincing explanation, within Advaita Vedanta, of why the cause must therefore be "the infinite unmanifest Absolute," while the effect is the finite, manifest material world). And the reverse direction also holds true.

In relation to the finiteness-transfiniteness principle, if there is finiteness (e.g., the finite number of the levels of truth or reality in Advaita Vedanta, that is, the three levels known as the transcendental, pragmatic, and apparent), there is transfiniteness (e.g, the transfinite nature of reality at the transcendental level, in accordance to Advaita Vedanta, such that it is "omnipresent, omnipotent, incorporeal, impersonal, transcendent," etc.). And the reverse direction also holds true.

In relation to the preciseness-vagueness principle, if there is preciseness (e.g., the precise identification, in Advaita Vedanta, of Brahman as "the infinite unmanifest Absolute"), there is vagueness (e.g., the unclarity in the conception of Brahman in Advaita Vedanta, because it "cannot be correctly described"). And the reverse direction also holds true.

In relation to the simpleness-complicatedness principle, if there is simpleness (e.g., the relatively simplistic analysis of the creation of the universe in a mystical way, in that the effect like the universe pre-existed in the cause like "the infinite unmanifest Absolute"), there is complicatedness (e.g., the relatively more complicated analysis of the Advaita Vedanta theory of creation by questioning its claims, as shown in the criticisms). And the reverse direction also holds true.

In relation to the openness-hiddenness principle, if there is openness (e.g., the open exploration of the illusion in the creation of the world, in accordance to Advaita Vedanta), there is hiddenness (e.g., the hidden bias in Advaita Vedanta, in its obsession with the spiritual quest for salvation in relation to "the infinite unmanifest Absolute"). And the reverse direction also holds true.

In relation to the denseness-emptiness principle, if there is denseness (e.g., the relatively denser concentration of spiritual concerns with "the reunuciation of enjoyments of objects" in the material world, in Advaita Vedanta), there is emptiness (e.g, the relatively less dense, or more empty, concentration of mundane concerns with sensual life enjoyments, in Advaita Vedanta). And the reverse direction also holds true.

In relation to the slowness-quickness principle, if there is slowness (e.g., the relatively slower readiness of a follower of Advaita Vedanta to let himself preoccupied with actions for materialist life success), there is quickness (e.g, the relatively quicker readiness of a followre of Advaita Vedanta to concentrate on meditation for the "liberation from the cycle of births and deaths"). And the reverse direction also holds true.

In relation to the expansion-contraction principle, if there is expansion (e.g., the relatively more developed ability of a follower of Advaita Vedanta to understand "the canonical texts consisting of [a] the Upanishads, [b] the Bhagavad Gita and [c] the Brahma Sutras"), there is contraction (e.g., the relatively less developed ability of a follower of Advaita Vedanta to understand non-Hindu schools of thought, like the theory of relativity in physical cosmology). And the reverse direction also holds true.

In relation to the theory-praxis principle, if there is theory (e.g, the theoretical construction of acosmism), there is praxis (e.g., the practical application of acosmism, as in the use of Advaita Vedanta for spiritual salvation). And the reverse direction also holds true.

In relation to the convention-novelty principle, if there is convention (e.g., the conventional wisdom about non-duality in Hindu philosophy, like the one in Advaita Vedanta), there is novelty (e.g., the alternative novel challenge to this conventional wisdom by Dvaita Vedanta, which allows duality). And the reverse direction also holds true.

In relation to the evolution-transformation principle, if there is evolution (e.g., the natural evolution of humans to live in awe of the universe or nature), there is transformation (e.g, the technical transformation of human relation with the universe by way of the invention of systems of thought like Advaita Vedanta, which searches for that blissful state where "there is no difference between Self and God"). And the reverse direction also holds true.

In relation to the symmetry-asymmetry principle, if there is symmetry (e.g., the co-existence of different Vedanta schools in the history of Hindu philosophy), there is asymmetry (e.g., the popularity of Advaita Vedanta among those who accept the teachings of Adi Shankara, but the acceptance of other sub-schools of Vedanta by those who follow different teachings of Dvaita Vedanta and others). And the reverse direction also holds true.

In relation to the softness-hardness principle, if there is softness (e.g., the concentration on meditation in Advaita Vedanta), there is hardness (e.g., "the reunuciation of enjoyments of objects" in the material world, in Advaita Vedanta). And the reverse direction also holds true.

In relation to the regression-progression principle, if there is regression (e.g., the regression made in Advaita Vedanta, as shown by the criticisms against its claims), there is progression (e.g., the progress made in Advaita Vedanta for the better understanding of the nature of suffering inherent in the obsession with the "I"). And the reverse direction also holds true.

And in relation to the same-difference principle, if there is similarity in outcome (e.g., the contribution to a happy life, regardless of whether this is done by way of a spiritual quest or by way of a material pursuit), there is difference in outcome (e.g., the contribution to a happy life by way of the spiritual quest for salvation, but the contribution to a happy life by way of the craving for enjoyments in the material world). And the reverse direction also holds true.

Idealism, the Mind, and the Problem of Universals

The Advaita Vedanta theory of creation (as discussed in the previous section), because of its claim about the illusion in the creation of the universe (which is not real, in its finite manfest nature of the material world), adopts an "idealistic" view of reality, for the existence of an underlying spiritual reality which "preceded the existence of insentient matter" (like the universe).

This then leads us to a hotly contested debate in metaphysics, that is, on what "reality" is, or what really "exists."

Precisely here, there are three opposing sides in the debate, namely, (a) idealism, (b) realism, and (c) nominalism—to be explicated hereafter, in that order.

Idealism and the Problem of Universals

The term "idealism" refers to "the philosophical theory that maintains that the ultimate nature of reality is based on the mind or ideas." (WK 2010w)

By contrast, the opposite of idealism (in the philosophy of mind) is "materialism" (as a version of realism), which does not need "the mind" or "consciousness" (of intelligent life) as the basis of reality and in fact "sees no incoherence in a scenario of there being a cosmos where no sentient subject ever develops"; in fact, materialism even accepts "a wholly unknown universe where neither any subject, nor any object of a subject's experience ever exists." (WK 2010w)

Unlike materialism—idealism, in the context of "the ancient philosophy of the Vedas" (to which Advaita Vedanta as discussed in the previous section belongs), refers to "the dynamic consciousness of living beings that emanates from the divine cosmic source." (WK 2010w)

Such terms like "divineness," "absoluteness," "eternity," "transcendentalness," "Godness," "Oneness," "wholeness," and so on, when used to describe the Brahman at the transcendental level of reality, are known as "universals" in metaphysics.

For the idealists, these universals do not really exist except being used as a limited way to describe the Brahman (or "the divine cosmic source"), which, unfortunately, as those in Advaita Vedanta so often insist, "cannot be correctly described" by universals or any other terms in the world.

By definition, "a universal is what particular things have in common, namely characteristics or qualities. In other words, universals are repeatable or recurrent entities that can be instantiated or exemplified by many particular things. For example, suppose there are two chairs in a room, each of which is green. These two chairs both share the quality of 'chairness,' as well as greenness or the quality of being green. Metaphysicians call this quality that they share a 'universal,' because it can be instantiated or exemplified by many particular things." (WK 2010x; H. Price 1953)

Normally, but not always, "universals" can be identified with the use of "the Ness-Ity-Hood Principle," in that universals often have endings like (a) "ness" (e.g., "absoluteness"), (b) "ity" (e.g., "eternity") and (c) "hood" (e.g., chairhood). (WK 2010x; F. Feldman 2005) Of course, there are exceptions to any rule (e.g., "charity").

In addition, there are three major types of "universals," namely, (a) "types or kinds" (e.g., "dogness"), (b) "properties" (e.g., "redness"), and (c) "relations" (e.g., "betweenness"). (WK 2010x; M. Loux 2001)

However, the "universal-particular" distinction should not be confused with the "abstract-concrete" distinction, even though the two are related (as summarized in *Table 2.3*).

On the one hand, the two distinctions are related, because, "[p]aradigmatically, universals are abstract (e.g., humanity), whereas particulars are concrete (e.g., the person of Socrates)." (WK 2010x)

On the other hand, the two distinctions are not always the same; for example, "one might hold that numbers [e.g., the number 5] are particular yet abstract objects." (WK 2010x & 2010y)

In general, but not always, an abstract object can be identified with the use of two criteria, namely, (a) that an abstract object (e.g., "justice") "lacks a location in space" and (b) an abstract object (e.g., "empty set") "lacks any causal powers" (or "cannot act on other objects"). (WK 2010y)

With this clarification of the two distinctions in mind—the problem with universals, for the idealists, is that they do not exist (or are not real), although they can be used as a limited way to describe the transcendental reality of the Brahman, which, unfortunately, as those in Advaita Vedanta so often insist, "cannot be correctly described" by universals or any other terms in the world.

Thus, for those who claim the illusion in the creation of the universe, universals about the universe (or anything else) do not exist (or are not real). At the transcendental level of reality, for Advaita Vedanta, for example, Atman (the mind) and the Brahman (God) are one and the same, and this Oneness "cannot be correctly described" by universals or any other terms in the world (as these terms, be they universals or not, do not really exist).

The Critical Reply from the Realists

But idealism has its enemies, one of which is known as "realism." Consider two of its proponents below, for illustration.

Firstly, Plato argues that universals exist and are not merely "mental" (as the idealists would claim); they are "non-mental entities." (WK 2010x) In fact, Plato is even more extreme, in holding that univerals exist in a metaphysical realm, which is independent of both the physical world and the mind.

For the realists, "universals must be posited as distinct entities in order to account for various phenomena. For example,…[for] Plato…, universals are required for certain general words to have meaning and for the sentences in which they occur to be true or false. Take the sentence 'Djivan Gasparyan is a musician.' The realist may claim that this sentence is only meaningful and expresses a truth because there is an individual, Djivan Gasparyan, who possesses a certain quality, musicianship. Thus it is assumed that the property is a universal which is distinct from the particular individual who has the property." (WK 2010z; M. MacLeod 2006)

And secondly, Berrand Russell (1912:42-43), another realist, criticized idealism in a different way, in that it is tautological, as he thus wrote in the following passage against the idealist argument defended by George Berkeley: "If we say that the things known must be in the mind, we are either un-duly limiting the mind's power of knowing [without learning from the outside world], or we are uttering a mere tautology. We are uttering a mere tautology if we mean by 'in the mind' the same as by 'before the mind', i.e., if we mean merely being apprehended by the mind. But if we mean this, we shall have to admit that what, in this sense, is in the mind, may nevertheless be not mental. Thus when we realize the nature of knowledge, Berkeley's argument is seen to be wrong in substance as well as in form, and his grounds for supposing that 'idea'—i.e. the objects apprehended—must be mental, are found to have no validity whatever. Hence his grounds in favour of the idealism may be dismissed." (WK 2010w)

The Compromise of the Nominalists

Yet, there is a third position, which is somewhere between idealism and realism—and is called "nominalism."

On the one hand, nominalism is against realism, because, for the nominalists, "only individuals or particulars exist and deny that universals are real (i.e. that they exist as entities or beings)." (WK 2010z)

On the other hand, nominalism is also critical to idealism, because it holds that, as is true in the version of nominalism known as "conceptualism" ("as a middle way between nominalism and realism"), universals can exist as a "concept" in the human mind (as opposed to the "divine cosmic" mind at the transcendental level of reality for the idealists like those in Advaita Vedanta). In fact, the term "nominalism" derives "from the Latin *nomen* ('name')." (WK 2010z)

Well known nominalists include, for instance, William of Ockham, D. C. Williams (1953), David Lewis (1983), H. H. Price (1953), and W. V. O. Quine (1961). (WK 2010z)

For illustration, consider the nominalist views of Peter Abelard and William of Ockham, as two of "the most influential and thorough nominalist[s]. Abelard's and Ockham's version of nominalism is sometimes called conceptualism, which presents itself as a middle way between nominalism and realism, asserting that there is something in common among like individuals, but that it is a concept in the mind, rather than a real entity existing independently of the mind." (WK 2010z)

For Ockham, "only individuals existed and...universals were only mental ways of referring to sets of individuals. 'I maintain,' he wrote, 'that a universal is not something real that exists in a subject...but that it has a being only as a thought-object in the mind [objectivum in anima].' As a general rule, Ockham argued against assuming any entities that were not necessary for explanations. Accordingly, he wrote, there is no reason to believe that there is an entity called 'humanity' that resides inside, say, Socrates. Nothing further is explained by saying that. This is in accord with the analytical method which has since come to be called Ockham's razor, the principle that the explanation of any phenomenon should make as few assumptions as possible." (WK 2010z)

However, there is a weakness in the conceptualist argument: "Critics argue that conceptualist approaches only answer the psychological question of universals. If the same concept is correctly and non-arbitrarily applied to two individuals, there must be some resemblance or shared property between the two individuals that justifies their falling under the same concept and that is just the metaphysical problem that universals were brought in to address, the starting-point of the whole problem." (WK 2010z; M. MacLeod 2006)

A More Impartial Perspective

These criticisms of idealism from the two vantage points of realism and nominalism are not to dismiss idealism outright but are to allow us to learn from all sides of the debate on the possibility and desirability of the contested beginnings of the universe (and the aftermath) from the perspective of the mind (with the problem of universals as a case study here).

Each side has its own assumptions and emphasis, with its own distinctive implications for understanding the world (e.g., Ockham's razor for conceptualism by Ockham, the divine cosmic mind for idealism in Advaita Vedanta, and the Ideal Forms for realism by Plato).

In addition, the analysis reveals something interesting about the ontological principles in existential dialectics, and good examples include the formalness-informalness principle, the absoluteness-relativeness principle, the predictability-unpredictability principle, the explicability-inexplicability principle, the finiteness-transfiniteness principle, the preciseness-vagueness principle, the simpleness-complicatedness principle, the openness-hiddenness principle, the denseness-emptiness principle, the slowness-quickness principle, the expansion-contraction principle, the convention-novelty principle, the evolution-transformation principle, the symmetry-asymmetry principle, the regression-progression principle, and the same-difference principle.

For instance, in relation to the formalness-informalness principle, if there is formalness (e.g., the formal requirement of completeness in a system of ideas for provability), there is informalness (e.g., the ambivalent nature of probability in regard to the Brahman in Advaita Vedanta, in that it cannot completely be known because it "cannot be correctly described," but the Atman, or self, and the Brahman can be converged into Oneness, under certain conditions for salvation as guided by a Guru). And the reverse direction also holds true.

In relation to the absoluteness-relativeness principle, if there is absoluteness (e.g., the absolute Realist viewpoint by Plato that universals exist in the metaphysical realm), there is relativeness (e.g., what is real for Plato in regard to the universals is not so for the idealists in regard to the illusion of the world). And the reverse direction also holds true.

In relation to the predictability-unpredictability principle, if there is predictability (e.g., the predictable tendency of an idealist to deny the existence of univerals in the external world), there is unpredictability (e.g., the more difficult task to predict what a particular idealist can accept in relation to an understanding of what the mind really refers to, which can mean different things to different idealists, like the "divine cosmic" mind in Advaita Vedanta). And the reverse direction also holds true.

In relation to the explicability-inexplicability principle, if there is explicability (e.g., the explanation, by those in Advaita Vedanta, that universals do not really describe the transcendental reality of the Brahman, because the Brahman "cannot be correctly described"), there is inexplicability (e.g., the lack of sufficient explanation, within Advaita Vedanta, of why the Brahman is supposed to be real, if it cannot even be correctly described). And the reverse direction also holds true.

In relation to the finiteness-transfiniteness principle, if there is finiteness (e.g., the finte number of categories in the classification of universals, as shown in the three known as the "types," "properties," and "relations"), there is transfiniteness (e.g., the transfinite number of all the possible types, properties, and relations for universals in the world). And the reverse direction also holds true.

In relation to the preciseness-vagueness principle, if there is preciseness (e.g., the precise identification of "lacking any causal powers" as one criterion for abstract objects), there is vagueness (e.g., the vagueness in the identification of "lacking any causal powers" as one criterion for abstract objects, because "it is not clear exactly what it is to have a causal power," for the critics). (WK 2010y) And the reverse direction also holds true.

In relation the simpleness-complicatedness principle, if there is simpleness (e.g., the relatively simplistic claim by the idealists that universals do not exist), there is complicatedness (e.g., the relativley more complicated analysis of the idealist claim by way of questioning its assumptions, as pointed out by the realists and the nominalists). And the reverse direction also holds true.

In relation to the openness-hiddenness principle, if there is openness (e.g., the open exploration, in idealism, of the relatonship between universals and the unreality of the world), there is hiddenness (e.g., the hidden bias in the idealist view, as shown in Advaita Vedanta, because of its spiritual preference for the alleged "divine cosmic source" which is prior to the creation of the universe). And the reverse direction also holds true.

In relation to the denseness-emptiness principle, if there is denseness (e.g., the relatively denser concentration of concerns, for a Platonic realist, with a metaphysical realm for universals which is independent of both the physical world and the mind), there is emptiness (e.g., the relatively more empty concentraiton of concerns, for a Platonic realist, with the cosmic mind as advocated by the idealists in Advaita Vedanta). And the reverse direction also holds true.

In relation to the slowness-quickness principle, if there is slowness (e.g., the relatively slower readiness of an idealist to accept universals as real), there is quickness (e.g., the relatively quicker readiness of an idealist in Advaita Vednata to accept the existence of the divine cosmic mind, which cannot be described by words and is prior to the creation of the world). And the reverse direction also holds true.

In relation to the expansion-contraction principle, if there is expansion (e.g., the relatively more developed ability of a conceptualist to be frugal "against assuming any entities that were not necessary for explanations," also kwown as Ockham's razor), there is contraction principle (e.g., the relatively less developed ability of a conceptualist to be complicated in assuming many entities and their properties like those about the Brahman by the idealists). And the reverse direction also holds true.

In relation to the convention-novelty principle, if there is convention (e.g., the conventional wisdom, as argued by Plato, about the reality of universals), there is novelty (e.g., the alternative novel challenge to the conventional wisdom about the Platonic reality of universals by the nominalists like Abelard and Ockham, who argued instead their existence as concepts in the human mind only). And the reverse direction also holds true.

In relation to the evolution-transformation principle, if there is evolution (e.g., the natural evolution of humans to communicate with each other in a primitive way in the state of nature, without the existence of elaborated systems of language yet), there is transformation (e.g., the technical transformation of human communication by way of the invention of metaphysics, like elaborated systems of universals). And the reverse direction also holds true.

In relation to the symmetry-asymmetry principle, if there is symmetry (e.g., the co-existence of different schools of thought about the problem of universals through the ages), there is aymmetry (e.g. the popularity of Platonic realism in ancient Greece, but the acceptance of conceptualism in the Middle Ages, especially with the works of Abelard and Ockham). And the reverse direction also holds true.

In relation to the regression-progression principle, if there is regression (e.g., the regression made in idealism, as shown in the criticisms as made by its criticcs, like the realists and the nominalists), there is progression (e.g., the progress made in idealism for the better understanding of why universals do not exist in the external world). And the reverse direction also holds true.

And in relation to the same-difference principle, if there is similarity in outcome (e.g., the contribution to the understanding of the world, be this done with the use of universals or the rejection of them), there is difference in outcome (e.g., the contribution to the understanding of the world by way of the use of universals as concepts, with implications for Ockham's razor; but the contribution to the understanding of the world by way of the rejection of universals as adequately desccribing the Brahman, which is prior to the creation of the universe, for those in Advaita Vedanta). And the reverse direction also holds true.

Its Contested Beginnings—and Nature

The contested beginnings of the universe (and the aftermath) can also be examined from a different perspective, this time, in relation to nature, as it can enrich the understanding of the topic in a distinctive way.

An excellent case study here concerns the Big Bang theory and the pre-Big Bang cosmological models (as summarized in *Table 2.4*).

The Big Bang Theory, and the Pre-Big Bang Alternatives

A good way to start the evaluation of the possibility and desirability of the contested beginnings of the universe from the perspective of nature has to do with the Big Bang theory, which is one of the most important theories in the history of physics.

Major Contributors

The term "Big Bang," as originally proposed by Georges Lemaître at M.I.T. in 1927 (for a new cosmology to understand the beginning of the universe), refers to "the cosmological model of the initial conditions and subsequent development of the Universe that is supported by the most comprehensive and accurate explanations from current scientific evidence and observation." (WK 2010aa)

Generally, "the term Big Bang generally refers to the idea that the Universe has expanded from a primordial hot and dense initial condition at some finite time in the past (best available measurements in 2009 suggest that the initial conditions occurred around 13.3 to 13.9 billion years ago), and continues to expand to this day." (WK 2010aa; B. Feuerbacher 2006; E. Wright 2009; E. Komatsu 2009; E. Menegoni 2009)

Needless to say, quite a number of different scentists had contributed to the development of the Big Bang theory. Consider a few below, for illustration.

For instance, back in 1912, "Vesto Slipher measured the first Doppler shift of a 'spiral nebula' (spiral nebula is the obsolete term for spiral galaxies), and soon discovered that almost all such nebulae were receding from Earth. He did not grasp the cosmological implications of this fact, and indeed at the time it was highly controversial whether or not these nebulae were 'island universes' outside our Milky Way." (WK 2010aa; V. Slipher 1913 & 1915)

In 1922, "Alexander Friedmann, a Russian cosmologist and mathematician, derived the Friedmann equations from Albert Einstein's equations of general relativity, showing that the Universe might be expanding in contrast to the static Universe model advocated by Einstein at that time." (WK 2010aa; A. Friedman 1999)

Then, in 1924, "Edwin Hubble's measurement of the great distance to the nearest spiral nebulae showed that these systems were indeed other galaxies. Independently deriving Friedmann's equations in 1927, Georges Lemaître, a Belgian physicist and Roman Catholic priest, proposed that the inferred recession of the nebulae was due to the expansion of the Universe." (WK 2010aa; G. Lemaître 1931)

A few years later, in 1931, "Lemaître went further and suggested that the evident expansion in forward time required that the Universe contracted backwards in time, and would continue to do so until it could contract no further, bringing all the mass of the Universe into a single point, a 'primeval atom' where and when the fabric of time and space comes into existence." (WK 2010aa)

During the 1920's and 1930's, Hubble (1929) used "the 100-inch (2,500 mm) Hooker telescope at Mount Wilson Observatory…to estimate distances to galaxies whose redshifts had already been measured, mostly by Slipher. In 1929, Hubble discovered a correlation between distance and recession velocity—now known as Hubble's law. Lemaître had already shown that this was expected, given the Cosmological Principle." (WK 2010aa; P. Peebles 2003; E. Christianson 1995)

By late 1940's, "two distinct possibilities emerged. One was Fred Hoyle's steady state model, whereby new matter would be created as the Universe seemed to expand. In this model, the Universe is roughly the same at any point in time. The other was Lemaître's Big Bang theory, advocated and developed by George Gamow, who introduced big bang nucleosynthesis (BBN)[23] and whose associates, Ralph Alpher and Robert Herman, predicted the cosmic microwave background radiation (CMB)." (WK 2010aa; F. Hoyle 1948; R. Alpher 1948)

It was before long, however, that "the observational evidence, most notably from radio source counts, began to favor the latter [the Big Bang theory]. The discovery and confirmation of the cosmic microwave background radiation in 1964 secured the Big Bang as the best theory of the origin and evolution of the cosmos. Much of the current work in cosmology includes understanding how galaxies form in the context of the Big Bang, understanding the physics of the Universe at earlier and earlier times, and reconciling observations with the basic theory." (WK 2010aa; A. Penzias 1965)

By the 1990's, "[h]uge strides in Big Bang cosmology have been made...as a result of major advances in telescope technology as well as the analysis of copious data from satellites such as COBE, the Hubble Space Telescope and WMAP. Cosmologists now have fairly precise and accurate measurements of many of the parameters of the Big Bang model, and have made the unexpected discovery that the expansion of the Universe appears to be accelerating." (WK 2010aa; N. Boggess 1992; D. Spergel 2006)

In retrospect, however, it is worth mentioning that, although Hoyle's model was rejected, "it was Hoyle who coined the phrase that came to be applied to Lemaître's theory, referring to it derisively as 'this big bang idea' during a BBC Radio broadcast in March 1949." (WK 2010aa; S. Singh 2010)

A Timeline of the Big Bang

The Big Bang theory also provides a timeline from the beginning to the present.

As of 2009, "[b]est available measurements...suggest that the initial conditions occurred between 13.3 and 13.9 billion years ago. It is convenient to divide the evolution of the universe since then into three phases," namely, (a) "the every early universe," (b) "the early universe," and (c) the phase of "structure formation." (WK 2010bb; E. Komatsu 2009; E. Menegoni 2009)

(a) The Very Early Universe

In this earliest period (which is the study of cosmogony), much is still unknown, in that "[t]he very early universe, which is still poorly understood, was the split second in which the universe was so hot that particles had energies higher than those currently accessible in particle accelerators on Earth. Therefore, while the basic features of this epoch have been worked out in the big bang theory, the details are largely based on educated guesses." (WK 2010bb)

With this caveat in mind—astrophysics have proposed three different epochs in the very early universe, that is, (a1) the "Planck epoch," (a2) the "grand unification epoch," and (a3) the "electroweak epoch." (WK 2010bb)

(a1) In the Planck epoch, the universe was "up to 10^{-43} seconds after the Big Bang," and "[i]f supersymmetry is correct, then during this time the four fundamental forces—electromagnetism, weak nuclear force, strong nuclear force and gravitation—all have the same strength, so they are possibly unified into one fundamental force. Little is known about this epoch, although different theories propose different scenarios." (WK 2010bb)

(a2) In the grand unification epoch, the universe was "'[b]etween 10^{-43} seconds and 10^{-36} seconds after the Big Bang.' As the universe expands and cools from the Planck epoch, gravitation begins to separate from the fundamental gauge interactions: electromagnetism and the strong and weak nuclear forces." (WK 2010bb)

However, "eventually, the grand unification is broken as the strong nuclear force separates from the electroweak force [the name for the unified forces of electromagnetism and the weak interaction]. This occurs as soon as inflation does." (WK 2010bb)

(a3) In the electroweak epoch, the universe was "'[b]etween 10^{-36} seconds and 10^{-12} seconds after the Big Bang.' The temperature of the universe is low enough (1028K) to separate the strong force from the electroweak force. This phase transition triggers a period of exponential expansion known as cosmic inflation. After inflation ends, particle interactions are still energetic enough to create large numbers of exotic particles, including W and Z bosons and Higgs bosons." (WK 2010bb)

This electroweak epoch can be further sub-divided into three smaller periods, namely, (a3.1) the "inflationary" period, (a3.2) the "reheating" period, and (a3.3) the "baryogenesis" period. (WK 2010bb)

(a3.1) In the "inflationary" period, the universe was "'[b]etween 10^{-36} seconds and 10^{-32} seconds after the Big Bang.' The temperature, and therefore the time, at which cosmic inflation occurs is not known for certain. During inflation, the universe is flattened (its spatial curvature is critical) and the universe enters a homogeneous and isotropic rapidly expanding phase in which the seeds of structure formation are laid down in the form of a primordial spectrum of nearly-scale-invariant fluctuations....One scenario suggests that prior to cosmic inflation, the universe was cold and empty, and the immense heat and energy associated with the early stages of the big bang was created through the phase change associated with the end of inflation. According to the ΛCDM model, dark energy is present as a property of space itself, beginning immediately following the period of inflation, as described by the Equation of state (cosmology). ΛCDM says nothing about the fundamental physical origin of dark energy but it represents the energy density of a flat universe. Observations indicate that it has existed for at least 9 billion years." (WK 2010bb)

(a3.2) In the "repheating" period, "the exponential expansion that occurred during inflation ceases and the potential energy of the inflaton field decays into a hot, relativistic plasma of particles....At this point, the universe is dominated by radiation; quarks, electrons and neutrinos form." (WK 2010bb)

(a3.3) In the "baryogenesis" period, the universe has more baryons than antibaryons, but "[t]here is currently insufficient observational evidence to explain why the universe contains far more baryons than antibaryons." (WK 2010bb)

(b) The Early Universe

In this second phase of the universe, "the evolution of the universe proceeded according to known high energy physics. This is when the first protons, electrons and neutrons formed, then nuclei and finally atoms. With the formation of neutral hydrogen, the cosmic microwave background was emitted." (WK 2010bb)

Astrophysicists have also proposed six different epochs in the early universe, namely, (b1) the "supersymmetry breaking" epoch, (b2) the "quark epoch," (b3) the "hadron epoch," (b4) the "lepton epoch," (b5) the "photon epoch," and (b6) the "dark ages," as described below.

(b1) In the "supersymmetry breaking" epoch—"if supersymmetry is a property of our universe, then it must be broken at an energy as low as 1 TeV, the electroweak symmetry scale. The masses of particles and their superpartners would then no longer be equal, which could explain why no superpartners of known particles have ever been observed." (WK 2010bb)

(b2) In the quark epoch—the universe was "'between 10^{-12} seconds and 10^{-6} seconds after the Big Bang'....The fundamental interactions of gravitation, electromagnetism, the strong interaction and the weak interaction have now taken their present forms, but the temperature of the universe is still too high to allow quarks to bind together to form hadrons." (WK 2010bb)

(b3) In the hadron epoch—the universe was "'between 10^{-6} seconds and 1 second after the Big Bang.' The quark-gluon plasma that composes the universe cools until hadrons, including baryons such as protons and neutrons, can form. At approximately 1 second after the Big Bang neutrinos decouple and begin traveling freely through space. This cosmic neutrino background, while unlikely to ever be observed in detail, is analogous to the cosmic microwave background that was emitted much later." (WK 2010bb)

(b4) In the lepton epoch—the universe was "'between 1 second and 10 seconds after the Big Bang.' The majority of hadrons and anti-hadrons annihilate each other at the end of the hadron epoch, leaving leptons and anti-leptons dominating the mass of the universe. Approximately 10 seconds after the Big Bang the temperature of the universe falls to the point where new lepton/anti-lepton pairs are no longer created and most leptons and anti-leptons are eliminated in annihilation reactions, leaving a small residue of leptons." (WK 2010bb)

(b5) In the photon epoch—the universe was "'between 10 seconds and 380,000 years after the Big Bang.' After most leptons and anti-leptons are annihilated at the end of the lepton epoch the energy of the universe is dominated by photons. These photons are still interacting frequently with charged protons, electrons and (eventually) nuclei, and continue to do so for the next 300,000 years." (WK 2010bb)

During this epoch, there were three further periods identified by astrophysicists, namely, (b5.1) "nucleosynthesis," (b5.2) "matter domination," and (b5.3) "recombination."

(b5.1) In the "nucleosynthesis" period—the universe was "'between 3 minutes and 20 minutes after the Big Bang.' During the photon epoch the temperature of the universe falls to the point where atomic nuclei can begin to form. Protons (hydrogen ions) and neutrons begin to combine into atomic nuclei in the process of nuclear fusion. However, nucleosynthesis only lasts for about seventeen minutes, after which time the temperature and density of the universe has fallen to the point where nuclear fusion cannot continue. At this time, there is about three times more hydrogen than helium-4 (by mass) and only trace quantities of other nuclei." (WK 2010bb; E. Wright 2009)

(b5.2) In the "matter domination period—the universe was "about 70,000 years after the Big Bang. At this time, the densities of non-relativistic matter (atomic nuclei) and relativistic radiation (photons) are equal. According to ΛCDM, at this stage, cold dark matter dominates, paving the way for gravitational collapse to amplify the tiny inhomogeneities left by cosmic inflation, making dense regions denser and rarefied regions more rarefied. However, because present theories as to the nature of dark matter are inconclusive, there is as yet no consensus as to its origin at earlier times." (WK 2010bb).

(b5.3) In the "recombination" period—the universe was about "'377,000 years after the big bang.' Hydrogen and helium atoms begin to form and the density of the universe falls. This is thought to have occurred about 377,000 years after the Big Bang. Hydrogen and helium are at the beginning ionized, i. e., no electrons are bounded to the nuclei, which are therefore electrically charged (+1 and +2 respectively). As the universe cools down, the electrons get captured by the ions, making them neutral. This process is relatively fast (actually faster for the helium than for the hydrogen) and is known as recombination. At the end of recombination, most of the atoms in the universe are neutral, therefore the photons can now travel freely: the universe has become transparent. The photons emitted right after the recombination can now travel undisturbed and are those that we see in the cosmic microwave background (CMB) radiation. Therefore the CMB is a picture of the universe at the end of this epoch." (WK 2010bb; G. Hinshaw 2009)

(b6) In the "dark ages"—"[b]efore decoupling occurs most of the photons in the universe are interacting with electrons and protons in the photon-baryon fluid. The universe is opaque or 'foggy' as a result. There is light but not light we could observe through telescopes. The baryonic matter in the universe consisted of ionized plasma, and it only became neutral when it gained free electrons during 'recombination,' thereby releasing the photons creating the CMB. When the photons were released (or decoupled) the universe became transparent. At this point the only radiation emitted is the 21 cm spin line of neutral hydrogen. There is currently an observational effort underway to detect this faint radiation, as it is in principle an even more powerful tool than the cosmic microwave background for studying the early universe." (WK 2010bb)

(c) The Phase of Structure Formation

And in the third phase of the universe, "[m]atter then continued to aggregate into the first stars and ultimately galaxies, quasars, clusters of galaxies and superclusters formed." (WK 2010bb)

Astrophysicists have also proposed six different periods in the phase of "strcture formation,, namely, (c1) the "quasars and reionization" period, (c2) the "stars" period, (c3) the "galaxies" period, (c4) the "groups, clusters, and superclusters" period, (c5) the "solar system" period, and (c6) the period at present, as described below.

(c1) In the "quasars and reionization" period—"[t]his lasted '150 million to 1 billion years.' The first quasars form from gravitational collapse. The intense radiation they emit reionizes the surrounding universe. From this point on, most of the universe is composed of plasma." (WK 2010bb)

(c2) In the "stars" period—"[t]he first stars, most likely Population III stars, form and start the process of turning the light elements that were formed in the Big Bang (hydrogen, helium and lithium) into heavier elements. However, as of yet there have been no observed Population III stars, which leaves their formation a mystery." (WK 2010bb)

(c3) In the "galaxies" period—"large volumes of matter collapse to form a galaxy. Population II stars are formed early on in this process, with Population I stars formed later. Johannes Schedler's project has identified a quasar CFHQS 1641+3755 at 12.7 billion light-years away, when the Universe was just 7 percent of its present age....The Hubble Ultra Deep Field shows a number of small galaxies merging to form larger ones, at 13 billion light years, when the Universe was only 5% its current age. Based upon the emerging science of nucleocosmochronology, the Galactic thin disk of the Milky Way is estimated to have been formed 8.3 ± 1.8 billion years ago." (WK 2010bb; L. Arany-Prado 2005)

(c4) In the "groups, clusters, and superclusters" peirod—"gravitational attraction pulls galaxies towards each other to form groups, clusters and superclusters." (WK 2010bb)

(c5) In the "solar system" period—"this lasted for '8 billion years.' Finally, objects on the scale of our solar system form. Our sun is a late-generation star, incorporating the debris from many generations of earlier stars, and formed roughly 5 billion years ago, or roughly 8 to 9 billion years after the big bang." (WK 2010bb)

(c6) In ther period at present—"the best current data estimate the age of the universe today as 13.7 billion years since the big bang. Since the expansion of the universe appears to be accelerating, superclusters are likely to be the largest structures that will ever form in the universe. The present accelerated expansion prevents any more inflationary structures entering the horizon and prevents new gravitationally bound structures from forming." (WK 2010bb)

Major Evidences

The timeline of the Big Bang so constructed is based on some kinds of observational evidence, which are known as "the four pillars of the Big

Bang theory"—namely, (a) "the Hubble-type expansion seen in the redshifts of galaxies," (b) "the detailed measurements of the cosmic microwave background," (c) "the abundance of light elements" from Big Bang nucleosynthesis, and (d) "the large scale distribution and apparent evolution of galaxies," as explained below. (WK 2010bb; M. Gladders 2007; CC 2010)

(a) The "Expansion of the Universe"

The first observational evidence has to do with the "expansion of the universe," in that "[t]he Universe began about ten billion years ago in a violent explosion; every particle started rushing apart from every other particle in an early super-dense phase. The fact that galaxies are receding from us in all directions is a consequence of this initial explosion and was first discovered observationally by Hubble....Projecting galaxy trajectories backwards in time means that they converge to a high density state—the initial fireball." (CC 2010)

In addition, "[t]he Copernican or cosmological principle states that the Universe appears the same in every direction from every point in space. It amounts to asserting that our position in the Universe—with respect to the very largest scales—is in no sense preferred. There is considerable observational evidence for this assertion, including the measured distributions of galaxies and faint radio sources, though the best evidence comes from the near-perfect uniformity of the relic cosmic microwave background radiation. This means that any observer anywhere in the Universe will enjoy much the same view as we do, including the observation that galaxies are moving away from them." (CC 2010)

(b) The "Origin of the Cosmic Background Radiation"

The second observational evidence has to do with the "origin of the cosmic background radiation," in that "[a]bout 100,000 years after the Big Bang, the temperature of the Universe had dropped sufficiently for electrons and protons to combine into hydrogen atoms, $p + e \rightarrow H$. From this time onwards, radiation was effectively unable to interact with the background gas; it has propagated freely ever since, while constantly losing energy because its wavelength is stretched by the expansion of the Universe. Originally, the radiation temperature was about 3000 degrees Kelvin, whereas today it has fallen to only 3K." (CC 2010)

In fact, "[o]bservers detecting this radiation today are able to see the Universe at a very early stage on what is known as the 'surface of last scattering'. Photons in the cosmic microwave background have been travelling towards us for over ten billion years, and have covered a distance of about a million billion billion miles." (CC 2010)

(c) The "Nucleosynthesis of the Light Elements"

The third observational evidence has to do with the "nucleosynthesis of the elements," in that, "[p]rior to about one second after the Big Bang, matter—in the form of free neutrons and protons—was very hot and dense. As the Universe expanded, the temperature fell and some of these nucleons were synthesised into the light elements: deuterium (D), helium-3, and helium-4. Theoretical calculations for these nuclear processes predict, for example, that about a quarter of the Universe consists of helium-4, a result which is in good agreement with current stellar observations. The heavier elements, of which we are partly made, were created later in the interiors of stars and spread widely in supernova explosions." (CC 2010)

(d) The "Formation of Galaxies and Large-Scale Structure"

And the fourth observational evidence has to do with the "formation of galaxies and large-scale structure," in that "[a]t about 10,000 years after the Big Bang, the temperature had fallen to such an extent that the energy density of the Universe began to be dominated by massive particles, rather than the light and other radiation which had predominated earlier. This change in the form of the main matter density meant that the gravitational forces between the massive particles could begin to take effects, so that any small perturbations in their density would grow. Ten billion years later we see the results of this collapse." (CC 2010)

However,"[t]he standard cosmology, then, provides a framework for understanding galaxy formation, but it does not tell us about the origin of the primordial fluctuations required at 10,000 years. We must seek answers to questions like these from earlier epochs in the history of the Universe." (CC 2010)

Major Problems

Impressive as the Big Bang theory may seem, in light of the timeline of the Big Bang and the four observational evidences—there remain some major problems in the theory.

Consider a few of them below, for illustration.

Firstly, a serious problem with the Big Bang theory is that it does not really explain the origin of the universe, since even its account of the "very early universe" starts with "up to 10^{-43} seconds after the Big Bang"; even then, "[t]he very early universe…is still poorly understood." (WK 2010bb)

Secondly, another serious problem with the Big Bang theory is that it does not explain some anomalous phenomena like "why the Universe has more matter than antimatter." (WK 2010aa)

For instance, "[i]t is generally assumed that when the Universe was young and very hot, it was in statistical equilibrium and contained equal numbers of baryons and antibaryons. However, observations suggest that the Universe, including its most distant parts, is made almost entirely of matter. An unknown process called 'baryogenesis' created the asymmetry." (WK 2010aa; E. Kolb 1988)

Thirdly, still another serious problem with the Big Bang theory is that it assumes some exotic entities which remain to be proved and explained, and good examples include "dark energy" and "dark matter."

In regard to "dark matter"—this exotic entity was assumed, because, "[d]uring the 1970s and 1980s, various observations showed that there is not sufficient visible matter in the Universe to account for the apparent strength of gravitational forces within and between galaxies. This led to the idea that up to 90% of the matter in the Universe is dark matter that does not emit light or interact with normal baryonic matter....In particular, the Universe today is far more lumpy and contains far less deuterium than can be accounted for without dark matter." (WK 2010aa)

Although "[t]he evidence for dark matter comes from its gravitational influence on other matter," unfortunately, "no dark matter particles have been observed in laboratories. Many particle physics candidates for dark matter have been proposed, and several projects to detect them directly are underway." (WK 2010aa; W. Yao 2006) Rolf-Dieter Heuer, Director-General of the European Organization for Nuclear Research on the Swiss-French border near Geneva, frankly acknowledged: "We don't know what dark matter is." (R. Evans 2010)

And in regard to "dark energy"—this exotic entity was also assumed, because "[m]easurements of the redshift–magnitude relation for type Ia supernovae have revealed that the expansion of the Universe has been accelerating since the Universe was about half its present age. To explain this acceleration, general relativity requires that much of the energy in the Universe consists of a component with large negative pressure, dubbed 'dark energy....Results from the WMAP team in 2008, which combined data from the CMB and other sources, indicate that the Universe today is 72% dark energy, 23% dark matter, 4.6% regular matter and less than 1% neutrinos." (WK 2010aa; G. Hinshaw 2009)

But the problem with "dark energy" (or "negative pressure") is that "the exact nature of dark energy remains one of the great mysteries of the Big Bang." (WK 2010aa)

And fourthly, still one more serious problem with the Big Bang theory is that it can be motivated with religious considerations (especially in relation to those religious traditions favoring creationism).

For instance, some cosmologists "complained that the beginning of time implied by the Big Bang imported religious concepts into physics; this objection was later repeated by supporters of the steady state theory. This perception was enhanced by the fact that the originator of the Big Bang theory, Monsignor Georges Lemaître, was a Roman Catholic priest." (WK 2010aa; H. Kragh 1996; PBS 1998)

This critique remains, even though "there have been a variety of reactions by religious groups as to its implications for their respective religious cosmologies. Some accept the scientific evidence at face value, while others seek to reconcile the Big Bang with their religious tenets, and others completely reject or ignore the evidence for the Big Bang." (WK 2010aa; E. Wright 2009a)

The Alternative Pre-Big Bang Theories

These major problems in the Big Bang theory have led some scholars to propose alternative cosmological models to account for pre-Big Bang phenomena.

Although there are quite a number of these alternative cosmological models, consider two major ones below, for illustration.

(a) Brane Cosmology

In "brane cosmology"—our universe is understood as "the visible, four-dimensional universe" which "is restricted to a brane inside a higher-dimensional space, called the 'bulk'....In the bulk model, other branes may be moving through this bulk. Interactions with the bulk, and possibly with other branes, can influence our brane and thus introduce effects not seen in more standard cosmological models." (WK 2010cc)

Brane cosmology can come in different models, and one of them is known as the "ekpyrotic" model by Neil Turok and Paul Steinhardt (2002), which "hypothesizes that the origin of the observable universe occurred when two parallel branes collided." (WK 2010cc; G. Musser 2002)

Another model of brane cosmology, known as the "cyclic" model also by Paul Steinhardt (2001) and Neil Turok, takes a step further and "describes a universe exploding into existence not just once, but repeatedly over time. The theory could potentially explain why a mysterious repulsive form of energy known as the 'cosmological constant,' and which is accelerating the expansion of the universe, is several orders of magnitude smaller than predicted by the standard Big Bang model." (WK 2010ee)

In general, there are two noticeable advantages in brane csomology, when compared with the Big Bang theory.

The first advantage is that brane cosmology does not need the idea of "singularity" at the beginning of the universe (as assumed in the theory of general relativity), and this is important to understand the origin of the universe before the Big Bang, because "general relativity must break down before the Universe reaches the Planck temperature." (WK 2010aa; S. Hawking 1973)

Instead, the "ekpyrotic" model in brane csomology as proposed by Turok and Steinhardt suggests that "the universe did not start in a singularity, but came about from the collision of two branes. This collision released a large amount of energy which would appear very similar to the classic big bang, but avoids the singularity. The universe then expands until it appears almost featureless and flat, and this allows another cycle to take place with the universe being mostly removed of all debris." (Wk 2010dd)

And the second advantage is that brane cosmology can explain the anomalous problem concerning the weakness of gravity when compared with other three fundamental forces of nature.

In other words, "the model can explain the weakness of gravity relative to the other fundamental forces of nature, thus solving the so-called hierarchy problem. In the brane picture, the other three forces (electromagnetism and the weak and strong nuclear forces) are localized on the brane, but gravity has no such constraint and so much of its attractive power 'leaks' into the bulk. As a consequence, the force of gravity should appear significantly stronger on small (subatomic or at least sub-millimetre) scales, where less gravitational force has 'leaked.' Various experiments are currently underway to test this." (WK 2010cc)

(b) The Chaotic Inflation Theory

In the "chaotic inflation" theory proposed by Andre Linde (1986 & 1986a) and Alan Guth, among others—our univrese is just one of different universes in the multiverse, because "universal inflation ends locally here and there in a random fashion, each end-point leading to a bubble universe expanding from its own big bang." (WK 2010aa; J. Khoury 2002)

In this sense, both the "chaotic inflation" theory and the "cyclic" model in brane cosmology treat "the Big Bang as an event in a much larger and older Universe, or multiverse, and not the literal beginning." (WK 2010aa)

More techically speaking,, "chaotic Inflation theory models quantum fluctuations in the rate of inflation. Those regions with a higher rate of inflation expand faster and dominate the universe, despite the natural tendency of inflation to end in other regions. This allows inflation to continue forever, to produce future-eternal inflation." (WK 2010ff; A. Linde 1986)

Victor Stenger (2010) summarized the idea of chaotic inflation in this concise way: "Within the framework of established knowledge of physics and cosmology, our universe could be one of many in a super-universe or multiverse. Linde…has proposed that a background space-time 'foam' empty of matter and radiation will experience local quantum fluctuations in curvature, forming many bubbles of false vacuum that individually inflate into mini-universes with random characteristics. Each universe within the multiverse can have a different set of constants and physical laws. Some might have life of a form different from ours; others might have no life at all or something even more complex or so different that we cannot even imagine it. Obviously we are in one of those universes with life." (WK 2010ff)

In this sense, "different regions of this inflationary universe (termed a multiverse) decayed to a true vacuum state at different times, with decaying regions corresponding to 'sub'-universes not in causal contact with each other and resulting in different physical laws in different regions which are then subject to 'selection' which determine each region's components based upon (dependent on) the survivability of the quantum components within that region. The end result will be a finite number of universes with physical laws consistent within each region of spacetime." (WK 2010ff)

This also means that a given universe does not have to rapidly expand forever, as it can also be static or even decline: "More recently past-eternal models have been proposed which adhere to the perfect cosmological principle and have features of the steady state cosmos. The universe may also decay." (WK 2010ff; A. Aguirre 2002 & 2003; J. Gribbin 2010)

A Broader Viewpoint

Yet, the pre-Big Bang cosmological models have their share of the problems.

For instance, because both the "chaotic inflation" theory and the "cyclic" model in brane cosmology treat "the Big Bang as an event in a much larger and older Universe, or multiverse, and not the literal beginning"—a critical question to ask here is, Is there a beginning of the cycling? (WK 2010aa)

Even Steinhardt himself could not answer and thus said: "We don't know yet how to make that into a scientifically decidable question." (NS 2007) In other words, within the two models, "there is no way to tell whether the cycling has been going on forever....The problem is that information about previous cycles tends to get scrambled. Even if the cycling had a beginning, there may be no way to detect it." (NS 2007)

Yet, Steinhardt is optimistic enough to add that the research on pre-Big Bang cosmological models have "really opened the door for people to come up with other imaginative ideas that take us back to the big bang and beyond." (NS 2007)

In any event, the introduction to the Big Bang theory and its pre-Big Bang alternatives not only pinpoints the opposing sides in the debate on the origin of the universe (from the perspective of nature) but also teaches us something valuable about the ontological principles in existential dialectics, and good examples include the absoluteness-relativeness principle, the predictability-unpredictability principle, the explicability-inexplicability principle, the finiteness-transfiniteness principle, the preciseness-vagueness principle, the simpleness-complicatedness principle, the openness-hiddenness principle, the denseness-emptiness principle, the order-chaos principle, the slowness-quickness principle, the expansion-contraction principle, the convention-novelty principle, the evolution-transformation principle, the symmetry-asymmetry principle, the softness-hardness principle, the regression-progression principle, and the same-difference principle.

For instance, in relation to the absoluteness-relativeness principle, if there is absoluteness (e.g., the absolute view, wihtin the Big Bang theory, that the universe started in a singularity), there is relativeness (e.g., what is right for the Big Bang theory in regard to the need for the idea of singularity is not so for the "ekpyrotic" model in brane csomology, in that "the universe did not start in a singularity, but came about from the collision of two branes"). And the reverse direction also holds true.

In relation to the predictability-unpredictability principle, if there is predictability (e.g., the predictable occurrence, in accordance to the chaotic inflation theory, of some universes which are "subject to 'selection' which determine each region's components based upon...the survivability of the quantum components within that region," in the larger context of the multiverse), there is unpredictability (e.g., the more difficult task to predict which particular universes will survive in the multiverse and end up with which particular physical laws). (WK 2010ff) And the reverse direction also holds true.

In relation to the explicability-inexplicability principle, if there is explicability (e.g, the explanation, in accordance to the Big Bang theory, of the beginning of the universe on the basis of the Big Bang, with its different epochs), there is inexplicability (e.g., the lack of sufficient explanation, if viewed within the Big Bang theory, of what happened prior to the Big Bang, as critically pointed out by those in pre-Big Bang theories). And the reverse direction also holds true.

In relation to the finiteness-transfiniteness principle, if there is finiteness (e.g., the finite number of epochs which have been identified for the timeline of the Big Bang), there is transfiniteness (e.g., the mind-boggling numbers of particles, stars, galaxies, quasars, etc., in the timeline of the Big Bang). And the reverse direction also holds true.

In relation to the preciseness-vagueness principle, if there is preciseness (e.g., the precise identification, within the Big Bang theory, that "the Universe today is 72% dark energy, 23% dark matter, 4.6% regular matter and less than 1% neutrinos"), there is vagueness (e.g,. the vagueness in this identification, within the Big Bang theory, of "dark matter" and "dark energy," which, unfortunately, are not exactly clear as to what they really are, since they are still poorly understood in our time and their "exact nature...remains one of the great mysteries of the Big Bang"). (WK 2010aa; G. Hinshaw 2009) And the reverse direction also holds true.

In relation to the simpleness-complicatedness principle, if there is simpleness (e.g., the relatively simple analysis of the beginning of the universe on the basis of the Big Bang, with its different epochs, especially in the context of the standard Big Bang theory), there is complicatedness (e.g., the relatively more complicated analysis of the beginning of the universe by questioning the claims of the standard Big Bang theory and pointing out its major problems, as pointed out by the critics). And the reverse direction also holds true.

In relation to the openness-hiddenness principle, if there is openness (e.g., the open exploration of the beginning of the universe "up to 10^{-43} seconds after the Big Bang" in the Planck epoch, in accordance to the Big Bang theory), there is hiddenness (e.g., the hiddenness of the pre-Big Bang period, such that the Big Bang theory cannot account for). And the reverse direction also holds true.

In relation to the denseness-emptiness principle, if there is denseness (e.g., the extreme density of the universe when it started at the point of singularity, in accordance to the Big Bang theory), there is emptiness (e.g., the relatively lesser density of the universe in our present era, about "13.7 billion years since the big bang," because "the expansion of the universe appears to be accelerating," farther and farther away from the point of singularity, in accordance to the Big Bang theory). And the reverse direction also holds true.

In relation to the order-chaos principle, if there is order (e.g., "the universe began in its most orderly state," because of the second law of thermodynamics), there is chaos (e.g., "the entropy of the universe—a measure of its disorder—increases with time," because of the second law of thermodynamics). (NS 2007) And the reverse direction also holds true.

In relation to the slowness-quickness principle, if there is slowness (e.g., the relatively slower speed of the motion of particles through space, when compared with the expansion of space itself), there is quickness (e.g., the relatively faster speed of the expansion of space itself, which can travel faster than the speed of light, and "[t]his does not violate the speed limit dictated by relativity, since the expansion of space itself is happening faster than light, rather than the motion of particles through that space," which is subject to the limit of the speed of light). (NS 2007) And the reverse direction also holds true.

In relation to the expansion-contraction principle, if there is expansion (e.g., some universes within the multiverse can "rapidly expand forever," in accordance to the chaotic inflation theory), there is contraction (e.g., some universes withiin the multiverse can even "decay" over time and do not survive the "selection" process, in accordance to the chaotic inflation theory). (WK 2010ff) And the reverse direction also holds true.

In relation to the convention-novelty principle, if there is convention (e.g., the conventional wisdom about the beginning of the universe in terms of the Big Bang), there is novelty (e.g., the alternative novel challenge to the conventional wisdom about the Big Bang, by those scholars favoring pre-Big Bang theories, like Steinhardt and Turok, for "a universe exploding into existence not just once, but repeatedly over time"). And the reverse direction also holds true.

In relation to the evolution-transformation principle, if there is evolution (e.g., the natural evolution of humans to observe with their naked eyes in the state of nature), there is transformation (e.g., the technical transformation of human ability to observe the world by way of the invention of telescopes, which Hubble in 1929 used to "discover…a correlation between distance and recession velocity—now known as Hubble's law"). (WK 2010aa) And the reverse direction also holds true.

In relation to the symmetry-asymmetry principle, if there is symmetry (e.g., the co-existence of different cosmological models in the history of cosmology to understand the beginning of the universe, be they about the Big Bang theory and its pre-Big Bang alternatives), there is asymmetry (e.g., the domiannce of the Big Bang theory, especially since "the discovery and confirmation of the cosmic microwave background radiation in 1964"—but the rise of pre-Big Bang theories in our time, especially in the last two decades). And the reverse direction also holds true.

In relation to the softness-hardness principle, if there is softness (e.g., "electrons and protons to combine into hydrogen atoms, $p + e \rightarrow H$"), there is hardnes (e.g., when antimatter and matter, or antiparticles and partices, meet, they annihilate each other, instead of combining to form something new). (CC 2010) And the reverse direction also holds true.

In relation to the regression-progression principle, if there is regression (e.g., the regression made by the Big Bang theory in regard to some serious problems, as pointed out by the critics), there is progression (e.g., the progress made by the Big Bang theory to help us understand better the beginning of the universe, at least "up to 10^{-43} seconds after the Big Bang" in the Planck epoch). And the reverse direction also holds true.

And in relation to the same-difference principle, if there is similarity in outcome (e.g., the contribution to the understanding of the beginning of the universe, be it by way of the Big Bang theory or the pre-Big Bang alternatives), there is difference in outcome (e.g., the contribution to the understanding of the beginning of the universe "up to 10^{-43} seconds after the Big Bang" by way of the Big Bang theory—but the contribution to the understanding of the beginning of the universe prior to the Big Bang by way of the pre-Big Bang cosmological models). And the reverse direction also holds true.

Its Contested Beginnings—and Society

The contested beginnings of the universe (and the aftermath) can also be examined from a different perspective, this time, in relation to society, as it can cast a different light on the subject matter.

Two case studies are arranged here, namely, (2.4.1) earth-centric cosmology, and the politics of the church, and (2.4.2) correlative cosmology, and empire building—to be analyzed hereafter, in that order.

Earth-Centric Cosmology, and the Politics of the Church

The history of cosmology has its own political controversy, with important implications for the advancement of knowledge about the universe.

For instance, in the case of the Big Bang theory (as discussed in the previous section), one major criticism is that the Big Bang theory is driven with the religious politics of creationism, as "the originator of the Big

Bang theory, Monsignor Georges Lemaître, was a Roman Catholic priest." (WK 2010aa; H. Kragh 1996; PBS 1998)

One consequence is that the Big Bang theory does not focus on pre-Big Bang period for the prospect of an eternal cycling of our universe (within a larger universe known as "multiverse"), because, in accordance to the doctrine of creation, "an infinite temporal regress of events cannot exist." (WK 2010b) For this reason, some pre-Big Bang cosmological models have emerged in the last two decades or so to take a closer look at the possibility of this multiverse with no beginning or end—contrary to what the Big Bang theory would like us to believe.

But there is another more well-known case of political controversy in the history of cosmology, which has to do with the "Galileo affair" in the 17th century, involving the Catholic Church and Galileo Galilei in regard to the Copernican astronomy.

The Galileo Affair

More specifically, the term "the Galileo affair" refers to "a sequence of events, beginning sometime around 1610, during which Galileo Galilei came into conflict with the Catholic Church over his support of Copernican astronomy." (WK 2010gg; M. Finocchiaro 1999)

It all started when "Galileo published his *Sidereus Nuncius* (*Starry Messenger*), describing the surprising observations that he had made with...[his] new telescope. These and other discoveries exposed major difficulties with the understanding of the heavens that had been held since antiquity, and raised new interest in radical teachings such as the heliocentric theory of Copernicus (published in *De revolutionibus orbium coelestium* in 1543)." (WK 2010gg)

But the Roman Catholic Church did not take this unorthodox cosmological understanding slightly, because "the theory...seemingly contradicted several passages of Scripture. Galileo's part in the controversies over theology, astronomy, and philosophy culminated in his trial and sentencing in 1633 on a grave suspicion of heresy." (WK 2010gg)

Galileo's Cosmological Observations

To understand this Galileo affair, it is important to put things in a more historical perspective, by first understanding Galileo's pathbreaking observations in cosmology at the time.

In 1609 and 1610, "Galileo began his telescopic observations...and by March of 1610 was able to publish a small book, *The Starry Messenger* (*Sidereus Nuncius*), relating some discoveries that had not been dreamed of in the philosophy of the time: mountains on the Moon, lesser moons in orbit around Jupiter, and the resolution of what had been thought cloudy masses in the sky (nebulae) into collections of stars too faint to see individually without a telescope. Other observations followed, including the phases of Venus and the existence of sunspots." (WK 2010gg)

All these observation seemed benign at first glance, because they did not "directly contradict...Christian doctrine. However, they caused difficulties for theologians and for natural philosophers (the name given to scientists at the time, before the modern scientific method was developed), as they contradicted the scientific and philosophical ideas of the time, which were based on those of Aristotle and Ptolemy, whose teachings were closely associated with the Catholic Church of the time (despite their work being pagan writings). In particular, the geocentric model [with the view that the Earth was the center of the universe] was contradicted by the phases of Venus [as observed by Galileo], which indicated it was both inside and outside the theorized 'celestial sphere' of the sun, and the observation of moons orbiting Jupiter instead of the Earth." (WK 2010gg)

Of course, not every one was hostile to Galileo's observations; for instance, "Jesuit astronomers, experts both in Church teachings, science, and in natural philosophy, were at first skeptical and hostile to the new ideas, however, within a year or two the availability of good telescopes enabled them to repeat the observations." (WK 2010gg)

However, many of those in authority did not accept Galileo's findings; for instance, "Christopher Clavius, the most distinguished astronomer of his age, never was reconciled to the idea of mountains on the Moon, and outside the Collegium many still disputed the reality of the observations." (WK 2010gg)

To make things worse, Galileo was also to blame, because he made enemies with some people at the time over other issues, like the dispute with "Christoph Scheiner, a prominent Jesuit," over the credit of being the first to discover "sunspots" (without knowing that "the Chinese had already been familiar with them for centuries")—and also with others "over the reasons that objects float or sink in water, siding with Archimedes against Aristotle." (WK 2010gg; M. Sharratt 1994)

Disputes with others like this were "unfriendly, and Galileo's blunt and sometimes sarcastic style, though not extraordinary in academic debates of the time, made him enemies." (WK 2010gg)

Galileo's Cosmological Observations vs. the Scripture

Things turned nasty in 1613, with the first suggestion of heresy against Galileo.

After all, at the time, "the majority of people subscribed to the Ptolemaic theory of geocentrism that the earth was the center of the universe and that all heavenly bodies revolved around the Earth. This theory accorded with the limited scientific knowledge available at the time, agreed with a literalist interpretation of Scripture in several places, such as Chronicles 16:30, Psalm 93:1, Psalm 96:10, Psalm 104:5, Ecclesiastes 1:5." (WK 2010gg)

Furthermore, "since in the Incarnation the Son of God had descended to the earth and become man, it seemed fitting that the Earth be the center around which all other celestial bodies moved. Heliocentrism, the theory that the Earth was a planet, which, along with all the others, revolved around the Sun, contradicted both geocentrism and the prevailing theological support of the theory." (WK 2010gg)

In this highly political environment in favor of the scripture, in 1913, Cosimo Boscaglia told "Galileo's patron, Cosimo II de' Medici" that the "telescopic discoveries [by Galileo] were valid, but the motion of the Earth was obviously contrary to Scripture. Galileo was defended on the spot by a Benedictine abbot, Benedetto Castelli, who was also a professor of mathematics and a former student of Galileo's. This exchange, reported to Galileo by Castelli, led Galileo to write a letter to Castelli, expounding his views on what he considered the most appropriate way of treating scriptural passages which made assertions about natural phenomena." (WK 2010gg; M. Sharratt 1994)

Then, "Tommaso Caccini, a Dominican friar, appears to have made the first dangerous attack on Galileo. Preaching a sermon in Florence at the end of 1614, he denounced Galileo, his associates, and mathematicians in general (a category that included astronomers). The biblical text for the sermon on that day was Joshua 10, in which Joshua makes the Sun stand still." (WK 2010gg)

By "late 1614 or early 1615, one of Caccini's fellow Dominicans, Niccolò Lorini, acquired a copy of Galileo's letter to Castelli, which he considered of sufficiently doubtful orthodoxy to bring to the attention of the Inquisition. In February 1615 he accordingly sent a copy to the Secretary of the Inquisition, Cardinal Paolo Emilio Sfondrati, with a covering letter critical of Galileo's supporters. On March 19, Caccini arrived at the Inquisition's offices in Rome to denounce Galileo for his Copernicanism and various other alleged heresies supposedly being spread by his pupils." (WK 2010gg; S. Drake 1978; M. Sharratt 1994)

Bellarmine's Intervention for a Compromise

The Inquisition's offices in Rome at first asked "Robert Cardinal Bellarmine, one of the most respected Catholic theologians of the time,...to adjudicate the dispute between Galileo and his opponents, including both religious zealots and secular university professors." (WK 2010gg)

Bellarmine was willing to seek a compromise in the dispute between Galileo and his opponents. For instance, in his letter to Paolo Antonio Foscarini, a Carmelite father, on April 12, 1615, with a copy sent to Galileo—Bellarmine suggested that "it is prudent for him and Galileo to limit themselves to treating heliocentrism as a merely hypothetical phenomenon and not a physically real one. Further on he says that interpreting heliocentrism as physically real would be 'a very dangerous thing, likely not only to irritate all scholastic philosophers and theologians, but also to harm the Holy Faith by rendering Holy Scripture false.' Moreover, while the topic was not inherently a matter of faith, the statements about it in Scripture were so by virtue of who said them—namely, the Holy Spirit. He conceded that if there were conclusive proof, 'then one would have to proceed with great care in explaining the Scriptures that appear contrary; and say rather that we do not understand them than that what is demonstrated is false.' However, demonstrating that heliocentrism merely 'saved the appearances' could not be regarded as sufficient to establish that it was physically real. Although he believed that the former may well have been possible, he had 'very great doubts' that the latter would be, and in case of doubt it was not permissible to depart from the traditional interpretation of Scriptures." (WK 2010gg; M. Finocchiaro 1989; R. Blackwell 1991)

In other words, "Bellarmine found no problem with heliocentrism so long as it was treated as a purely hypothetical calculating device and not as a physically real phenomenon, but he did not regard it as permissible to advocate the latter unless it could be conclusively proved. This put Galileo in a difficult position, because he believed that the available evidence strongly favoured heliocentrism, and he wished to be able to publish his arguments, but he did not have the necessary conclusive proof." (WK 2010gg; M. Sharratt 1994)

The Inquisition's Examination

In the end, Galileo refused to back down and carried out an "aggressive campaign to prevent the condemnation of Copernicanism." (WK 2010gg)

In 1615, "the Inquisition had taken a deposition from Gianozzi Attavanti..., as part of its investigation into the denunciations of Galileo by Lorini and Caccini. In this deposition, Attavanti confirmed that Galileo had advocated the Copernican doctrines of a stationary Sun and a mobile Earth, and as a consequence the Tribunal of the Inquisition would have eventually needed to determine the theological status of those doctrines." (WK 2010gg; F. Beretta 2005)

So, in 1616, "the Inquisition asked a commission of theologians, known as qualifiers, about the propositions of the heliocentric view of the universe....On February 24 the Qualifiers delivered their unanimous report: the idea that the Sun is stationary is 'foolish and absurd in philosophy, and formally heretical since it explicitly contradicts in many places the sense of Holy Scripture...'; while the Earth's movement 'receives the same judgement in philosophy and...in regard to theological truth it is at least erroneous in faith.'" (WK 2010gg; A. Fantoli 2005; E. McMullin 2005)

Then, "Pope Paul V instructed Bellarmine to deliver this result to Galileo, and to order him to abandon the Copernican opinions; should Galileo resist the decree, stronger action would be taken. On February 26, Galileo was called to Bellarmine's residence, and accepted the orders. On March 5, the decree was issued by the Congregation of the Index, prohibiting, condemning, or suspending various books which advocated the truth of the Copernican system." (WK 2010gg; S. Drake 1978)

So, "Galileo...saw heliocentrism formally declared an idea that could not be held as truth, for lack of evidence. It was consequently termed heretical by the Qualifiers, since it contradicted the literal meaning of the Scriptures....Copernicus' *De Revolutionibus*, though not formally banned, was removed from circulation pending revisions, and in fact was not fully cleared until the 19th century. Though Galileo was personally safe, and his works had not been banned, there was now much doubt (felt by other astronomers as far away as Germany) whether it was possible to do serious work in Copernican astronomy." (WK 2010gg)

Galileo on Trial, and the Aftermath

With this order from Pope Paul V, Galileo stayed quiet for some years until 1632, when he "published his book, *Dialogue Concerning the Two Chief World Systems*, with formal authorization from the Inquisition for a book which presented a balanced view of both Copernican's and church's theories. However, in the book, the Copernican theory clearly receives better treatment, instead of a balanced view. Because of this, Galileo was ordered to appear before the Inquisition for trial." (WK 2010gg)

In 1633, "Galileo was ordered to Rome to stand trial on suspicion of heresy..., 'for holding as true the false doctrine taught by some that the sun is the center of the world,' against the 1616 condemnation, since 'it was decided at the Holy Congregation...on 25 Feb 1616 that...the Holy Office would give you an injunction to abandon this doctrine, not to teach it to others, not to defend it, and not to treat of it; and that if you did not acquiesce in this injunction, you should be imprisoned.' A panel of theologians, consisting of Melchior Inchofer, Agostino Oreggi and Zaccaria Pasqualigo, reported on the *Dialogue*. Their opinions were strongly argued in favour of the view that the *Dialogue* taught the Copernican theory." (WK 2010gg; M. Finocchiaro 1989; M. Sharratt 1994)

Not surprsingly, "Galileo was found guilty, and the sentence of the Inquisition was in three essential parts," as shown below.

- "Galileo was found 'vehemently suspect of heresy,' namely of having held the opinions that the Sun lies motionless at the centre of the universe, that the Earth is not at its centre and moves, and that one may hold and defend an opinion as probable after it has been declared contrary to Holy Scripture. He was required to 'abjure, curse, and detest' those opinions." (WK 2010gg; A. Fantoli 2005; M. Finocchiaro 1989)
- "He was originally ordered to be imprisoned, however, due to his advanced age (70 at the time of the trial), the sentence was later commuted to house arrest for the rest of his life." (WK 2010gg)
- "His offending *Dialogue* was banned; and in an action not announced at the trial, publication of any of his works was forbidden, including any he might write in the future." (WK 2010gg; S. Drake 1978; M. Sharratt 1994)

Later, however, "Galileo was allowed to return to his villa at Arcetri near Florence, where he spent the remainder of his life under house arrest....His standing would remain questioned at every turn. In March 1641, Vincentio Reinieri, a follower and pupil of Galileo, wrote him at Arcetri that an Inquisitor had recently compelled the author of a book printed at Florence to change the words 'most distinguished Galileo' to 'Galileo, man of noted name.'" (WK 2010gg; S. Drake 1978)

One had to wait until 1835 for a more favorable view of the Galileo affair by the Church, when "Copernicus's *De Revolutionibus* and Galileo's *Dialogue* were then subsequently omitted from the...edition of the *Index* [*of Forbidden Books*]." (WK 2010gg)

The vindication finally came in 1992, when Pope John Paul II thus praised Galileo in *L'Osservatore Romano*: "Thanks to his intuition as a brilliant physicist and by relying on different arguments, Galileo, who practically invented the experimental method, understood why only the sun could function as the centre of the world, as it was then known, that is to say, as a planetary system. The error of the theologians of the time, when they maintained the centrality of the Earth, was to think that our understanding of the physical world's structure was, in some way, imposed by the literal sense of Sacred Scripture." (WK 2010gg)

And in 2000, "Pope John Paul II issued a formal apology for all the mistakes committed by some Catholics in the last 2,000 years of the Catholic Church's history, including the trial of Galileo among others." (WK 2010gg; R. Carroll 2000)

In Retrospect

However, this major turnaround in the Galileo affair (from the verdict against him in 1633 to the papal apology in 2000) does not settle, once and for all, the final words on the Galileo affair for everyone.

Even today, there are some who want to defend the decision by the Catholic Church against Galileo, and some good examples include prominent philosophers like Ernst Bloch, C.F. Von Weizsacker, and Paul Feyerabend.

For instance, Feyerabend thus wrote: "The Church at the time of Galileo kept much more closely to reason than did Galileo himself, and she took into consideration the ethical and social consequences of Galileo's teaching too. Her verdict against Galileo was rational and just, and the revision of this verdict can be justified only on the grounds of what is politically opportune." (WK 2010gg) After all, Galileo did not have conclusive proof of heliocentrism at the time. (WK 2010gg; M. Sharratt 1994)

And Cardinal Joseph Ratzinger (1994) used Feyerabend's quotation to then conclude: "It would be foolish to construct an impulsive apologetic on the basis of such views." (WK 2010gg)

The introduction to this recurrent controversy concerning the Galileo affair allows us to see the opposing sides of the debate, such that the very understanding of the universe is not to the extent that its respective ideologues would like us to believe. Each side has its own reasons and motives to make assertions in the way that it does, and these reasons and motives vary from one era to another and from one authority to another.

In addition, the analysis of the Galileo affair can also shed some light on the ontological principles in existential dialectics, and good examples include the absoluteness-relativeness principle, the predictability-unpredictability principle, the explicability-inexplicability principle, the preciseness-vagueness principle, the simpleness-complicatedness principle, the openness-hiddenness principle, the denseness-emptiness principle, the slowness-quickness principle, the expansion-contraction principle, the theory-praxis principle, the convention-novelty principle, the evolution-transformation principle, the symmetry-asymmetry principle, the softness-hardness principle, the regression-progression principle, and the same-difference principle.

For instance, in relation to the absoluteness-relativeness principle, if there is absolute viewpoint (e.g., the absolute view of the Catholic Church, at the time of Galileo, that "the earth was the center of the universe," on the basis of "a literalist interpretation of Scripture"), there is relative counterpart (e.g., what is correct for the Catholic Church in regard to geocentrism is not so for Galileo in regard to helocentrism). And the reverse direction also holds true.

In relation to the predictability-unpredictability principle, if there is predictability (e.g., the predictable tendency of the Catholic Church to interpret the Scripture in a way so as to defend it), there is unpredictability (e.g., the more difficult task to predict what particular interpretation that the Catholic Church would creatively use to defend the Scripture in a given case is, since it varies from one era to another and from one authority to another, as shown by the shift of the church's position in regard to the Galileo affair over the centuries). And the reverse direction also holds true.

In relation to the explicability-inexplicability principle, if there is explicability (e.g., the explanation by Bellarmine "to treat…heliocentrism as a merely hypothetical phenomenon and not a physically real one," because Galileo "did not have the necessary conclusive proof"), there is inexplicability (e.g., the lack of sufficient explanation by Bellarmine of why "it was not permissible to depart from the traditional interpretation of Scriptures," just because heliocentrism was not conclusively proved—or why geocentrism must be true, only because heliocentrism was not conclusively proved at the time). And the reverse direction also holds true.

In relation to the preciseness-vagueness principle, if there is preciseness (e.g., the precise identification of the Earth as the center of the universe in geocentrism), there is vagueness (e.g, the vagueness in the geocentric view that the Earth is the center of the universe, as this can mean different things for different peoples, because one geocentric view is that "the stars, sun, and planets appear to revolve around the Earth each day, with the stars circling around the pole and those stars nearer the equator rising and setting each day and circling back to their rising point," and another geocentric view is that "the Earth is solid and stable it is not moving—but is at rest"). (WK 2010c) And the reverse direction also holds true.

In relation to the simpleness-complicatedness principle, if there is simpleness (e.g., the relatively simplistic geocentric view of the Earth as the center of the universe), there is complicatedness (e.g., the relatively complicated helocentric view that the Earth is not the center of the universe, but part of the solar system, with the Sun as the center and other objects revolving around it). And the reverse direction also holds true.

In relation to the openness-hiddenness principle, if there is openness (e.g., the open defense of geocetrism by the Catholic Church against the helocentric view by Galileo), there is hiddenness (e.g.,the hidden bias in the Catholic Church because of its unquestioned faith in the Scripture). And the reverse direction also holds true.

In relation to the denseness-emptiness principle, if there is denseness (e.g., the relatively denser concentration of individuals with faith in geocentrism within the Catholic Church at the time of Galileo), there is emptiness (e.g., the relatively less dense, or more empty, concentration of individuals like Galileo and other astronomers working with good telescopes to have faith in geocentrism). And the reverse direction also holds true.

In relation to the slowness-quickness principle, if there is slowness (e.g., the relatively slower readiness of those in the Catholic Church to accept heliocentrism unless it is conclusively proved), there is quickness (e.g., the relatively quicker readiness of those like Galileo to accept heliocentrism, because "the available evidence strongly favoured heliocentrism," even though "he did not have the necessary conclusive proof"). And the reverse direction also holds true.

In relation to the expansion-contraction principle, if there is expansion (e.g., the relatively more developed ability of those in the Catholic Church to be good in biblical hermeneutics), there is contraction (e.g., the relatively less developed ability of those in the Catholic Church to be good in working with telescopes for scientific discovery). And the reverse direction also holds true.

In relation to the theory-praxis principle, if there is theory (e.g., the theoretical construction of the Scripture), there is praxis (e.g., the practical application of the Scripture to reject heliocentrism in the Galileo affair). And the reverse direction also holds true.

In relation to the convention-novelty principle, if there is convention (e.g., the conventional wisdom about geocentrism at the time of Galileo), there is novelty (e.g., the alternative novel challenge to geocentrism by Galileo, by way of his original heliocentric observations). And the reverse direction also holds true.

In relation to the evolution-transformation principle, if there is evolution (e.g., the natural evolution of humans to observe things with their naked eyes in the state of nature), there is transformation (e.g., the technical transformation of human observation by the invention of telescopes). And the reverse direction also holds true.

In relation to the symmetry-asymmetry principle, if there is symmetry (e.g., the co-existence of different views of the universe in human history, like geocentrism and heliocentrism), there is asymmetry (e.g., the popularity of geocentrism in antiquity until the 16th century or so, but the increasing acceptance of heliocentrism from the late 16th century onward). And the reverse direction also holds true.

In relation to the softness-hardness principle, if there is softness (e.g., the praise of those who conformed to the geocentric view of the Catholic Church at the time of Galileo), there is hardness (e.g., the punishment of those who rejected the geocentric view of the Catholic Church at the time of Galileo, who was put under house arrest, for example, for his heliocentrism). And the reverse direction also holds true.

In relation to the regression-progression principle, if there is regression (e.g., the regression made by heliocentrism at the time of Galileo, which "not only...irritate[d]...scholastic philosophers and theologians, but also...harm[ed] the Holy Faith by rendering Holy Scripture false" and was not even backed up with "necessary conclusive proof," as pointed out by Bellarmine), there is progression (e.g., the progress made by heliocentrism, like the one by Galileo, to help us understand better why the Earth is not the center of the universe). And the reverse direction also holds true.

And in relation to the same-difference principle, if there is similarity in outcome (e.g., the contribution to our understanding of the world, be this so by way of heliocentrism or geocentrism), there is difference in outcome (e.g., the contribution to our understandiing of the world by geocentrism from a religious standpoint in accordance to the Scripture, but the contribution to our understanding of the world by heliocentrism from a scientific standpoint in accordance to observations with telescopes). And the reverse direction also holds true.

Correlative Cosmology, and Empire Building

Another way to understand the possibility and desirability of the contested beginnings of the universe (and the aftermath) from the perspective of society has to do with the relationship between "correlative cosmology" and empire building.

Correlative Cosmology in Ancient China

The term "correlative cosmology" here refers to the ancient Chinese cosmology as "a framework of conceptions and relations," which is "an immense system of correlation-building, based on interlaced pairs (correlated to *Ying-Yang*...), fours (correlated to the four directions), fives (correlated to the Five Phases or *Wuxing*...), eights (correlated to the Eight Trigrams), and so on." (A. Wang 2000)

This complicated system of "correlation-building" provides "an orderly system of correspondence among various domains of reality in the universe, correlating categories of the human world, such as the human body, behavior, morality, the sociopolitical order, and historical changes, with categories of the cosmos, including time, space, the heavenly bodies, seasonal movement, and natural phenomena." (A. Wang 2000)

In this sense, "correlative cosmology" links the natural world with the human world, as one of correspondence, in the sense that it constitutes what Schwartz once called "a kind of anthropocosmology in which entities, processes, and classes of phenomena found in nature correspond to or 'go together with' various entities, processes, and classes of phenomena in the human world." (A. Wang 2000)

However, some scholars like A. C. Graham argue that "correlative cosmology" had also appeared in some non-Chinese civilizations. (A. Wang 2000)

Be so as this may, our concern here is with the ancient Chinese version.

Sifang and Wuxing Versions

Correlative cosmology in ancient China does not consist of only one version, as it had evolved into different ones over time, and two good examples here for illustration concern "Sifang" and "Wuxing." (A. Wang 2000)

The origin of *Wuxing* can be traced back to an earlier version of correlative cosmology known as *Sifang*, which refers to "the archaic cosmology of the Four Quarters" during the Bronze Age of Chinese history (viz., during the Shang and Western Zhou eras, ca. 1700-771 B.C.). (A. Wang 2000)

Although the two, *Wuxing* and *Sifang*, share "a certain symbolic resemblance and some continuity," there are "fundamental differences between the two in their structures, functions, and meanings." (A. Wang 2000)

For instance, *Wuxing* emerged "from the fourth to third centuries B.C., during the transition from the Bronze Age to the imperial era. These early forms of *Wuxing* absorbed certain structural and symbolic features of *Sifang*, and became one of many systems of classification that coexisted in mantic and ritual practices of the time. At this stage, *Wuxing* was loosely defined and unstematically used. Later, in the course of the formation of centralized empires, it was elevated to become the core of correlative cosmology, the predominant system of classification, and a shared discourse among different interest groups in their political interactions." (A. Wang 2000)

The Close Relationship Between Cosmology and Politics

But correlative cosmology so understood, even when contextualized within the history of *Wuxing* and its historical predecessor *Sifang*, remains vague, in terms of its particular meanings.

Consider a few interpretations of correlative cosmology below, for illustration.

Firstly, one interpretation is that "the leading critics of Chinese history and culture in the early twentieth century—such as Liang Qichao and Gu Jiegang—deconstructed *Wuxing* cosmology by revealing the political context within which it was formed and the political motivations of the forgers of *Wuxing* texts or textual fragments." (A. Wang 2000)

But a weakness of this interpretation is that these critics "treated cosmology only as a product of textual forgeries carried out by a few Qin and Han compliers motivated by utilitarian and political concerns. Such a conclusion reduced a profound cultural phenomenon—a cosmology that prevailed in all of Chinese society and persisted throughout history—to a mere lie invented by a few cunning flatterers and usurpers to cover up the real matter of power." (A. Wang 2000)

Secondly, another interpretation is that "historians of philosophy and science have studied *Wuxing* cosmology primarily as a mode of thinking….While varying in methodology and interests, they have commonly developed a theme suggested by Sima Qian (b. 145 B.C.), attributing the creation of *Wuxing* cosmology to an individual thinker active at the eve of the imperial era, Zou Yan (305-240 B.C.), and his school of philosophy, which has been retrospectively labeled *Yin-Yang jia* (the school of *Yin and Yang*). This school of philosophy is a combination of magic and science according to Feng Youlan, or a school of 'naturalists' according to Needham....[Later] Schwartz sees Zou Yan as a pioneer of Han Confucianism, initiating the fusion of cosmology with Confucian values." (A. Wang 2000).

But a weakness of this interpretation is that "treating cosmology as a school of philosophy or a mode of thinking represents another kind of reductionism: it gives ontological or analytical priority to the products of the mind, to ideas, meanings, and thoughts,...neglecting or obscuring its immense sybmolic manifestation in everyday cultural practice and material production." (A. Wang 2000)

And thirdly, still another interpretation is that, as suggested by Aihe Wang (2000), it treats correlative cosmology "as a single, but highly dynamic, process of cultural-political change that is constituted by fragmentary events, by conflicts and disputes, and by contests for power and control."

In this interpretation, correlative cosmology in ancient China was transformed, especially in relation to the two political eras—"from the hegemonic states of the Bronze Age (the Shang and Western Zhou, ca. 1700-771 B.C.) to the unified empires of the Iron Age (Qin and Han, 221 B.C.-220 A.D.)." (A. Wang 2000)

In this sense, "cosmology was constructive to power while being at the same time constantly transformed by the political process"; for instance, "[t]he ruling clans of the Bronze Age drew legitimacy through a cosmological system known as *Sifang* (the Four Quarters), in which the king and his ancestral line were believed to be the conduit of divine authority." (A. Wang 2000)

Later, there was a shift from *Sifang* to *Wuxing*; in "about 400 B.C., the shift to *Wuxing* (commonly knonw as the Five Elements, in which the cosmic energies of wood, fire, earth, metal, and water constantly interact) both paved the way for and was subseqently refined by the politics of a unified, imperial order." (A. Wang 2000)

But a weakness of this interpertation is that it too treats cosmology in a reductionistic way, although differently from the first two—this time, in understanding cosmology as a cultural fabric in its constant interaction with the political realm.

By contrast, my methodological holism calls for a treatment of cosmology from the multiple perspectives of the mind, nature, society, and culture in the most comprehensive way of examining reality.

From a Larger Picture of Things

These criticisms of the political treatment of correlative cosmology are not to dismiss the close relationship between cosmology and politics but to show us the opposing sides of the debate, such that the business of interpreting a cosmology is not to the extent that its respective ideologues would like us to believe. Each interpretation has its merit, just as it has its own drawback.

More importantly, the analysis of correlative cosmology and its political motives can also reveal something refreshing about the ontological principles in existential dialectics, and good examples include the absoluteness-relativeness principle, the predictability-unpredictability principle, the explicability-inexplicability principle, the finiteness-transfiniteness principle, the preciseness-vagueness principle, the simpleness-complicatedness principle, the openness-hiddenness principle, the denseness-emptiness principle, the slowness-quickness principle, the expansion-contraction principle, the theory-praxis principle, the convention-novelty principle, the evolution-transformation principle, the symmetry-asymmetry principle, the softness-hardness principle, the regression-progression principle, and the same-difference principle.

For instance, in relation to the absoluteness-relativeness principle, if there is absoluteness (e.g., the absolute view, in ancient Chinese correlative cosmology, of the universe as "an orderly system of correspondence among various domains of reality in the universe, correlating categories of the human world…with categories of the cosmos," without any transcendental entity like God outside it), there is relativeness (e.g., what is obvious for the Chinese correlative cosmology in regard to the absolute existence of the universe without the need of any transcendental entity like God is not so for Christian cosmology in regard to the need of a transcendental entity like God, who created the universe). And the reverse direction also holds true.

In relation to the predictability-unpredictability principle, if there is predictability (e.g., the predictable tendency of a correlative cosmologist to correlate everything in the natural world with its corresponding equivalent in the human world), there is unpredictability (e.g., the more difficult task to predict which particular correlation a correlative cosmologist would refer to in a particular case, since it varies from one version of correlative cosmology to another, from one historical era to another, from one expert to another, from one category of correlation to another, and so on). And the reverse direction also holds true.

In relation to the explicability-inexplicability principle, if there is explicability (e.g., the explanation, by those in Chinese correlative cosmology, of the need of a "unified, imperial order," on the basis of the universe as "an orderly system of correspondence…correlating categories of the human world…with categories of the cosmos"), there is inexplicability (e.g, the lack of convincing explanation, by those in Chinese correlative cosmology, of why it must be true that the universe is "an orderly system of correspondence" like that). And the reverse direction also holds true.

In relation to the finiteness-transfiniteness principle, if there is finitenes (e.g., the finite number of "correlations" in the universe as defined in Chinese correlative cosmology, such as those based on "pairs," "fours," "fives," "eights," and the like), there is transfiniteness (e.g., the amazingly huge number of things that these "correlations" can possibly refer to in the entire universe). And the reverse direction also holds true.

In relation to the preciseness-vagueness principle, if there is preciseness (e.g., the precise identification of the historical origin of *Wuxing* in *Sifang*, which is "the archaic cosmology of the Four Quarters" during the Bronze Age of Chinese history), there is vagueness (e.g., the vagueness in understanding the historical origin itself, since it is subject to different historical interpretations, not only because of the "archaic" nature of *Sifang* at the time but also because of the "loosely defined and unstematically used" *Wuxing* at the time). And the reverse direction also holds true.

In relation to the simpleness-complicatedness principle, if there is simpleness (e.g., the relatively simplistic analysis of Chinese correlative cosmology as "a kind of anthropocosmology in which entities, processes, and classes of phenomena found in nature correspond to or 'go together with' various entities, processes, and classes of phenomena in the human world"), there is complicatedness (e.g., the relatively more complicated analysis of Chinese correlative cosmology in terms of different conflicting interpretations as pointed out by the critics, who do not share a consensus). And the reverse direction also holds true.

In relation to the openness-hiddenness principle, if there is openness (e.g., the open exploraton of the correlations between things in the natural world and those in the human world, in accordance to Chinese correlative cosmology), there is hiddenness (e.g., the hidden bias in interpreting the validity of correlative cosmology, because of the different political motives closely linked with the interpretations). And the reverse direction also holds true.

In relation to the denseness-emptiness principle, if there is denseness (e.g., the relatively denser concentration of those who accepted the "correlations" in the universe on the basis of "fours" as in *Sifang*, during the Bronze Age of ancient Chinese history), there is emptiness (e.g., the relatively less dense, or more empty, concentration of those who accepted the "correlations" in the universe on the basis of "fives," during the Bronze Age of ancient Chinese history). And the reverse direction also holds true.

In relation to the slowness-quickness principle, if there is slowness (e.g., the relatively slower readiness of those "leading critics" like Liang Qichao and Gu Jiegang—who "deconstructed *Wuxing* cosmology" on the basis of revealing "the political motivations of the forgers of *Wuxing* texts"—to link correlative cosmology with empire building as proposed by A. Wang), there is quickness (e.g., the relatively quicker readiness of those like A. Wang to interpret correlative cosmology as a cultural fabric in its constant interaction with the political realm, for the service of empire building). And the reverse direction also holds true.

In relation to the expansion-contraction principle, if there is expansion (e.g., the relatively more developed ability of those who specialize in correlative cosmology to think about reality in terms of "correlations" between the natural world and the human world), there is contraction (e.g., the relatively less developed ability of those who specialize in correlative cosmology to think logically, not correlatively, in a more systematic way as in Western symbolic logic). And the reverse direction also holds true.

In relation to the theory-praxis principle, if there is theory (e.g., the theoretical construction of correlative cosmology), there is praxis (e.g.,. the practical application of correlative cosmology, as in *Wuxing*, by those in power for "the politics of a unified, imperial order" in "about 400 B.C."). And the reverse direction also holds true.

In relation to the convention-novelty principle, if there is convention (e.g., the conventional wisdom about *Sifang* cosmology during the Bronze Age of ancient Chinese history), there is novelty (e.g., the alternative novel challenge to the conventional wisdom about *Sifang* cosmology by its successor known as *Wuxing* cosmology for a different system of correlations based on fives). And the reverse direction also holds true.

In relation to the evolution-transformation principle, if there is evolution (e.g., the natural evolution of humans to observe things in the world in the state of nature), there is transformation (e.g., the technical transformation of human observations about the world by way of the invention of correlative cosmology for "an immense system of correlation-building"). And the reverse direction also holds true.

In relation to the symmetry-asymmetry principle, if there is symmetry (e.g., the co-existence of different versions of correlative cosmology in Chinese history, be they about *Wuxing*, *Sifang*, and the like), there is asymmetry (e.g., the popularity of *Sifang* durng the Bronze Age, but the acceptance of *Wuxing* "from the fourth to third centuries B.C., during the transition from the Bronze Age to the imperial era"). And the reverse direction also holds true.

In relation to the softness-hardness principle, if there is softness (e.g., the refinement and acceptance of correlative cosmology by those in *Wuxing*), there is hardness (e.g., the deconstruction and rejection of correlative cosmology by those critics like Liang Qichao and Gu Jiegang). And the reverse direction also holds true.

In relation to the regression-progression principle, if there is regression (e.g., the problems in *Wuxing* cosmology, as pointed out by the critics, with their different interpretations), there is progression (e.g., the progress made in correlative cosmology to help us understand some correlations in category between the natural world and the human world). And the reverse direction also holds true.

And in relation to the same-difference principle, if there is similarity in outcome (e.g., the contribution to the better understanding of correlative cosmology, regardless of the existence of different interpretations), there is difference in outcome (e.g., the contribution to the better understanding of correlative cosmology by way of identifying its forgers, but the contribution to the better understanding of correlative cosmology by way of linkage its role to empire building). And the reverse direction also holds true.

Its Contested Beginnings—and Culture

The contested beginnings of the universe (and the aftermath) can also be explored from a different perspective, this time, in relation to culture, as it can reveal something different about the subject matter.

Two case studies are in order here, namely, (2.5.1) theology, and the cosmological argument, and (2.5.2) hylomorphism, ethics, and the teleology of the universe—to be analyzed hereafter, respectively.

Theology, and the Cosmological Argument

The first case study here concerns the relationship between theology and the cosmological argument.

The term "the cosmological argument" refers to "an argument for the existence of a First Cause (or instead, an Uncaused cause) to the universe, and by extension is often used as an argument for the existence of an 'unconditioned' or 'supreme' being, usually then identified as God." (WK 2010hh)

The cosmological argument is also known as "an argument from universal causation, an argument from first cause, the causal argument or the argument from existence. The basic premise…is that something caused the Universe to exist, and this First Cause must be God. It has been used by various theologians and philosophers over the centuries, from the ancient Greek Plato and Aristotle to the medieval St. Thomas Aquinas and the 20th century Frederick Copleston." (WK 2010hh)

One way to summarize the cosmological argument can be shown below. (WK 2010hh)

- "Every finite and contingent being has a cause."
- "Nothing finite and contingent can cause itself."
- "A causal chain cannot be of infinite length."
- "Therefore, a First Cause (or something that is not an effect) must exist."

Surely, one can question the premises and conclusion of the argument (as will be clear later)—but, before that, one must first understand the different variants of the cosmological argument.

Three Variants of the Cosmological Arguement

The cosmological argument can come in all shapes and sizes, and three good illustrations concern, namely, (a) "the argument from causation, *in esse*" (in existence), (b) "the argument from causation, *in fieri*" (in becoming), and (c) "the argument from contingency"—to be introduced hereafter, in that order (and summarized in *Table 2.5*). (WK 2010hh)

(a) The Argument from Causation, *in Esse* (in Existence)

The first variant of the cosmological argument is one about causation in existence (*in esse*).

The term "in esse" here means "in existence," which "is more akin to the light from a candle or the liquid in a vessel. George Hayward Joyce [1922]…explained that '...where the light of the candle is dependent on the candle's continued existence, not only does a candle produce light in a room in the first instance, but its continued presence is necessary if the illumination is to continue. If it is removed, the light ceases. Again, a liquid receives its shape from the vessel in which it is contained; but were the pressure of the containing sides withdrawn, it would not retain its form for an instant.'" (WK 2010hh)

Aristotle, for instance, defended the argument from causation, *in esse*, in terms of "the idea of a First Cause, often referred to as the 'Prime Mover' or 'Unmoved Mover' (πρῶτον κινοῦν ἀκίνητον or primus motor) in his work *Metaphysics*." (WK 2010hh)

For Aristotle (and his teacher Plato), "the underlying essence of the Universe always was in existence and always would be (which in turn follows Parmenides' famous statement that 'nothing can come from nothing'). Aristotle posited an underlying *ousia* (essence or substance) of which the Universe was composed, and it was this *ousia* that the Prime Mover organized and set into motion. The Prime Mover did not organize matter physically, but was instead a being who constantly thought about thinking itself, and who organized the Cosmos by being itself the object of 'aspiration or desire.' The Prime Mover was, to Aristotle, a 'thinking on thinking,' an eternal process of pure thought.'" (WK 2010hh; MEP 1967)

(b) The Argument from Causation, *in Fieri* (in Becoming)

The second variant of the cosmological argument has to do with causation, *in fieri* (in becoming), which refers to "the process of becoming" and is "similar to building a house. Once it is built, the builder walks away, and it stands on its own accord....It may require occasional maintenance, but that is beyond the scope of the first cause argument." (WK 2010hh)

In general, "the modern slants on the cosmological argument, including the Kalam argument, tend to lean very strongly towards an *in fieri* argument." (WK 2010hh)

(c) The Argument from Contingency

And the third variant of the cosmological argument involves the factor of "contingency," which, unlike the first two variants, refers to "the possibility of a Universe that has no beginning in time" or the assumption that "the Universe could, under different circumstances, conceivably *not* exist (contingency)." (WK 2010hh)

But since the universe turns out to exist after all, "there must be something to explain why the Universe exists," in that "its existence must have a cause." (WK 2010hh)

Thomas Aquinas, for example, defended the argument from contingency, in "claiming that there must be something to explain why the Universe exists. Since the Universe could, under different circumstances, conceivably not exist (contingency), its existence must have a cause—not merely another contingent thing, but something that exists by necessity (something that must exist in order for anything else to exist). In other words, even if the Universe has always existed, it still owes its existence to an Uncaused Cause, although Aquinas used the words '...and this we understand to be God.'" (WK 2010hh)

In other words, "[c]ontingent beings, therefore, are insufficient to account for the existence of contingent beings: there must exist a necessary being whose non-existence is an impossibility, and from which the existence of all contingent beings is derived." (WK 2010hh)

Gottfried Leibniz also supported the argument from contingency, with his work on the "principle of sufficient reason" in 1714, when he wrote: "There can be found no fact that is true or existent, or any true proposition, without there being a sufficient reason for its being so and not otherwise, although we cannot know these reasons in most cases....Why is there something rather than nothing? The sufficient reason...is found in a substance which...is a necessary being bearing the reason for its existence within itself." (WK 2010hh)

Main Criticisms against the Cosmological Argument

These variants of the cosmological argument, however, have drawn a lot of criticisms from their critics.

Consider a few of them below, for illustration.

Firstly, a major criticism against the cosmological argument is that it questionably exempts the First Cause from any further causation.

In other words, "[o]ne objection to the argument is that it leaves open the question of why the First Cause is unique in that it does not require a cause. Proponents argue that the First Cause is exempt from having a cause, while opponents argue that this is special pleading or otherwise untrue. The problem with arguing for the First Cause's exemption is that it raises the question of why the First Cause is indeed exempt." (WK 2010hh; B. Reichenbach 2006; A. Cline 2008)

Secondly, another major criticism against the cosmological argument is that it is empirically over-generalized beyond our world of experience.

In other words, "the premise of causality [in the cosmological argument] has been arrived at via a posteriori (inductive) reasoning, which is dependent on experience. David Hume highlighted this problem of induction and argued that causal relations were not true a priori (deductively)....Even though causality applies to the known world, it does not necessarily apply to the universe at large. In other words, it is unwise to draw conclusions from an extrapolation of causality beyond experience. The rules of causality only make sense in the context of time, which obviously did not exist before the creation of the universe, thus it could be considered nonsensical to speak of pre-universal 'causes,' specifically a First Cause, when discussing the origins of the universe." (WK 2010hh; B. Reichenbach 2006)

Thirdly, still another major criticism against the cosmological argument is that, "even if one accepts the argument as a proof of a First Cause, it does not identify that First Cause with God." (WK 2010hh)

In other words, "the argument does not go on to ascribe to the First Cause some of the basic attributes commonly associated with, for instance, a theistic God, such as immanence or omnibenevolence. Rather, it simply argues that a First Cause (e.g. the Big Bang, God, or an unarticulated First Cause) must exist." (WK 2010hh; L. Craig 2008; A. Cline 2008)

Any attempt to extract such attributes to the First Cause requires a separate proof, which is not in the cosmological argument.

And fourthly, still one more major criticism against the cosomological argument is that modern physics has shown that things in nature can move themselves without the Prime Mover.

In other words, "[t]he argument for a Prime Mover is based on a foundation of Aristotelian metaphysics. Some physicists feel that the development of the laws of thermodynamics in the 19th century and quantum physics in the 20th century have weakened a purely scientific expression of the cosmological argument. Modern physics has many examples of bodies being moved without any known moving body, apparently undermining the first premise of the Prime Mover argument: every object in motion must be moved by another object in motion." (WK 2010hh)

For instance, "[p]hysicist Michio Kaku [1994] directly addresses the cosmological argument in his book *Hyperspace*, saying that it is easily dismissed by the law of conservation of energy and the laws governing molecular physics. He gives an example—'gas molecules may bounce against the walls of a container without requiring anything or anyone to get them moving.' According to Kaku, these molecules could move forever, without beginning or end. So, there is no need for a First Mover to explain the origins of motion." (WK 2010hh)

Privileging Neither Side

These criticisms against the cosmological argument do not mean to favor one side of the debate on the contested beginnings of the universe but to reveal the opposing sides of the debate, such that the possibility and desirability of the contested beginnings of the universe are not to the extent that the respective spokespersons would like us to believe.

In addition, the analysis of theology and the cosmological argument also teaches us a valuable lesson about the ontological principles in existential dialectics, and good examples include the formalness-informalness principle, the absoluteness-relativeness principle, the partiality-totality principle, the predictability-unpredictability principle, the explicability-inexplicability principle, the finiteness-transfiniteness principle, the preciseness-vagueness principle, the simpleness-complicatedness principle, the openness-hiddenness principle, the denseness-emptiness principle, the slowness-quickness principle, the expansion-contraction principle, the theory-praxis principle, the convention-novelty principle, the evolution-transformation principle, the symmetry-asymmetry principle, the regression-progression principle, and the same-difference principle.

For instance, in relation to the formalness-informalness principle, if there is formalness (e.g., the formal requirement of consistency in classical logic, in that there is either A or non-A, but not both), there is informalness (e.g., the non-formal acceptance of both A and non-A, like the argument by Aquinas that the universe can be both contingent and necessary, since "Aquinas' argument is both *in fieri* and *in esse*...plus an additional argument from contingency"). (WK 2010hh) And the reverse direction also holds true.

In relation to the absoluteness-relativeness principle, if there is absoluteness (e.g., the absolute view in the cosmological argument that the First Cause is exempt from any further causation), there is relativeness (e.g., what is reasonable for those in the cosmological argument in regard to the uncaused First Cause is not so for the critics who question this exemption). And the reverse direction also holds true.

In relation to the partiality-totality principle, if there is partial viewpoint (e.g., the partial viewpoint of a contingent entity that it is caused by something else, like "the light of the candle" which "is dependent on the candle's continued existence"), there is totalistic viewpoint (e.g., the holistic view of all things in the universe, such that there can be the First Cause for all contingent entties, so that the whole is not the sum of its parts, in accordance to the cosmological argument). And the reverse direction also holds true.

In relation to the predictability-unpredictability principle, if there is predictability (e.g., the predictable tendency of certain contingent entities to have their existence dependent on something else, like "a liquid"which "receives its shape from the vessel in which it is contained"), there is unpredictability (e.g., the more difficult task to predict something outside the realm of experience like the First Cause before the creation of the universe, because"it is unwise to draw conclusions from an extrapolation of causality beyond experience" and "it could be considered nonsensical to speak of pre-universal 'causes,' specifically a First Cause"). And the reverse direction also holds true.

In relation to the explicability-inexplicability principle, if there is explicability (e.g., the explanation, in accordance to the cosmological argument, that objects move because of some moving body as their cause), there is inexplicability (e.g., the lack of convincing explanation by those in the cosmological argument to rebuke the counter-argument by Kaku, in that "the law of conservation of energy and the laws governing molecular physics" show that "gas molecules may bounce against the walls of a container without requiring anything or anyone to get them moving"). And the reverse direction also holds true.

In relation to the finiteness-transfiniteness principle, if there is finiteness (e.g., the finite number of the First Cause, which cannot be further caused, in accordance to the cosmological argument), there is transfiniteness (e.g., the transfinite number of causes in the universe after the First Cause, in accordance to the cosmological argument). And the reverse direction also holds true.

In relation to the preciseness-vagueness principle, if there is preciseness (e.g., the precise identification of the First Cause as that which causes all contingent things in the universe, in accordance to the cosmological argument), there is vagueness (e.g.., the vagueness in the attribtutes of the First Cause, since, for the critics, any attempt to extract such attributes to the First Cause requires a separate proof, which is not in the cosmological argument). And the reverse direction also holds true.

In relation to the simpleness-complicatedness principle, if there is simpleness (e.g., the relatively simple analysis of the cosmological arguemnt in terms of its three variants), there is complicatedness (e.g., the relatively complicated analysis of the cosmological argument by pointing out various problems as shown by the critics). And the reverse direction also holds true.

In relation to the openness-hiddenness principle, if there is openness (e.g., the open exploration of the relationship between the First Cause and the contingency of things, in accordance to the cosmological argument), there is hiddenness (e.g.,the hidden bias in the cosmological argument, because of its assumption of a suprasensible being, a Prime Mover, a God, or something like that in the First Cause). And the reverse direction also holds true.

In relation to the denseness-emptiness principle, if there is denseness (e.g., the relatively denser concentration of concerns with the First Cause for those in the cosmological argument), there is emptiness (e.g., the relatively less dense, or more empty, concentration of concerns with the infinite regress of causation for those in the cosmological argument). And the reverse direction also holds true.

In relation to the slowness-quickness principle, if there is slowness (e.g., the relatively slower readiness of those in the cosmological argument to accept that the universe has no beginning), there is quickness (e.g., the relatively quicker readiness of those in the Big Bang theory to accept that "the question 'What was there before the Universe?' makes no sense," because "the concept of 'before' becomes meaningless when considering a situation without time"). (WK 2010hh) And the reverse direction also holds true.

In relation to the expansion-contraction principle, if there is expansion (e.g., the relatively more developed ability of those in the cosmological argument to manipulate deduction for the search of necessity, even though the premise of causality in the world of experience is inductive and uncertain), there is contraction (e.g., the relatively less developed ability of those in the cosmological argument to use more induction for the understanding of the world with a high dose of uncertainty). And the reverse direction also holds true.

In relation to the theory-praxis principle, if there is theory (e.g., the theoretical construction of the cosmological argument), there is praxis (e.g., the practical application of the cosmological argument for the proof of the existence of God by Christian theologians like Aquinas). And the reverse direction also holds true.

In relation to the convention-novelty principle, if there is convention (e.g.,the conventional wisdom about the cosmological argument as defended by Aristotle in antiquity and by Aquinas in the medieval era), there is novelty (e.g., the alternative novel challenge to this conventional wisdom about the cosmology argument, by those in the Big Bang theory in the modern times, who instead hold that "the question 'What was there before the Universe?' makes no sense"). And the reverse direction also holds true.

In relation to the evolution-transformation principle, if there is evolution (e.g., the natural evolution of humans in the state of nature to understand the world from the realm of experience in a primitive way), there is transformation (e.g., the technical transformation of the human mode of thinking about the world by way of the systematic invention of different variants of the cosmological argument for the First Cause). And the reverse direction also holds true.

In relation to the symmetry-asymmetry principle, if there is symmetry (e.g., the co-existence of different views about the existence of the universe over the ages), there is asymmetry (e.g., the popularity of the cosmological argument by Aquinas in the medieval era, but the acceptance of the Big Bang theory in the modern era by scientists). And the reverse direction also holds true.

In relation to the regression-progression principle, if there is regression (e.g., the regression made by the cosmological argument, as shown by the critics in terms of its major problems), there is progression (e.g., the progress made by the cosmological argument in raising anew the unresolved question concernng the First Cause of the universe). And the reverse direction also holds true.

And in relation to the same-difference principle, if there is similarity in outcome (e.g., the contribution to the better understanding of the origins of the universe, be this so by way of the cosmological argument or by the Big Bang theory), there is difference in outcome (e.g., the contribution to the better understanding of the origins of the universe, by way of the cosmological argument in terms of the First Cause—but the contribution to the better understanding of the origins of the universe, by way of the Big Bang theory in terms of a Big Bang, and nothing before that). And the reverse direction also holds true.

Hylomorphism, Ethics, and the Teleology of the Universe

The analysis of the cosmological argument (in the previous section) raises an important question concerning any teleology that the Prime Mover (the First Cause) had designed for the universe before setting it in motion.

Aristotle offers his answer, by way of his treatment of hylomorphism in the context of an ethical theory based on a teleological understanding of the universe (as organized by the First Cause before setting it in motion).

The term "hylomorphism" here is from "Greek ὑλο-" ("hylo-") for 'wood, matter" and "Greek μορφή" ("morphē") for "form"—and refers to "a philosophical theory developed by Aristotle, which analyzes substance into matter and form. More precisely, substances are conceived of as forms inhering in matter." (WK 2010ii)

Matter and Form

For Aristotle, "the term 'matter' (*hyle*) has a somewhat different meaning than the term 'matter' in modern English. In modern English, the term 'matter' often refers to a specific kind of substance, namely physical substance. In contrast, for Aristotle, 'matter' is a relative term. For Aristotle, the question is not 'Is X matter?' but, rather, 'What is the matter of X?' Aristotle defines X's matter as the 'constituents' of X, as 'that out of which' X is made." (WK 2010ii)

In this sense, "something can be matter without being physical. For example, letters are the matter of syllables." (WK 2010ii; H. Mendell 2004)

By contrast, unlike "matter" (which is "potentiality")—"form" is "actuality," in that it gives matter its shape, flavor, color, and the like.

In other words, for Aristotle, "if bronze is a bronze sphere's matter, then roundness is its form. Bronze is potentially a bronze sphere. It becomes actually a bronze sphere when given roundness. Thus, roundness is an 'actuality' of the bronze sphere—or, rather, part of the bronze sphere's actuality: it is the shape that a thing needs in order to actually be a bronze sphere. Forms need not be shapes. According to Aristotle's theory of perception, the senses perceive an object by receiving its form. The senses receive such things as colors and flavors. Thus, forms include such properties as colors and flavors, not just shapes." (WK 2010ii)

When a form becomes perfect, it gives matter its fullest realization (or its "entelechy"); so, "[t]he perfection of the form of a thing is its entelechy in virtue of which it attains its fullest realization of function"; for instance, "the entelechy of the body is the soul." (WK 2010jj)

Body-Soul Hylomorphism

When applied to living things, the Aristotelian hylomorphism makes a distinction between the body and the soul, corresponding to the distinction between matter and form.

For Aristotle, "a soul" is "that which makes a living thing alive. Life is a property of living things, just as knowledge and health are. Therefore, a soul is a form—that is, a property or set of properties—belonging to a living thing. Furthermore, Aristotle says that a soul is related to its body as form to matter." (WK 2010ii)

In this way, "there is no problem in explaining the unity of body and soul, just as there is no problem in explaining the unity of wax [matter] and its shape [form]. Just as a wax object consists of wax with a certain shape, so a living organism consists of a body with the property of life, which is its soul. On the basis of his hylomorphic theory, Aristotle rejects the Pythagorean doctrine of reincarnation, ridiculing the notion that just any soul could inhabit just any body." (WK 2010ii)

This Aristotelian body-soul hylomorphism was later exploited by medieval Christian theologians for the idea about "the transubstantiation of the Eucharist's bread and wine to the body and blood of Jesus"; later, "[t]he most remarkable adaptation is probably that of Aquinas, who distinguished the spiritual world with its subsistent forms (*formae separatae*) from the material with its inherent forms which exist only in combination with matter." (WK 2010jj)

Ethics, and the Teleology of the Universe

In any event, the Aristotelian body-soul hylomorphism has implications for an ethical theory based on a teleological view of the universe.

The reason is straightforward, in that the universe, for Aristotle, exists to realize its potential, as first organized by the Prime Mover (the First Cause) and then set into motion.

The universe, for Aristotle, is "inherently purposeful. Basically, Aristotle claims that potentiality exists for the sake of actuality. Thus, matter exists for the sake of receiving its form, and an organism has sight for the sake of seeing. Now, each thing has certain potentialities as a result of its form. Because of its form, a snake has the potential to slither. Hence, we can say that the snake *ought to* slither. The more a thing achieves its potential, the more it succeeds in achieving its purpose." (WK 2010ii; T. Irwin 1990) And when a form gives matter its fullest realization of functionality, it is known as "entelechy." (WK 2010jj)

This then means that, "[b]ecause of his form, a human being has certain abilities. Hence, his purpose in life is to exercise those abilities as well and as fully as possible. Now, the most characteristic human ability, which is not included in the form of any other organism, is the ability to think. Therefore, the best human life is a life lived rationally." (WK 2010ii)

Major Criticisms against the Body-Soul Hylomorphism

However, the Aristotelian body-soul hylomorphism has received a lot of criticisms by its critics.

Consider a few examples below, for illustration.

Firstly, a major criticism against the Aristotelian body-soul hylomorphism is that a dead body does not remain the same after its death, but a matter (like bronze) remains constant (even when it no longer has a certain form with matter over time).

This is a problem for Aristotle, who argues that "a living thing's (proximate) matter is its body, which needs a soul in order to be alive. Similarly, a bronze sphere's (proximate) matter is bronze, which needs roundness in order to be a sphere. Now, bronze remains the same bronze after ceasing to be a sphere. Therefore, it seems that a body should remain the same body after death. However, Aristotle implies that a body is no longer the same body after death. Moreover, Aristotle says that a body that has lost its soul is no longer potentially alive. But if a living thing's matter is its body, then that body should be potentially alive by definition." (WK 2010ii; C. Shields 2007)

Secondly, another major criticism against the Aristotelian body-soul hylomorphism is that it is not clear what substantial (necessary) forms a soul is to have, as opposed to accidental (non-essential) formes.

For instance, "[s]ome…thinkers argued that X's soul is X's only substantial form, responsible for all of the features of X's body. In contrast, other…thinkers argued that a living being contains at least two substantial forms—(1) the shape and structure of its body, and (2) its soul, which makes its body alive." (WK 2010ii; R. Cross 1998; A. Kenny 1993)

Thirdly, still major criticism against the Aristotelian body-soul hylomorphism is that it contradicts what Aristolte has to say about the "intellect."

For instance, "Aristotle says that the intellect (*nous*), the ability to think, has no bodily organ (in contrast with other psychological abilities, such as sense-perception and imagination). In fact, he says that it is not mixed with the body and suggests that it can exist apart from the body. This seems to contradict Aristotle's claim that the soul is a form or property of the body. Thus, interpreters of Aristotle have faced the problem of explaining how the intellect fits into Aristotle's hylomorphic theory of the soul. To complicate matters further, Aristotle distinguishes between two kinds of intellect or two parts of the intellect. These two intellectual powers are traditionally called the 'passive intellect' and the 'active intellect' or 'agent intellect.'"(WK 2010ii)

There are three answers to this contradiction in Aristotle's work, but none of them succeeds in resolving the contradiction.

The first unsatisfactory answer is that "a person's ability to think (unlike his other psychological abilities) belongs to some incorporeal organ distinct from his body. This would amount to a form of dualism. However, according to some scholars, it would not be a full-fledged Cartesian dualism. This interpretation creates what Robert Pasnau has called the 'mind-soul problem': if the intellect belongs to an entity distinct from the body, and the soul is the form of the body, then how is the intellect part of the soul?" (WK 2010ii; V. Caston 2006; C. Shields 1993; R. Pasnau 2001)

The second unsatisfactory answer is that "the passive intellect is a property of the body. In contrast, the agent intellect is a substance distinct from the body. Some proponents of this interpretation think that each person has his own agent intellect, which presumably separates from the body at death. Others interpret the agent intellect as a single divine being, perhaps the Unmoved Mover, Aristotle's God." (WK 2010ii; V. Caston 1999; J. Vella 2008; T. McEvilley 2002)

The third unsatisfactory answer is that "an individual form is capable of having properties of its own. According to this interpretation, the soul is a property of the body, but the ability to think is a property of the soul itself. If that is the case, then the soul is the body's form and yet thinking need not involve any bodily organ." (WK 2010ii; C. Shields 1988)

Fourthly, still one more major criticism against the Aristotelian body-soul hylomorphism is that it is not a necessary way to understand the world, because there exist alternative good explanations in modern science; for instance, a good alternative explanation is simply that "the order we observe is the result of a conglomeration of different properties and forces" as understood in modern science like physics, without the messy body-soul hylomorphism. (WK 2010jj)

Fifthly, still another more major criticism against the Aristotelian body-soul hylomorphism is that it does not really explain much of anything.

For instance, Rene Descartes once rightly criticized the very idea of "substantial forms" in the body-soul hylomorphism: "They were introduced by philosophers solely to account for the proper action of natural things, of which they were supposed to be the principles and bases....But no natural action at all can be explained by these substantial forms, since their defenders admit that they are occult, and that they do not understand them themselves. If they say that some action proceeds from a substantial form, it is as if they said it proceeds from something they do not understand; which explains nothing." (WK 2010jj)

And sixthly, still another major criticism against the Aristotelian body-soul hylomorphism is that its teleological view of the universe (that form gives matter its fullest realization, as in "entelechy") is questionable.

For instance, in contemporary neo-Darwinian evolutionary theory, it is no longer true that phenotypes necessarily have any evolutionary purpose, like being better in functionality (as naively assumed in its classical version). This is something that I already went to greath lengths to explain in *Beyond Nature and Nurture* (2006).

Putting Things in Perspective

Again, these criticisms against the Aristotelian body-soul hylomorphism do not mean to dismiss it outright but to let us learn from the two opposing sides of the debate on the teleology of the universe, such that the possibility and desirability of the contested beginnings of the universe (and the aftermath) from the perspective of culture are not to the extent that the respective spokespersons would like us to believe.

Furthermore, the analysis of hylomorphism, ethics, and the teleology of the universe reveals something insightful about the ontological principles in existential dialectics, and good examples include the absoluteness-relativeness principle, the predictability-unpredictability principle, the explicability-inexplicability principle, the finiteness-transfiniteness principle, the preciseness-vagueness principle, the simpleness-complicatedness principle, the openness-hiddenness principle, the denseness-emptiness principle, the slowness-quickness principle, the expansion-contraction principle, the theory-praxis principle, the convention-novelty principle, the evolution-transformation principle, the symmetry-asymmetry principle, the regression-progression principle, and the same-difference principle.

For instance, in relation to the absoluteness-relativeness principle, if there is absoluteness (e.g., the absolute teleological view of the universe by Aristotle, in that the universe is "inherently purposeful," in the sense that form gives matter its actuality), there is relativeness (e.g., what is true for the Aristotelian teleological view of the universe in regard to the full realization of matter is not necessarily so for the contemporary neo-Darwinian evolutionary theory in regard to the sub-optimality of evolution in nature). And the reverse direction also holds true.

In relation to the predictability-unpredictability principle, if there is predictability (e.g., the predictable tendency of those in the body-soul hylomorphism to argue that a body needs a soul to be alive), there is unpredictability (e.g., the more difficult task to predict what would happen to the "intellect" after the death of the body, because Aristotle also said, in a contradictory way, that the intellect "can exist apart from the body" and that "the soul is a form or property of the body"). And the reverse direction also holds true.

In relation to the explicability-inexplicability principle, if there is explicability (e.g, the explanation by Aristotle that "a soul is related to its body as form to matter"), there is inexplicability (e.g., the lack of convincing explanation, by Aristotle, of how exactly the body-soul hylomorphism really explains much of anything, as the critics like Descartes are quick to point out that "no natural action at all can be explained by these substantial forms" and, if "their defenders...say that some action proceeds from a substantial form, it is as if they said it proceeds from something they do not understand, which explains nothing"). And the reverse direction also holds true.

In relation to the finiteness-transfiniteness principle, if there is finiteness (e.g., the finite number of kinds of form to be classified, like "substantial"forms vs. "accidental" forms), there is transfiniteness (e.g., the huge number of different particular things in the universe which could possibly be classified under "substantial" forms or under "accidental"forms). And the reverse direction also holds true.

In relation to the preciseness-vagueness principle, if there is preciseness (e.g., the precise identification of the soul as the substantial form of the body), there is vagueness (e.g., the vagueness in the identification of the soul as the substantial form of the body, since "[s]ome...thinkers argued that X's soul is X's only substantial form, responsible for all of the features of X's body," but "other...thinkers argued that a living being contains at least two substantial forms" like "the shape and structure of its body" and "its soul, which makes its body alive"). And the reverse direction also holds true.

In relation to the simpleness-complicatedness principle, if there is simpleness (e.g., the relatively simple analysis by Aristotle of the body-soul hylomorphism), there is complicatedness (e.g., the relatively more complicated analysis of the body-soul hylomorphism as shown in the criticisms by the critics). And the reverse direction also holds true.

In relation to the openness-hiddenness principle, if there is openness (e.g., the open exploration by Aristotle of the relationship between the body and the soul, in that "a soul is related to its body as form to matter"), there is hiddenness (e.g., the hidden bias in the Aristotelian body-soul hylomorphism, because of his teleological view of the universe as originally organized by the Prime Mover and subsequently set in motion). And the reverse direction also holds true.

In relation to the denseness-emptiness principle, if there is denseness (e.g., the relatively denser concentration of the followers of the Aristotelian hylomorphism in the medieval era, especially when applied for "the Christian doctrines such as the transubstantiation of the Eucharist's bread and wine into the body and blood of Jesus"), there is emptiness (e.g., the relatively less dense, or more empty, concentration of the followers of the Aristotelian hylomorphism in ancient China). And the reverse direction also holds true.

In relation to the slowness-quickness principle, if there is slowness (e.g., the relatively slower readiness of the followers of the Aristotelian hylomorphism to understand the world on the basis of "different properties and forces" as understood in modern science like physics, without the messy body-soul hylomorphism), there is quickness (e.g., the relatively quicker readiness of the followers of the Aristotelian hylomorphism to understand the world on the basis of form and matter). And the reverse direction also holds true.

In relation to the expansion-contraction principle, if there is expansion (e.g., the relatively more developed ability of the followers of the Aristotelian hylomorphism to deal with philosophical, especially metaphysical, problems), there is contraction (e.g., the relatively less developed ability of the followers of the Aristotelian hylomorphism to deal with technical and scientific problems like those in modern physics in understanding the world). And the reverse direction also holds true.

In relation to the theory-praxis principle, if there is theory (e.g., the theoretical construction of hylomorphism by Aristotle), there is praxis (e.g., the practical application of hylomorphism for "the Christian doctrines such as the transubstantiation of the Eucharist's bread and wine into the body and blood of Jesus"). And the reverse direction also holds true.

In relation to the convention-novelty principle, if there is convention (e.g., the conventional wisdom about hylomorphism and the teleology of the universe in Greek antiquity), there is novelty (e.g., the altenative novel challenge to this conventional wisdom about hylomorphism, by contemporary neo-Darwinian evolutionary theory that things can come into existence without implying any evolutionary purpose for optimal functionality). And the reverse direction also holds true.

In relation to the evolution-transformation principle, if there is evolution (e.g., the natural evolution of humans in the state of nature to understand the world by observations in their relatively primitive life), there is transformation (e.g,. the technical transformtion of human understanding of the world by way of the invention of metaphysics like body-soul hylomorphism). And the reverse direction also holds true.

In relation to the symmetry-asymmetry principle, if there is symmetry (e.g., the co-existence of different views about the existence of the world in human history), there is asymmetry (e.g., the popularity of hylomorphism in Greek antiquity and medieval Christendom, but the rise of modern science like physics to explain the working of the universe, without the old idea of "substantial forms," etc.). (WK 2010jj) And the reverse direction also holds true.

In relation to the regression-progression principle, if there is regression (e.g., the regression made by the Aristotelian teleological view of the universe, as shown in the criticisms by the critics), there is progression (e.g., the progress made by the Aristotelian teleological view of the universe to help us understand the virtue of realizing the potential of things in the world). And the reverse direction also holds true.

And in relation to the same-difference principle, if there is similarity in outcome (e.g, the contribution to the understanding of the world, regardless of whether it is by way of a teleology of the universe or an impartial scientific viewpoint), there is difference in outcome (e.g., the contribution to the understanding of the world by way of the Aristotelian emphasis on the importance of full actualization of potential—but the contribution to the understanding of the world by way of an impartial scientific analysis of "different properties and forces" as in modern science, for instance, without any normative appeal to a universal telos). And the reverse direction also holds true.

The Questionability of Its Contested Beginnings

This comprehensive analysis of the possibility and desirability of the contested beginnings of the universe (and the aftermath), from the perspectives of the mind, nature, society, and culture, is vital for the critical understanding that reality is not to the extent that the spokespersons from both sides of the debate would like us to believe.

Each view of a contested beginning of the universe and its aftermath has its persuasion, just as it has its own fair share of questionability.

For instance, those who argue for initial cosmologies (that there was once a beginning of the universe) need to answer the follow-up question of why the universe had to begin like that or what really happened before that.

Conversely, those who argue for eternal cosmologies (that there was no beginning of the universe at all, if traced further back to space and time) also need to answer the follow-up question of why there is this infinite regress of causation or what exactly this infinity is supposed to refer to.

The important point here is to not favor any one side of the debate at the expense of the other but to teach us the opposing sides of the debate, such that reality is not to the extent that each respective side would like us to believe.

But this is only one part of a larger story, since the other half concerns the evolution of the universe and its speculative ends. Since this chapter deals with the contested beginnings of the universe and the aftermath, the next chapter will therefore deal with the evolution of the universe and its speculative ends.

With this in mind, let's now turn to Chapter Three for the study of the universe and its speculative ends—for which we now turn to chapter three.

Table 2.1. The Universe and Its Contested Beginnings

- **Its Contested Beginnings—and the Mind**
  —Ex: acosmism, and the illusion in the creation of the universe
  —Ex: idealism, the mind, and the problem of universals

- **Its Contested Beginnings—and Nature**
  —Ex: the Big Bang theory, and the Pre-Big Bang alternatives

- **Its Contested Beginnings—and Society**
  —Ex: earth-centric cosmology, and the politics of the church
  —Ex: correlative cosmology, and empire building

- **Its Contested Beginnings—and Culture**
  —Ex: theology, and the cosmological argument
  —Ex: hylomorphism, ethics, and the teleology of the universe

Notes: The examples in the categories are solely illustrative (not exhaustive), and the comparison is relative (not absolute), nor are they necessarily mutually exclusive. And some can be easily re-classified elsewhere. As generalities, they allow exceptions.
Source: A summary of Ch.2 of *BCOS*

Table 2.2. The Three Levels of Truth (or Reality) in Advaita Vedanta

- **"The Transcendental or the Pāramārthika Level"**
  —This is the level of truth (or reality) where "Brahman is the only reality and nothing else." (WK 2010u)

- **"The Pragmatic or the Vyāvahārika Level"**
  —This is the level of truth (or reality) where "both Jiva (living creatures or individual souls) and Iswara are true; here, the material world is also true." (WK 2010u)

- **"The Apparent or the Prāthibhāsika Level"**
  —This is the level of truth (or reality) where "material world reality is actually false, like illusion of a snake over a rope or a dream." (WK 2010u)

Source: From *Sec. 2.2.1* of *BCOS*. See book for citations.

Table 2.3. The Universal and the Abstract

• **The Universal-Particular Distinction**

—Normally, but not always, "universals" can be identified with the use of "the Ness-Ity-Hood Principle," in that universals often have endings like (a) "ness" (e.g., "absoluteness"), (b) "ity" (e.g., "eternity") and (c) "hood" (e.g., chairhood). (WK 2010x; F. Feldman 2005) Of course, there are exceptions to any rule (e.g., "charity").

—In addition, there are three major types of "universals," namely, (a) "types or kinds" (e.g., "dogness"), (b) "properties" (e.g., "redness"), and (c) "relations" (e.g., "betweenness"). (WK 2010x; M. Loux 2001)

• **The Abstract-Concrete Distinction**

—However, the "universal-particular" distinction should not be confused with the "abstract-concrete" distinction, even though the two are related.

—On the one hand, the two distinctions are related, because, "[p]aradigmatically, universals are abstract (e.g., humanity), whereas particulars are concrete (e.g., the person of Socrates)." (WK 2010x)

—On the other hand, the two distinctions are not always the same; for example, "one might hold that numbers [e.g., the number 5] are particular yet abstract objects." (WK 2010x & 2010y)

—In general, but not always, an abstract object can be identified with the use of two criteria, namely, (a) that an abstract object (e.g., "justice") "lacks a location in space" and (b) an abstract object (e.g., "empty set") "lacks any causal powers" (or "cannot act on other objects"). (WK 2010y)

• **The Opposing Sides in the Debate**

—Realism

—Idealism

—Nominalism (with Conceptualism as a sub-version)

Source: From *Sec. 2.2.2* of *BCOS*. See book for citations.

Table 2.4. The Big Bang Theory, and the Pre-Big Bang Alternatives

- **The Big Bang Theory**
  - —*Timeline of the Big Bang*
    - •Ex: the very early universe
    - •Ex: the early universe
    - •Ex: the phase of structure formation
  - —*Major Evidences*
    - •Ex: the "expansion of the universe"
    - •Ex: the "origin of the cosmic background radiation"
    - •Ex: the "nucleosynthesis of the light elements"
    - •Ex: the "formation of galaxies and large-scale structure"
  - —*Major Problems*
    - •Ex: it does not really explain the origin of the universe.
    - •Ex: it does not explain some anomalous phenomena like "why the Universe has more matter than antimatter."
    - •Ex: it assumes some exotic entities which remain to be proved and explained, and good examples include "dark energy" and "dark matter."
    - •Ex: it can be motivated with religious considerations (especially in relation to those religious traditions favoring creationism).

- **The Pre-Big Bang Alternatives**
  - —Ex: brane Cosmology
  - —Ex: the Chaotic Inflation Theory

Source: From *Sec. 2.3.1* of *BCOS*. See book for citations, like WK (2010aa), WK (2010bb), WK (2010cc), WK (2010ee), WK (2010ff), etc.

Table 2.5. The Cosmological Argument, and Its Problems

• **Three Variants of the Cosmological Argument**
—Ex: "the argument from causation, *in esse*" (in existence)
—Ex: "the argument from causation, *in fieri*" (in becoming)
—Ex: "the argument from contingency"

• **Main Problems with the Cosmological Argument**
—Ex: it questionably exempts the First Cause from any further causation.
—Ex: it is empirically over-generalized beyond our world of experience.
—Ex: "even if one accepts the argument as a proof of a First Cause, it does not identify that First Cause with God." (WK 2010hh)
—Ex: modern physics has shown that things in nature can move themselves without the Prime Mover.

Source: From *Sec. 2.5.1* of *BCOS*. See book for citations, like WK (2010hh), etc.

• Part Three •

The Speculative Ends

CHAPTER 3
THE UNIVERSE AND ITS SPECULATIVE ENDS

> It is even harder to realize that this present universe has evolved from an unspeakable unfamiliar early condition, and faces a future extinction of endless cold or intolerable heat. The more the universe seems comprehensible, the more it also seem pointless.
> —Steven Weinberg (1993: 154)

The Wonder of Its Speculative Ends

If there are the contested beginnings of the universe, there are likewise its speculative ends. After all, different thinkers over the ages had given different answers to such profound questions about whether the universe has a beginning and an end (as summarized in *Table 1.1* and *Table 1.2*).

With this controversial history in mind—this chapter here is to examine the evolution of the universe and its speculative ends (just as Chapter Two did the same to its contested beginnings and the aftermath)

If the contested beginnings of the universe and the aftermath have their persuasion, they also have their own questionability. By the same logic, if the evolution of the universe and its speculative ends have their wonder, they likewise have their own controversy.

A good way to explore the dual faces of the evolution of the universe and its speculative ends is by way of the evaluation of the extent to which the evolution of the universe and its speculative ends are in fact both possible and desirabe.

With this aim in mind, this chapter provides a comprehensive analysis of the evolution of the universe and its speculative ends from the four perspectives of the mind, nature, society, and culture (in accordance to my sophisticated methodological holism, as explained in *Sec. 1.9*).

Thus, this chapter is organized in four main sections, in relation to (3.2) its speculative ends—and the mind, (3.3) its speculative ends—and nature, (3.4) its speculative ends—and society, and (3.5) its speculative ends—and culture, to be examined in what follows, respectively (and summarized in *Table 2.1*).

Its Speculative Ends—and the Mind

The evolution of the universe and its speculative ends can be analyzed from the perspective of the mind, as it can show us something distinctive about the subject matter.

Two case studies are in order here, namely, (3.2.1) involution, the mind, and the cosmic return to the Absolute, and (3.2.2) the esoteric seven, and the spiritual evolution of the universe—to be addressed hereafter, respectively.

Involution, the Mind, and the Cosmic Return to the Absolute

To understand "the cosmic return to the absolute" in the evolution of the universe (in accordance to esoteric cosmology), it is useful to start with the term "involution," which has a special meaning in esoteric cosmology, in that it refers to a spiritual "part of a complex sequence of cosmic cycles, called Round. When the universe attains a stage of sufficient density, the individual spirit is able to descend and participate in the evolution. Involution thus refers to the incarnation of spirit in an already established matter, the necessary prerequiste of evolution." (WK 2010kk)

In this sense, "involution" is the spiritual part of the "evolution" of the universe in its later stage. It is in this later stage of the evolution of the universe known as "involution" when the "self-consciousness" of intelligent life like humans is attained, and "the building of the vehicles through which the spirit in man manifests." (WK 2010kk)

By contrast, the early stage of the evolution of the universe is "to slowly carry life...deeper into denser and denser matter for the building of forms, till the nadir of materiality is reached"; then, the later stage of evolution takes hold: "From that point, life begins to ascend into higher Worlds. This succeeding period of existence, during which the individual human being develops self-consciousness into divine omniscience, is called 'spiritual evolution.'" (WK 2010kk)

Thus, if the evolution of the universe is temporal (and temporary), its cosmic cycle will eventually return to the original point of the Absolute (which is non-temporal), when self-conciousness rediscovers its original divine nature and proceeds to its final realization.

Three Aspects of Involution

A more detailed illustration of involution in esoteric cosmology can be given by way of the philosophy of spiritual evolution by Sri Aurobindo (1872-1950) in India.

Following the Indian (Hindu) spiritual tradition, Sri Aurobindo regarded "involution" as "the process by which the Omnipresent Reality, i.e., the Absolute Brahman, extends Itself to create a universe of separate forms from out of Its own Force/Energy." (WK 2010kk)

The evolution of the universe is therefore treated as "the movement forward by which the created universe evolves from its initial state of inconscience (i.e.,...matter), evolves animated life forms and mental beings (i.e., humans), and continues to evolve spiritual properties, and in that process rediscovers its Source. Such an Evolution of animated forms is only possible because at each stage of development, the developing entity contains within itself the conception of what it may become. Thus, the evolution of animated life out of matter supposes a previous involution of that animated capacity. This is akin to a seed that already has the essence of the tree that will emerge from it." (WK 2010kk)

The evolution of the universe is thus "the development of all entities in the cosmos, including humans, in order to attain their fulfillment, including the discovery of spiritual Delight, which was, and always is, the experience of the Source Creator. The evolution is the progressive development from the original inconscience of matter into life (movement, sensation, desire, etc., and living physical beings), and from thence to mind (in conscious animals and most especially humans—the self-conscious thinking animal), and from thence to spiritualized mind, culminating in the Supermind or Truth Consciousness (as supramental individuals, and finally the supramental, i.e., a divine life on earth)," as summarized below (and also in *Table 3.2*). (WK 2010kk)

•"inconscience" (e.g., "matter")
•"life" (e.g., "movement," "sensation," "living physical beings")
•"mind" (e.g., "conscious animals," "self-conscious" humans)
•"spiritualized mind" (e.g., spiritual seekers)
•"supermind" (e.g., "truth consciousness," "divine life")

The understanding of the development of these entities in the evolution of the universe reveals to us the reason of evolution in the first place, in that the rationale of involution is "Delight—the Delight of Being (the Spirit or Absolute) moving to Delight of Becoming (temporal existence, the cosmos). Being throws itself forward into a multiplicity of forms, becoming lost in the inconscience of matter, and then through evolution it partakes in the Delight of rediscovering the Spirit which had been hidden in the interim." (WK 2010kk)

For Sri Aurobindo, involution can be further understood in terms of its three major aspects, namely, (a) "Sat" (for "existence"), (b) "Chit-Tapas" (for "consciousness"), and (c) "Ananda" (for "bliss"). (WK 2010kk)

These three aspects can be combined to form a new word, that is, "Sat-chit-ananda" (or simply "Satchitananda"). (WK 2010ll)

(a) Sat (for Existence)

The first aspect of involution is "Sat," which refers to "the vital state of that which is, was and always will be." (WK 2010kk)

Even though "Sat" (as "existence) is only "a beginnining"—"because it is pregnant with possibility, it is inexorably tied to Ananda or the recognition of Being and then the subsequent realization of bliss which is divine inner knowing. Each flows out of the other and then back again. It could be said that Sat only exists through Ananda or Consciousness; however, these levels of differentiation cannot grasp the true nature of either of these three qualities since they are interdependent." (WK 2010kk)

(b) Chit-Tapas (for Consciousness)

The second aspect of involution is "Chit-Tapas or Consciousness-Force," which "refers to the pure energy of Consciousness by which creation ultimately comes about; the infinite divine self-awareness which is also the infinite all-effective Will. It is also one of the seven planes of existence, according to the Vedic cosmology and the seven lokas of Hindu thought." (WK 2010kk)

For Sri Aurobindo, "following the Tantric dichotomy of Shiva and Shakti), *Chit* or Consciousness is not an inert and passive principle; but contains the potential spiritual Energy, *Tapas*, which in Creation becomes the dynamic and creative principle or Force, called *Shakti*. *Chit-Tapas* or *Chit-Shakti* is therefore the universal Consciousness-Force, the divine Energy; the Mother." (WK 2010kk)

(c) Ananda (for Bliss)

The third (and last) aspect of involution is "Ananda" (for "bliss"), which refers to the "delight" which "is the reason for creation, by which the Absolute extends its Delight of Being into multiplicity, losing itself in the inconscience and then through Delight rediscovering Itself through individuals realising their Divine nature and proceeding to spiritual realisation." (WK 2010kk)

This means that "the universe was created so that the Delight of the Infinite Spirit can manifest in all the forms of creation. When we discover our higher nature, the soul and spirit, we experience the delight for which we were came into being and of which we are a part." (WK 2010kk)

This cosmic cycle of being into multiplicity, losing in inconscience, rediscovering its divine nature, and proceeding to spritual development points to a return to the original Absolute, which has to do with "emanation." (WK 2010kk)

The Emanationist View of the Cosmic Return

This Indian (Hindu) idea of involution in the context of the evolution of the universe points to a cosmic return to its original Absolute, which is "based on an emanationist view of reality." (WK 2010mm)

In accordance to emanation, "Creation proceeds as an outpouring or even a transformation in the original Absolute or Godhead. The Supreme Light or Consciousness descends through a series of stages, gradations, worlds or hypostases, becoming progressively more material and embodied, before finally turning around to return to the One, retracing its steps through spiritual knowledge, contemplation and ascent." (WK 2010mm)

In the process, the cosmic return reveals a spiritual evolution towards "Godhead" (or the Absolute), which is "a common vision" unifying many esoteric cosmologists like "Sri Aurobindo [1977] and Pierre Teilhard de Chardin," both of whom "describe a progression from inanimate matter to a future state of Divine consciousness. Teilhard de Chardin refers to this as the Omega Point, and Sri Aurobindo as the Supermind." (WK 2010mm; M. Heindel 1909)

The view by Pierre Teilhard de Chardin, however, is a bit different from that by Sri Aurobindo, since "Teilhard...presented a teleological view of planetary and cosmic evolution, according to which the formation of atoms, molecules and inanimate matter is followed by the development of the biosphere and organic evolution, then the appearance of man and the noosphere as the total envelope of human thought. According to Teilhard evolution does not cease here but continues on to its culmination and unification in the Omega Point, which he identifies with Christ." (WK 2010mm)

The Problems with the Idea of the Cosmic Return

Yet, this idea of the cosmic return to the original Absolute for spiritual realization is not immune from criticisms.

Consider a few criticisms below, for illustration.

Firstly, a major criticism against the idea of the cosmic return for spiritual realization is that involution may well be a contingent (or non-necessary) outcome in the evolution of the universe, without the teleological assumption (that its proponents would like us to believe).

Or at least, those in esoteric cosmology have yet to prove that involution is necessary (with its teleological cosmic return), not contingent on evolutionary circumstances.

Secondly, another maor criticism against the idea of the cosmic return for spiritual realization is that it is not clear what this original Absolute really refers to—and the problem is all the more compounded when it is said to be non-temporal and divine, not to be completely humanly understood.

It is one thing to say something abstruse (or mysterious), but what is abstruse (or mysterious) is not the same as being true.

Thirdly, still another major criticism against the idea of the cosmic return for spiritual realization is that it is not clear why on earth the Absolute has to waste time and energy to go through the cosmic cycle from being into multiplicity, losing in inconscience, rediscovering its divine nature, and proceeding to spiritual development, only in order to confirm the nature of being, consciousness, and delight in involution.

Doesn't the Absolute have something else more worthwhile to do than wasting time and energy with this recurrent cosmic cycle in relation to the temporal material universe?

And fourthly, still one more major crticism against the idea of the cosmic return for spiritual realization is that it is puzzling indeed for the Absolute to have to go through this cosmic cycle so as to remind Itself of the delight in involution.

Does the Absolute suffer from an existential crisis, which needs the confirmation of the delight in existence, time and again, through involution in the cosmic cycle?

A More Detached View

Surely, these criticisms against the idea of the cosmic return to the Absolute for spiritual realization do not suggest to dismiss it but to make it easier for us to evaluate the two opposing sides of the debate, such that the possibility and desirability of the evolution of the universe (and its speculative ends) from the perspective of the mind are not to the extent that the respective spokespersons would like us to believe.

In addition, the analysis of the idea of the cosmic return to the Absolute for spiritual realization shows us something refreshing about the ontological principles in existential dialectics, and good examples include the absoluteness-relativeness principle, the predictability-unpredictability principle, the explicability-inexplicability principle, the finiteness-transfiniteness principle, the preciseness-vagueness principle, the simpleness-complicatedness principle, the openness-hiddenness principle, the denseness-emptiness principle, the change-constancy principle, the slowness-quickness principle, the expansion-contraction principle, the theory-praxis principle, the convention-novelty principle, the evolution-transformation principle, the symmetry-asymmetry principle, the regression-progression principle, and the same-difference principle.

For instance, in relation to the absoluteness-relativeness principle, if there is absoluteness (e.g., the absolute view by Sri Aurobindo that there is "the Omnipresent Reality" as "the Absolute Brahman," which is a non-temporal precursor to the existence of the universe), there is relativeness (e.g., what is true for Sri Aurobindo in regard to the Absolute Brahman is not so for those in the Big Bang theory, who do not accept the scientific legitimacy of any entity prior to the Big Bang). And the reverse direction also holds true.

In relation to the predictability-unpredictability principle, if there is predictability (e.g., the predictable tendency of a cosmic cycle to go through the development of all entities from "inconscience" to "divine life," in accordance to the idea of involution), there is unpredictability (e.g., the more difficult task to predict, if understood solely from the vantage point of involution, how many more rounds of the cosmic cycle there will be in future history). And the reverse direction also holds true.

In relation to the explicability-inexplicability principle, if there is explicability (e.g., the explanation of the cosmic return to the Absolute for spiritual realization on the basis of "emanation" as "an outpouring or even a transformation in the original Absolute or Godhead," in conjunction with involution, in accordance to those in esoteric cosmology), there is inexplicability (e.g., the lack of convincing explanation, by those in esoteric cosmology, of why this emanation is necessary and, also, why it is true at all). And the reverse direction also holds true.

In relation to the finiteness-transfiniteness principle, if there is finiteness (e.g., the finite number of the categories of entities in the evolution of the universe—that is, "matter," "life," "mind," "spiritualized mind," and "supermind," in accordance to Sri Aurobindo), there is transfiniteness (e.g., the amazingly huge number of particular things in the universe which can possibly be classified under each of the finite categories of "matter," "life," "mind," "spiritualized mind," and "supermind," in accordance to Sri Aurobindo). And the reverse direction also holds true.

In relation to the preciseness-vagueness principle, if there is preciseness (e.g., the precise identification of "the Omnipresent Reality" as "the Absolute Brahman," which is a non-temporal precursor to the existence of the universe, for Sri Aurobindo), there is vagueness (e.g., the vagueness in this precise identification of "the Omnipresent Reality," since it is not clear what exactly it really refers to, as it is not "temporal" but "divine," which, unfortunately, cannot be totally grasped by humans). And the reverse direction also holds true.

In relation to the simpleness-complicatedness principle, if there is simpleness (e.g., the relatively simple analysis of the cosmic cycle for spiritual realization, in accordance to esoteric cosmology), there is complicatedness (e.g., the relatively more complicated analysis of the cosmic cycle for spiritual realization by questioning its assumptions and claims, as shown in the criticisms). And the reverse direction also holds true.

In relation to the openness-hiddenness principle, if there is openness (e.g., the open exploration of the relationship between the evolution of the universe and its involution for spiritual realization), there is hiddenness (e.g., the hidden bias in the idea of the cosmic return to the Absolute in involution, because of its esoteric ideology of emanation). And the reverse direction also holds true.

In relation to the denseness-emptiness principle, if there is denseness (e.g., the relatively denser concentration of believers in involution in India), there is emptiness (e.g., the relatively less dense, or more empty, concentration of believers in involution among contemporary astrophysicists, especially those in the Big Bang theory). And the reverse direction also holds true.

In relation to the change-constancy principle, if there is change (e.g., the development of different entities in the evolution of the universe over time, in accordance to esoteric cosmology), there is constancy (e.g., the recurrence of the cosmic cycle in the universe, in accordance to esoteric cosmology). And the reverse direction also holds true.

In relation to the slowness-quickness principle, if there is slowness (e.g., the relatively slower readiness of such entities as life in physical beings and animals to become self-conscious), there is quickness (e.g., the relatively quicker readiness of such entities as humans to become self-conscious). And the reverse direction also holds true.

In relation to the expansion-contraction principle, if there is expansion (e.g., the relatively more developed ability of spiritual seekers in esoteric cosmology to develop their mind for spiritual realization), there is contraction (e.g., the relatively less developed ability of spiritual seekers in eosteric cosmology to cope with materialist development). And the reverse direction also holds true.

In relation to the theory-praxis principle, if there is theory (e.g., the theoretical construction of emanation in esoteric cosmology), there is praxis (e.g., the practical application of emanation in esoteric cosmology for the development of Christianity by Pierre Teilhard de Chardin). And the reverse direction also holds true.

In relation to the convention-novelty principle, if there is convention (e.g., the conventional wisdom about involution in the Indian, particularly Hindu, spiriutal tradition), there is novelty (e.g., the alternative novel challenge to this conventional wisdom about involution, by Sri Aurobindo, who revised and expanded the Indian, particularly Hindu, tradition for his vision of the "supermind" for "divine life"). And the reverse direction also holds true.

In relation to the evolution-transformation principle, if there is evolution (e.g., the natural evolution of different entities in the history of the universe, in accordance to Sri Aurobindo), there is transformation (e.g., the technical transformation of the existence of entities in the world, like that of self-conscious humans, by way of the invention of esoteric cosmology for a more systematic spritual realization). And the reverse direction also holds true.

In relation to the symmetry-asymmetry principle, if there is symmetry (e.g., the co-existence of different spiritual schools of thought in India, regardless of whether it is about the original Vedas in antiquity or else), there is asymmetry (e.g., the popularity of the original Vedas in ancient India and also later, but the more acceptance of the version by Sri Aurobindo among his followers in modern time). And the reverse direction also holds true.

In relation to the regression-progression principle, if there is regression (e.g., the serious problems in the idea of the cosmic return to the Absolute for spiritual realization, as shown in the criticisms which question its assumptions and claims), there is progression (e.g., the progress made by the idea of the cosmic return to the Absolute for a better understanding of the virtue of seeking spiritualtiy for the human mind in the evolution of the universe). And the reverse direction also holds true.

And in relation to the same-difference principle, if there is similarity in outcome (e.g., the contribution to the enhancement of life, regardless of whether it is done by spirituality or by materiality), there is difference in outcome (e.g., the contribution to the enhancement of life, by way of spirituality as in esoteric cosmology, but the contribution to the enhancement of life, by way of materiality as in capitalism). And the reverse direction also holds true.

The Esoteric Seven, and the Spiritual Evolution of the Universe

The analysis of the cosmic return to the Absolute (in the previous section) points to a teleology of epigenesis (after the creation of the universe) for spiritual evolution, with its (speculative) eventual ending towards a divine union with the Absolute—for the mind.

So, a more detailed analysis of this epigenesis for spiritual evolution is needed.

With this aim in mind, a good starting point concerns the magical number seven in esoteric cosmology—as the second case study of the possibility and desirability of the evolution of the universe (and its specualtive ends) from the perspective of the mind.

The Magical Number Seven

A well known branch of esoteric cosmology is "theosophy," which literally means "god-wisdom," from "Greek: θεοσοφία" ("theosophia"), and it is thus no wonder that "the Oxford English Dictionary defines theosophy as…'any system of speculation which bases the knowledge of nature upon that of the divine nature.'" (WK 2010nn)

Theosophy can be dated back to antiquity, when it was used "in Neoplatonism" in ancient Greece and also developed in the Indian (Hindu) tradition; later, it attracted the attention of "a group of Renaissance philosophers [like] Cornelius Agrippa, Paracelsus, Robert Fludd, and, especially, Jacob Boehme." (WK 2010nn)

In modern times, theosophy received a big boost when "Helena Petrovna Blavatsky (1831–1891), Henry Steel Olcott (1832–1907), and William Quan Judge (1851–1896), established the Theosophical Society in 1875." (WK 2010nn)

In general, "theosophists believe that religion, philosophy, science, the arts, commerce, and philanthropy, among other 'virtues,' lead people ever closer to 'the Absolute.' Planets, solar systems, galaxies, and the cosmos itself are regarded as conscious entities, fulfilling their own evolutionary paths." (WK 2010nn)

What unify them all has to do with "the spiritual units of consciousness in the universe," known as "the Monads, which may manifest as angels, human beings or in various other forms." (WK 2010nn)

(a) Seven Root Races

But one thing which stands out in theosophy is its obsession with the esoteric number seven, which is used to explain a lot of different things, including the Monads, human or not.

For instance, "theosophical writings propose that human civilizations, like all other parts of the universe, develop cyclically through seven stages. Blavatsky [1977] posited that the whole humanity, and indeed every reincarnating human monad, evolves through a series of seven 'Root Races,'" as summarized below. (WK 2010nn & 2010oo)

- •In the 1st stage—"ethereal" things (like immaterial "pure spirits")
- •In the 2nd stage—"physical bodies…lived in Hyperborea"
- •In the 3rd stage—first "truly human[s]…existed on the lost continent of Lemuria" (with developed "sexual characteristics")
- •In the 4th stage—"modern humans finally developed on the continent of Atlantis"
- •In the 5th stage—current humans in "a time of reawakening humanity's psychic gifts"
- •In the 6th and 7th stages—the future emergence of "more advanced peoples…set to develop on the reemerging Lemurian continent."

The use of the esoteric number seven is also extended to the analysis of the evolution of the spirit in the universe over time (as described below).

(b) Seven Vestures of the Soul

The obession with the number seven also shows up when "theosophists opine that the most material of the vestures of the Soul are interpenetrated by the particles of the more subtle vesture." (WK 2010nn)

For instance, for these theosophists, the "most material body…is, as science is aware, mostly space at its atomic level (as all matter is known to be), and these interstitial spaces are inhabited by those subtler particles of the Astral Body or Linga sarira, and so on for the other more energy-like envelopes of the Soul." (WK 2010nn)

What is important "about this interpenetration of each sheath…is that we see the inner person as a fluid and unbroken continuity, although varying in density/flexibility and energy….And since matter is merely the material counterpart of consciousness (ultimately our aspect being pure consciousness), this interpenetration of sheaths allows for consciousness to interpenetrate Man's nature and explains how we are sensitive to what we think of as external stimuli, through the five senses." (WK 2010nn)

To understand better this interpenetration of sheaths, theosophists distinguish seven kinds of bodies in relation to the spirit/matter forms (which are called "the septenary"), as summarized below. (WK 2010nn)

- "The first body"—refers to "sthula-sarira" (in Sanskrit), or "a gross body, impermanent because of its wholly compounded character. The physical body is usually considered as the lowest substance-principle. The physical form is the result of the harmonious co-working on the physical plane of forces and faculties." (WK 2010nn)
- "The second body"—refers to "linga-Sarira" (in Sanskrit), or "the model-body or astral body, only slightly more ethereal than the physical body. It is the astral model around which the physical body is built, and from which the physical body flows or develops as growth proceeds." (WK 2010nn)
- "The third body"—refers to "prana" (in Sanskrit), or "the breath of life. This life or prana works on, in, and around us, pulsating unceasingly during the term of physical existence. Prana is 'the radiating force or Energy of Atma—as the Universal Life and the One Self....Prana or Life permeates the whole being of the objective Universe; and is called a 'principle' only because it is an indispensable factor and the deus ex machina of the living man.'" (WK 2010nn)
- "The fourth body"—refers to "kāma" (in Sanskrit), or "desire,...the driving, impelling force. Born from the interaction of atman, buddhi, and manas, kama per se is a colourless force, good or bad according to the way the mind and soul use it. It is the seat of the living electrical impulses, desires, and aspirations, considered in their energetic aspect." (WK 2010nn)
- "The fifth body"—refers to "manas" (in Sanskrit), or thinking, as "the seat of mentation and egoic consciousness; in humanity Manas is the human person, the reincarnating ego, immortal in essence, enduring in its higher aspects through the entire manvantara. When embodied, manas is dual, gravitating toward buddhi in its higher aspects and in its lower aspects toward kama. The first is intuitive mind, the second the animal, ratiocinative consciousness, the lower mentality and passions of the personality." (WK 2010nn)
- "The sixth body"—refers to "Buddhi" (in Sanskrit), or "the pure, universal spirit, hence an inseparable garment or vehicle of atman, which is, in its essence, of the highest plane of akasa or alaya. In man

buddhi is the spiritual soul, the faculty of discriminating, the channel through which streams divine inspiration from the atman to the ego, and therefore that faculty which enables us to discern between good and evil: spiritual conscience. The qualities of the buddhic principle when awakened are higher judgment, instant understanding, discrimination, intuition, love that has no bounds, and consequent universal forgiveness." (WK 2010nn)

- "The seventh body"—refers to "Atman" (in Sanskrit), or "pure consciousness, that cosmic self which is the same in every dweller on this globe and on every one of the planetary or stellar bodies in space. It is the feeling and knowledge of 'I am,' pure cognition, the abstract idea of self. It does not differ at all throughout the cosmos except in degree of self-recognition. It may also be considered as the First Logos in the human microcosm. During incarnation the lowest aspects of atman take on attributes, because it is linked with buddhi, as the buddhi is linked with manas, as the manas is linked with kama, etc." (WK 2001nn)

But where will the evolution of these seven vestures of the soul (or simply "the septenary") end in the long run?

An Internal Split in the Theosophical Movement

The answer to this above question also depends on which group in the theosohical movement one wants to ask. Precisely here shows an internal split in the theosophical movement.

For illustration, consider the modern disagreement between the Theosophical Society (with its relatively Eastern orientation, focusing more on the Hindu spiritual tradition in regard to the cosmic return to the Absolute) and the Anthroposophical Society (with its relatively Western orientation, focusing more on the second coming of Christ in the teleological end of spiritual evolution in the universe).

It all started when Helena Blavatsky, in 1875, "founded the Theosophical Society in New York City together with Henry Steel Olcott....Blavatsky was a world traveler who eventually settled in India where, with Olcott, she established the headquarters of the Society in Bangalore." (WK 2010nn)

After Blavatsky's death in 1891, however, the Theosophical Society started its internal schism into different groups, each with different views. For example, "Annie Besant became leader of the society based in Adyar, India, while William Quan Judge split off the American Section of the Theosophical Society in New York which later moved to Point Loma,

Covina, and Pasadena, California under a series of leaders: Katherine Tingley, Gottfried de Purucker, Colonel Arthur L. Conger, James A. Long, Grace F. Knoche, and in March 2006 Randell C. Grubb." (WK 2010nn)

Later, "Rudolf Steiner created a successful branch of the Theosophical Society in Germany. He focused on a Western esoteric path that incorporated the influences of Christianity and natural science, resulting in tensions with Annie Besant…; these were seriously exacerbated by Steiner refusing members of the Order of the Star of the East membership in the Theosophical Society's German Section." (WK 2010nn)

One major reason is that "Steiner was vehemently opposed to The Order of the Star of the East's proclamation that the young boy, Jiddu Krishnamurti, was the incarnation of Maitreya (who was believed to have 'over-shadowed' Jesus Christ)….Krishnamurti later repudiated this role and left the Society to pursue an independent career of spiritual teaching." (WK 2010nn)

Thus, in 1913, Steiner decided to have a formal break with the Theosophical Society and "founded his own Anthroposophical Society; the great majority of German-speaking theosophists joined the new society, which grew rapidly. Steiner later became most famous for his ideas about education, resulting in an international network of 'Steiner Schools,' also known as Waldorf schools. Other influences of anthroposophical thought include biodynamic agriculture, anthroposophic medicine and the acting techniques of Michael Chekhov." (WK 2010nn)

Unlike the Theosophical Society—the Anthroposophical Society is more Western-orientated, in that Steiner, for instance believed that "Christ is a being who unifies all religions. He believed that Christ is not any particular religious faith, but instead is the central force in human evolution. This Christ Being is, according to Steiner, not only the Redeemer of the Fall from Paradise, but also the unique pivot and meaning of earth's evolutionary processes and of human history, which he believed to be manifested in all religions and cultures." (WK 2010)

Other organizations followed and further divided the theosophical movement, and good examples include "the Agni Yoga, 'I AM' Activity, The Bridge to Freedom and The Summit Lighthouse. These various offshoots dispute the authenticity of their rivals. Thus followers of the United Lodge of Theosophists will claim that only 'the Writings of HPB, William Quan Judge and Robert Crosbie can be trusted to contain unadulterated concepts and ethical direction.'" (WK 2010nn)

Troublesome Afterthoughts

The internal split of the theosophical movement speaks volumes of the disagreement among its followers on the final fate of the spiritual evolution of the universe.

This then raises a larger critical question concerning the validity of theosophy (and, for that matter, anthroposophy) as a spiritual philosophy and religion. Consider a few criticisms below, for illustration.

Firstly, the internal schism of the theosophical movement makes it easier for the critics to caution any acceptance of the viewpoints as propounded by the followers of the movement, because each group will offer different opinions.

For instance, some like Steiner elevate the status of "Christ" as "the central force in human evolution," while others like those in the Indian section of the Theosophical Society focus on the cosmic return to the Absolute for divine life.

Secondly, the use of the esoteric number seven in theosophy can be regarded as a form of superstition, which then raises the alarm flag that its analyses of "the seven root races" and "the seven vestures of the soul" are arbitrary, so as to conform to the superstitious obsession with the number seven in esoteric cosmology.

And thirdly, the claims made by the proponents of theosophy are controversially biased against different non-white races and ethnic groups. This is true, in spite of the official call for "the Universal Brotherhood of Humanity" in theosophy. (WK 2010nn)

The reason is that Blavatsky wrote in quite a number of passages about the inferiority of some non-white races and ethnic groups. For instance, she once wrote that "the Semites, especially the Arabs, are later Aryans—degenerate in spirituality and perfected in materiality." (WK 2010nn)

In another passage, Blavatsky explicitly mentioned some "inferior races," in her own words: "No amount of culture, nor generations of training amid civilization, could raise such human specimens as the Bushmen, the Veddhas of Ceylon, and some African tribes, to the same intellectual level as the Aryans....The 'sacred spark' is missing in them and it is they who are the only inferior races on the globe, now happily—owing to the wise adjustment of nature which ever works in that direction—fast dying out." (WK 2010oo)

She also used the words "semi-human stocks" to refer to "the wild men of Borneo, the Veddhas of Ceylon,...most of the remaining Australians, Bushmen, Negritos, Andaman Islanders, etc." (WK 2010oo)

It is no wonder that Blavtsky's work on "the root races" has been used by some Nazi groups for the ideology of white supramacy. A good case in point is "Guido von List (and his followers such as Lanz von Liebenfels)," who "later took up some of Blavatsky's theories, mixing them with nationalism to formulate Ariosophy, a precursor of nazism. Ariosophy emphasized intellectual expositions of racial evolution. The Thule Society was one of several German occult groups drawing on Ariosophy to preach Aryan supremacy. It provides a direct link between occult racial theories and the racial ideology of Hitler and the emerging Nazi party." (WK 2010mm; J. Spielvogel 1986)

A Lesson to Learn

In any event, these criticisms against the theosophical movement do not mean to discredit it outright but to show us the two opposing sides of the debate, such that the possibility and desirability of the evolution of the universe (and its speculative ends) from the perspective of the mind are not to the extent that the respective spokespersons would like us to believe.

Furthermore, the analysis of the magical number seven and the spiritual evolution of the universe can also teach us an important lesson about the ontological principles in existential dialectics, and good examples include the absoluteness-relativeness principle, the predictability-unpredictability principle, the explicability-inexplicability principle, the finiteness-transfiniteness principle, the preciseness-vagueness principle, the simpleness-complicatedness principle, the openness-hiddenness principle, the denseness-emptiness principle, the slowness-quickness principle, the expansion-contraction principle, the theory-praxis principle, the convention-novelty principle, the evolution-transformation principle, the symmetry-asymmetry principle, the regression-progression principle, and the same-difference principle.

For instance, in relation to the absoluteness-relativeness principle, if there is absoluteness (e.g., the absolute view about the esoteric number seven in theosophy, which has been used to analyze the seven root races and the septenary, for instance), there is relativeness (e.g., what is approriate for those in theosophy to appeal to the esoteric number seven is not necessarily so for the critics, who question its superstition). And the reverse direction also holds true.

In relation to the predictability-unpredictability principle, if there is predictability (e.g., the predictable tendency of those in theosophy like Blavatsky to analyze the spiritual evolution of the universe in terms of "seven root races" and the "seven vestures of the soul"), there is unpredictability (e.g., the more difficult task to predict correctly when exactly the seventh race, for instance, will actually emerge, if looked within Blavatsky's analysis alone). And the reverse direction also holds true.

In relation to the explicability-inexplicability principle, if there is explicability (e.g., the explanation by Blavatsky of the existence of the "inferior races" like "the Bushmen, the Veddhas of Ceylon, and some African tribes," on the basis of their "adjustment of nature which ever works in that direction"), there is inexplicability (e.g., the lack of convincing explanation, by Blavatsky, of why the superior Aryans are able to succeed, also by their "adjustment of nature," albeit in a different direction instead). And the reverse direction also holds true.

In relation to the fìniteness-transfiniteness principle, if there is finiteness (e.g., the finite number of seven root races in Blavatsky's analysis of the spiritual evolution of human monads), there is transfiniteness (e.g., the amazingly huge number of particular things which could possibly be classified under the seven root races in Blavatksy's analysis). And the reverse direction also holds true.

In relation to the preciseness-vagueness principle, if there is preciseness (e.g., the precise identification of the seven bodies in the literature on the septenary in theosophy), there is vagueness (e.g., the vagueness in the analysis of each of the seven bodies, since it is not clear what such abstract terms like "pure consciousness" really refer to, since they can mean different things for different theosophical interpreters in different contexts). And the reverse direction also holds true.

In relation to the simpleness-complicatedness principle, if there is simpleness (e.g., the relatively simple analysis of the races and the soul from the vantage point of the esoteric number seven, in theosophy), there is complicatedness (e.g., the relatively more complicated analysis of the races and the soul by challenging its assumptions and claims, as pointed out by the critics). And the reverse direction also holds true.

In relation to the openness-hiddenness principle, if there is openness (e.g., the relatively open exploration of the universal brotherhood of humanity in theosophy), there is hiddeness (e.g., the hidden bias in the claims by Blavatsky, one of the founders of the Theosophical Society, on different races, with the Aryans as the superior one and many of the non-white folks as the "inferior" ones). And the reverse direction also holds true.

In relation to the denseness-emptiness principle, if there is denseness (e.g., the relatively denser concentration of believers in the Theosophical Society with a relatively Eastern orientation in India), there is emptiness (e.g., the relatively less dense, or more empty, concentration of believers in the Theosophical Society with a relatively Eastern orientation in Germany, especially among those in the Anthroposophical Society).

In relation to the slowness-quickness principle, if there is slowness (e.g., the relatively slower readiness of someone like Steiner to accept a special boy like Jiddu Krishnamurti as "the incarnation of Maitreya…who was believed to have 'over-shadowed' Jesus Christ"), there is quickness (e.g., the relatively quicker readiness of those in The Order of the Star in the East, the Indian section of the Theosophical Society, to accept a special boy like Jiddu Krishnamurti as "the incarnation of Maitreya…who was believed to have 'over-shadowed' Jesus Christ"). And the reverse direction also holds true.

In relation to the expansion-contraction principle, if there is expansion (e.g., "in the last several decades" of the 20th century and the early 21st, "there was a steady increase in membership in India," in regard to the Theosophical Society), there is contraction (e.g,, "outside India, the membership has been dropping," in regard to the Theosophical Society, in the same period of the the last several decades of the 20th century and the early 21st). (WK 2010nn) And the reverse direction also holds true.

In relation to the theory-praxis principle, if there is theory (e.g., the theoretical construction of the "seven root races" by Blavatsky in the Theosophical Society), there is praxis (e.g,. the practical application of the analysis of the "seven root races" by some Nazi groups for the ideology of white supramacy). And the reverse direction also holds true.

In relation to the convention-novelty principle, if there is convention (e.g., the conventional wisdom about the relatively Eastern orientation in the original Theosophical Society), there is novelty (e.g., the alternative novel challenge to the conventional wisdom about the relativley Eastern orientation in the original Theosophical Society, by Steiner for his Anthroposophical Society with a relatively more Western orientation). And the reverse direction also holds true.

In relation to the evolution-transformation principle, if there is evolution (e.g., the natural evolution of humans to interact with nature in a primitive way in the state of nature), there is transformation (e.g., the techical transformation of human understanding of nature by way of the invention of a systematic philosophy and religion known as theosophy, with its analyses of the "seven root races" and the "seven vestures of the soul"). And the reverse direction also holds true.

In relation to the symmetry-asymmetry principle, if there is symmetry (e.g., the co-existence of different groups in the theosophical movement over time, regardless of whether they are located in India, Germany, or in the U.S.), there is asymmetry (e.g., the popularity of the Order of the Star in the East, the eastern section of the Theosophical Society, in India—but the more acceptance of the Anthroposophical Society in Germany). And the reverse direction also holds true.

In relation to the regression-progression principle, if there is regression (e.g., the regression made by the Theosophical Society, like the problems about its claims concerning the "inferiority" of non-Western races and ethnic groups, as pointed out by the critics), there is progression (e.g., the progress made by the Theosophical Society for the possibility of the Universal Brotherhood of Humanity). And the reverse direction also holds true.

And in relation to the same-difference principle, if there is similarity in outcome (e.g., the contribution to the better understanding of the spiriutal evolution of the universe, regardless of whether this is so by way of the cosmic return to the Absolute or the second coming of Christ), there is difference in outcome (e.g., the contribution to the better understanding of the spiriutal evolution of the universe, by way of the cosmic return to the Absolute as in the eastern section of the Theosophical Society—but the contribution to the better understanding of the spiriutal evolution of the universe, by way of the second coming of Christ, as in the Anthroposophical Society). And the reverse direction also holds true.

Its Speculative Ends—and Nature

The evolution of the universe and its speculative ends can also be analyzed from the perspective of nature, as it can reveal something insightful about the subject matter.

A good case study here concerns the scientific debate about the ultimate fate of the universe.

The Scientific Debate about the Ultimate Fate of the Universe

The Big Bang theory (as already discussed in *Sec. 2.3*) serves as a good scientific starting point to discuss the ultimate fate of the universe, because, in accordance to the theory, "the universe started with a Big Bang," so now a legitimate question to ask is: What then is the ultimate fate of this expansion of the universe? (WK 2010qq)

The Current Scientific Basis

The current scientfic basis to answer this question centers around "an important parameter" known as "the density parameter, Omega (Ω)," which is defined "as the average matter density of the universe divided by a critical value of that density." (WK 2010qq)

There are three possible values of Omega (Ω), namely, depending on "whether Ω is equal to, less than, or greater than 1. These are called, respectively, the flat, open and closed universes." (WK 2010qq)

These three values can define "the overall geometry of the universe, and not to the local curving of spacetime caused by smaller clumps of mass (for example, galaxies and stars). If the primary content of the universe is inert matter, as in the dust models popular for much of the 20th century, there is a particular fate corresponding to each geometry. Hence cosmologists aimed to determine the fate of the universe by measuring Ω, or equivalently the rate at which the expansion was decelerating." (WK 2010qq)

By the turn of the 21st century, there is increasing evidence that the universe is "open," which means that Ω is less than 1, so the universe is accelerating its expansion, with no end in sight. (WK 2010qq)

This is so, because "observations of supernovae in distant galaxies have been interpreted as consistent with a universe whose expansion is accelerating. Subsequent cosmological theorizing has been designed so as to allow for this possible acceleration, nearly always by involving dark energy, which in its simplest form is just a positive cosmological constant. In general dark energy is a catch-all term for any hypothesised field with negative pressure, usually with a density that changes as the universe expands." (WK 2010qq)

The Shape of the Universe

To understand the nature of "dark energy" (as indicated above) can shed some light on the shape of the universe, which in turn can then reveal the ultimate fate of the universe.

There are three possible shapes (or geometries) of the universe, in relation to the value of Ω (as indicted above), and they are, namely, (a) "flat universe," (b) "closed universe," and (c) "open universe," as summarized below (and also in *Table 3.3*).

(a) Flat Universe

The first scenario is for Ω=1, this means that "the average [matter] density of the universe exactly equals the critical density," so that there is no sufficient matter in the universe for its gravity to cause its eventual collapse in the distant future, nor sufficient lack of it to make the universe expand forever either. (WK 2010qq)

In this scenario, "the geometry of the universe is flat: as in Euclidean geometry, the sum of the angles of a triangle is 180 degrees and parallel lines continuously maintain the same distance." (WK 2010qq)

This is the fate of the universe which Einstein initially preferred, because "he and his contemporaries believed in a static universe" at the time, much to his regret later. (WK 2010qq)

To make this work on paper, Einstein fooled himself to believe that "his equations could easily be solved in such a way as to allow the universe to be expanding now, and to contract in the far future, he added to those equations what he called a cosmological constant, essentially a constant energy density unaffected by any expansion or contraction, whose role was to offset the effect of gravity on the universe as a whole in such a way that the universe would remain static" (and would not collapse in the end, out of its own gravity).

Unfortunately, the bad news is that, "after Hubble announced his conclusion that the universe was expanding, Einstein wrote that his cosmological constant was 'the greatest blunder of my life.'" (WK 2010qq

(b) Closed Universe

The second scenario is $\Omega > 1$, which means that the average matter density of the universe is greater than the critical density, so that there is sufficient matter in the universe for its gravity to cause its eventual collapse in the distant future

In this scenario, "the geometry of space is closed like the surface of a sphere. The sum of the angles of a triangle exceeds 180 degrees and there are no parallel lines; all lines eventually meet. The geometry of the universe is, at least on a very large scale, elliptic." (WK 2010qq)

In other words, "in a closed universe lacking the repulsive effect of dark energy, gravity eventually stops the expansion of the universe, after which it starts to contract until all matter in the universe collapses to a point, a final singularity termed the 'Big Crunch,' by analogy with Big Bang." (WK 2010qq)

However, this scenario is now ruled out, because astrophysicists now believe that "the universe has a large amount of dark energy (as suggested by recent findings)" and therefore "the expansion of the universe can continue forever."

This then brings us to the third scenario, an "open" universe.

(c) Open Universe

The third scenario is $\Omega < 1$, which means that the average matter density of the universe is less than the critical density, so that there is less matter in the universe than what is needed for its gravity to cause its eventual collapse in the distant future, such that all forms of matter in the universe are far enough from each other to allow the universe expand indefinitely.

Nowadays, astrophysicists believe that "the critical density is estimated to be approximately five atoms (of monatomic hydrogen) per cubic metre, whereas the average density of the Universe is believed to be 0.2 atoms per cubic metre. Therefore, the Universe will expand forever." (WK 2010rr; M. Rees 2000)

In this scenario, "the geometry of space is open, i.e., negatively curved like the surface of a saddle. The angles of a triangle sum to less than 180 degrees, and lines that do not meet are never equidistant; they have a point of least distance and otherwise grow apart. The geometry of such a universe is hyperbolic." (WK 2010qq)

Even in the absence of dark energy, "a negatively curved universe expands forever, with gravity barely slowing the rate of expansion"; however, "[w]ith dark energy, the expansion not only continues but accelerates. The ultimate fate of an open universe is either universal heat death, the 'Big Freeze,' or the 'Big Rip,' where the acceleration caused by dark energy eventually becomes so strong that it completely overwhelms the effects of the gravitational, electromagnetic and weak binding forces." (WK 2010qq)

Some astrophysicists even took the trouble to estimate, roughly speaking, the future of the "open" universe in accordance to the following time frame.

- "50 Billion Years"—This is the time "until the Earth and the Moon become tidelocked, with each showing only one face to the other, assuming both survive the Sun's expansion." (WK 2010tt)
- "400 Billion Years"—This is the time until "all the Solar System's actinide elements will have decayed to less than 1% their current value, leaving bismuth as the heaviest traceable element." (WK 2010tt)
- "10^{12} (1 Trillion) Years"—This is the time "until star formation ends in galaxies as galaxies are depleted of the gas clouds they need to form stars." (WK 2010tt; F. Adams 1997)
- "2×10^{12} (2 Trillion Years)"—This is the time "until all galaxies outside the Local Supercluster are no longer detectable in any way, assuming

that dark energy continues to make the Universe expand at an accelerating rate." (WK 2010tt; L. Krauss 2000)

- "10^{13} (10 Trillion) to 2×10^{13} (20 Trillion) Years"—This is the "lifetime of the longest-lived stars, low-mass red dwarfs." (WK 2010tt; F. Adams 1997)
- "10^{14} (100 Trillion) Years"—This is the "high estimate for the time until star formation ends in galaxies. Once star formation ends and the least massive red dwarfs exhaust their fuel, the only stellar-mass objects remaining will be stellar remnants (white dwarfs, neutron stars and black holes.) Brown dwarfs will also remain." (WK 2010tt; F. Adams 1997)
- "10^{15} Years"—This is the "estimated time until planets are detached from their orbits. Whenever two objects pass close to each other, the orbits of their planets can be disrupted and the planets can be ejected from orbit around their parent objects. Planets with closer orbits take longer to be ejected in this manner on average because a passing object must make a closer pass to the planet's primary to eject the planet." (WK 2010tt; F. Adams 1997)
- "10^{19} to 10^{20} Years"—This is "the estimated time until brown dwarfs and stellar remnants are ejected from galaxies. When two objects pass close enough to each other, they exchange orbital energy with lower-mass objects tending to gain energy. The lower-mass objects can gain enough energy in this manner through repeated encounters to be ejected from the galaxy. This process will cause the galaxy to eject the majority of its brown dwarfs and stellar remnants." (WK 2010tt; F. Adams 1997 & 1999)
- "10^{20} Years"—This is the "estimated time until the Earth's orbit around the Sun decays via emission of gravitational radiation, if the Earth is neither first engulfed by the red giant Sun a few billion years from now nor ejected from its orbit by a stellar encounter before then." (WK 2010tt; F. Dyson 1979; K. Schröder 2008; I. Sackmann 1993)
- "10^{32} Years"—This is "the smallest possible value for the proton half-life consistent with experiment." (WK 2010tt; SLAC 2008)
- "3×10^{34} Years"—This is "the estimated time for all nucleons in the observable universe to decay, if the proton half-life takes its smallest possible value." (WK 2010tt; R. Irion 2000)
- "10^{36} Years"—This is "the mean half-life of a proton according to some theories." (WK 2010tt; D. Page 1976)
- "10^{41} Years"—This is "the largest possible value for the proton half-life." (WK 2010tt; F. Adams 1997 & 1999)

- "3×10^{43} Years"—This is "the estimated time for all nucleons in the observable universe to decay, if the proton half-life takes the largest possible value, 10^{41} years." (WK 2010tt; R. Irion 2000)
- "10^{65} Years"—This is the "estimated time for rigid objects like rocks to rearrange their atoms and molecules via quantum tunnelling, assuming that the proton does not decay. On this timescale all matter is liquid." (WK 2010tt; F. Dyson 1979)
- "2×10^{66} Years"—This is "the estimated time until a black hole with the mass of the Sun decays by the Hawking process." (WK 2010tt; D. Page 1976)
- "1.7×10^{106} Years"—This is "the estimated time until a supermassive black hole with a mass of 20 trillion solar masses decays by the Hawking process." (WK 2010tt; F. Dyson 1979)
- "10^{1500} Years"—This is "the estimated time until all matter decays to 56Fe," that is, "the most common isotope of iron." (WK 2010tt & 2010uu)
- "10^{10^{26}} Years"—This is the "low estimate for the time until all matter collapses into black holes, assuming no proton decay." (WK 2010tt; F. Dyson 1979)
- "10^{10^{50}} Years"—This is the "estimated time for a Boltzmann Brain to appear in the vacuum via a spontaneous entropy decrease." (WK 2010tt; A. Linde 2007)
- "10^{10^{76}} Years"—This is the "high estimate for the time until all matter collapses into neutron stars or black holes, again assuming no proton decay." (WK 2010tt; F. Dyson 1979)
- "10^{10^{10^{10^{2.08}}}} Years"—This is the "scale of an estimated Poincaré recurrence time for the quantum state of a hypothetical box containing a black hole with the mass within the presently visible region of our universe. This time assumes a statistical model subject to Poincaré recurrence. A much simplified way of thinking about this time is in a model where our universe's history repeats itself arbitrarily many times due to properties of statistical mechanics, this is the time scale when it will first be somewhat similar (for a reasonable choice of 'similar') to its current state again." (WK 2010tt)
- "10^{10^{10^{10^{10^{1.1}}}}} Years"—This is the "scale of an estimated Poincaré recurrence time for the quantum state of a hypothetical box containing a black hole with the estimated mass of the entire universe, observable or not, assuming a certain inflationary model with an inflaton whose mass is 10^{-6} Planck masses." (WK 2010tt)

Surely, there are other estimates, but the ones listed here in the time frame are illustrative (not exhaustive).

The End of the Universe

The reason for this rather detailed estimation of the future of the "open" universe (in accordance to the time frame) is that many astrophysicsts nowadays believe that the universe will have the fate as described in the third scenario, namely, the "open" universe.

Even here, however, there are several possibilities about the fate of the "open" universe, and good examples include (a) the "big freeze," (c) the "heat death," (c) the "big rip," and (d) "multiverse"—to be summarized below, in that order.

(a) The Big Freeze

The first possibility of the end of an "open" universe is called the "Big Freeze," which is "a scenario under which continued expansion results in a universe that asymptotically approaches absolute zero temperature....This scenario is currently the most commonly accepted theory within the scientific community." (WK 2010qq)

(b) The Heat Death

The second possibility of the end of an "open" universe is called the "Heat Death, which states that the universe goes to a state of maximum entropy in which everything is evenly distributed, and there are no gradients—which are needed to sustain information processing, one form of which is life." (WK 2010qq)

Currently, many astrophyscists do not favour this possibility, because "the Heat Death scenario is compatible with any of the three spatial models, but requires that the universe reach an eventual temperature minimum." (WK 2010qq)

(c) The Big Rip

The third possibility of the end of an "open" universe is called the "Big Rip," which refers to "the special case of phantom dark energy," such that "the density of dark energy increases with time, causing the rate of acceleration to increase, leading to a steady increase in the Hubble constant." (WK 2010qq)

This then means that "all material objects in the universe, starting with galaxies and eventually (in a finite time) all forms, no matter how small, will disintegrate into unbound elementary particles and radiation, ripped apart by the phantom energy force and shooting apart from each other. The end state of the universe is a singularity, as the dark energy density and expansion rate becomes infinite." (WK 2010qq)

But the problem with this third possibility is that the very idea of "phantom dark energy" is still very speculative.

(d) Multiverse (or No Complete End)

Finally, the fourth possibility of the end of an "open" universe is that the universe will not completely end, because "our uni-'verse' is merely one Big Bang among an infinite number of simultaneously expanding Big Bangs that are spread out over endless distances (open space)." (WK 2010qq)

In this scenario of "no complete end" of the universe—"each 'verse' may be either matter or antimatter, with an equal number in existence at any given time. As the 'verses' expand they collide and matter and antimatter annihilate, releasing energy. Heat death of a finite universe would be predicted as entropy increases, however, the infinite size of the multiverse and the infinite number of 'verses' could mean that new 'verses' would be formed as old 'verses' were annihilated." (WK 2010qq)

To put it in a different way, "a chain reaction multiverse would be analogous to a fireworks display (each explosion representing a Big Bang) that starts in one neighborhood and is followed by fireworks displays in surrounding neighborhoods and then in neighborhoods further out. The chain reaction of Big Bangs would continue to expand as Big Bang fuel is consumed. If the multiverse is open and the fuel is infinite then the chain reaction would expand forever." (WK 2010qq)

So, in the end, "the multiverse as a whole may never end completely," because some verses would be born somewhere and some others would be destroyed elsewhere. (WK 2010qq)

However, to avoid any conceptual misunderstanding of the term "multiverse," two qualifications are needed here.

The first qualification is that all these different "verses," by definition, do not interact with each other; otherwise, "if two 'universes' could affect one another, they would be part of a single universe. Thus, although some fictional characters [in sci-fi and fantasy works] travel between parallel fictional 'universes,' this is, strictly speaking, an incorrect usage of the term *universe*. The disconnected universes are conceived as being physical, in the sense that each should have its own space and time, its own matter and energy, and its own physical laws—that also challenges the definition of parallelity as these universes don't exist synchronously (since they have their own time) or in a geometrically parallel way (since there's no interpretable relation between spatial positions of the different universes)." (WK 2010b; G. Ellis 2004; M. Munitz 1959)

In addition, "[s]uch physically disconnected universes should be distinguished from the metaphysical conception of alternate planes of consciousness [in esoteric cosmology], which are not thought to be physical places and are connected through the flow of information." (WK 2010b)

And the second qualification is that the very idea of "multiverse" remains speculative. For instance, a related version of "multiverse" concerns "the many-worlds hypothesis," in accordance to which "a parallel universe is born with every quantum measurement; the universe 'forks' into parallel copies, each one corresponding to a different outcome of the quantum measurement." (WK 2010b)

But this usage of the term "multiverse" is "speculative and may be considered unscientific; no experimental test in one universe could reveal the existence or properties of another non-interacting universe." (WK 2010b)

Remaining Problems

In any event, not everyone agrees with the four likely possibilities of the end of the universe (as summarized above), even if they are based on science.

Consider a few criticisms below, for illustration.

Firstly, a major criticism against the discourse on the ultimate fate of the universe is that it is not clear that the future can be so neatly outlined in these possibilities.

For instance, some point out other factors which have been neglected in the literature on the ultimate fate of the univrese, and a good example concerns "the vaccum metastability event." (WK 2010qq)

A "vaccum metastability event" refers to the possibility that, when a "vacuum is not in its lowest energy state (a false vacuum), it could tunnel into a lower energy state....This has the potential to fundamentally alter our universe; in more audacious scenarios even the various physical constants could have different values, severely affecting the foundations of matter, energy, and spacetime. It is also possible that all structures will be destroyed instantaneously, without any forewarning." (WK 2010qq; I. Moss 1982; M. Stone 1976 & 1977; P. Frampton 1976 & 1977; S. Coleman 1977; C. Callan 1977)

Secondly, another major criticism against the discourse on the ultimate fate of the univrse is that many of these possibilities assume the existence of "dark energy," which, however, is still not well understood.

In other words, "each possibility described so far is based on a very simple form for the dark energy equation of state. But as the name is meant to imply, we know almost nothing of the real physics of the dark energy." (WK 2010qq)

This ignorance can spell trouble for those who use it for theoretical convenience without much critical thinking; for instance, "if the theory of inflation is true, the universe went through an episode dominated by a different form of dark energy in the first moments of the big bang; but inflation ended, indicating an equation of state much more complicated than those assumed so far for present-day dark energy. It is possible that the dark energy equation of state could change again resulting in an event that would have consequences which are extremely difficult to parametrize or predict. It is also possible the universe may never have an end and continue in its present state forever." (WK 2010qq)

Thirdly, still one more major criticism against the discourse on the ultimate fate of the univrse is that it may well be motivated by religious beliefs.

For instance, "many religions have an end-of-the-universe prediction. The theological study of the ultimate fate of the universe and/or ultimate destiny of mankind is known as eschatology. Many religious beliefs are cataclysmic, and some theists do not view the various scientific theories about the end of the universe as contradicting their religious beliefs. Alternatively, many religions, particularly Hinduism and Buddhism, believe in an unending cycle of apocalyptic destruction and re-creation." (WK 2010qq)

And fourthly, still another major criticism against the discourse on the ultimate fate of the univrse is that it takes no account of the role of what I origianlly called "post-humans" in altering the universe (and, for that matter, the multiverse) in the distant future, as this is something that I went to great lengths to explain in many of my books, in regard to my different visions of the post-humans.

Chapter Four will discuss this issue in the relation to "the post-human odyssey" in the context of post-cosmology.

From an Overall Standpoint

Once more, these criticisms against the literature on the ultimate fate of the universe do not mean to dismiss it outright but to show us the opposing sides of the debate, such that the possibility and desirability of the evolution of the universe (and its speculative ends) from the perspective of nature are not to the extent that the respective spokespersons would like us to believe.

In addition, the analysis of the ultimate fate of the universe can reveal something valuable about the ontological principles in existential dialectics, and good examples include the absoluteness-relativeness principle, the partiality-totality principle, the predictability-unpredictability principle, the explicability-inexplicability principle, the finiteness-transfiniteness principle, the preciseness-vagueness principle, the simpleness-complicatedness principle, the openness-hiddenness principle, the denseness-emptiness principle, the order-chaos principle, the slowness-quickness principle, the expansion-contraction principle, the theory-praxis principle, the convention-novelty principle, the evolution-transformation principle, the symmetry-asymmetry principle, the regression-progression principle, and the same-difference principle.

For instance, in relation to the absoluteness-relativeness principle, if there is absoluteness (e.g., the absolute view of the three possibilities of the ultimate end of the "open" universe, with the Big Bang theory as its theoretical basis), there is relativeness (e.g., what is correct for the three possibilities of the ultimate theory of the "open" universe is not necessarily so for the theory of multiverse, in that it will not have a complete end). And the reverse direction also holds true.

In relation to the partiality-totality principle, if there is partial viewpoint (e.g., the partial view, from the perspective of each "verse," that it was born and it will end), there is totalistic counterpart (e.g., the holistic view of the larger "multiverse," in that there can be a cosmic cycle, for it to not have a complete end, with uncertain consequences, so the view of the whole is not the sum of the views of the individual "verses"). And the reverse direction also holds true.

In relation to the predictability-unpredictability principle, if there is predictability (e.g., the predictable fate of the universe in that it will expan indefinitely, based on the current scientific evidences), there is unpredictability (e.g., the more difficult task to predict correctly, based on the same current scientific evidences, how and when exactly the universe will end, and whether or not it will end completely, strictly speaking). And the reverse direction also holds true.

In relation to the explicability-inexplicability principle, if there is explicability (e.g,. the explanation of the ultimate fate of the universe by scientists in terms of three possible scenarios, with the Big Bang theory as the scientific basis and the assumption of some entities like "dark energy," "dark matter," etc.), there is inexplicability (e.g., the lack of sufficient explanation of what these assumed entities like "dark energy" and "dark matter" really are, since, "as the name is meant to imply, we know almost nothing of the real physics of the dark energy"). And the reverse direction also holds true.

In relation to the finiteness-transfiniteness principle, if there is finiteness (e.g., the finite number of the possibilities concerning the ultimate fate of the universe, as shown in the three scenarios), there is transfiniteness (e.g., the transfinite number of time for the universe to end, in accordance to any of the three scenarios). And the reverse direction also holds true.

In relation to the preciseness-vagueness principle, if there is preciseness (e.g., the precise identification of "dark energy" as crucial to understand the ultimate fate of the universe), there is vagueness (e.g., the vagueness in the idenfication of "dark energy," because, "as the name is meant to imply, we know almost nothing of the real physics of the dark energy"). And the reverse direction also holds true.

In relation to the simpleness-complicatedness principle, if there is simpleness (e.g., the relatively simple outline of the future fate of the universe on the basis of the three scenarios), there is complicatedness (e.g., the relatively more complicated analysis of the ultimate fate of the universe by questioning the assumptions and claims in regard to the three scenarios, as shown in the problems as pointed out by the critics). And the reverse direction also holds true.

In relation to the openness-hiddenness principle, if there is openness (e.g., the open exploration of the ultimate fate of the universe, on the basis of the Big Bang theory and the assumpton of some entities like "dark energy" and "dark matter" in physics), there is hiddenness (e.g., the hidden nature of "dark matter" and "dark energy," such that the ultimate fate of the universe remains uncertain). And the reverse direction also holds true.

In relation to the denseness-emptiness principle, if there is denseness (e.g., the reltively denser concentration of believers in the "open" universe as the ultimate fate of the universe nowadays), there is emptiness (e.g., the relatively less dense, or more empty, concentration of believers in the "flat" universe as the ultimate fate of the universe nowadays). And the reverse direction also holds true.

In relation to the order-chaos principle, if there is order (e.g., the relatively orderly patterns of the universe, such that its ultimate fate can be scientifically discussed in terms of three major scenarios), there is chaos (e.g., the relatively chaotic nature of other aspects of the universe, like "the vaccum metastability event," with "the potential to fundamentally alter our universe" or that "all structures will be destroyed instantaneously, without any forewarning"). And the reverse direction also holds true.

In relation to the slowness-quickness principle, if there is slowness (e.g., the universe will take a long long time to end in the scenario of the "open" universe, if measured from the standpoint of average human lifespan), there is quickness (e.g., the universe is accelerating very fast in its expansion, if measured from the standpoint of average human running speed). And the reverse direction also holds true.

In relation to the expansion-contraction principle, if there is expansion (e.g., "new 'verses' would be formed" somewhere, in accordance to the multiverse theory), there is contraction (e.g., "old 'verses'" would be "annihilated" somewhere, in accordance to the multiverse theory). And the reverse direction also holds true.

In relation to the theory-praxis principle, if there is theory (e.g., the theoretical construction of different scenarios about the ultmate fate of the universe), there is praxis (e.g., the practical application of the theories concerning the ultimate fate of the universe for "countless sci-fi and fantasy works," which "use the threatened destruction of the universe as their plot device, usually with an evil supervillain and/or the incompetence of humanity as the cause, and usually with human ingenuity saving the day"). (WK 2010rr) And the reverse direction also holds true.

In relation to the convention-novelty principle, if there is convention (e.g., the conventional wisdom about the "static universe" which Einstein initially accepted), there is novelty (e.g., the alternative novel challenge to the conventional wisdom about the "static universe," by Hubble who "announced his conclusion that the universe was expanding").

In relation to the evolution-transformation principle, if there is evolution (e.g., the natural evolution of humans to observe the universe with awe and incomprehension in the state of nature), there is transformation (e.g., the technical transformation of human observation at the universe by way of the invention of telescopes and the scientific construction of the Big Bang theory). And the reverse direction also holds true.

In relation to the symmetry-asymmetry principle, if there is symmetry (e.g., the co-existence of different scenarios about the ultimate fate of the universe in the history of cosmology), there is asymmetry (e.g., the acceptance of the static universe by Einstein and his contemporaries in the old days, but the acceptance of the expanding universe by many astrophysicists nowadays). And the reverse direction also holds true.

In relation to the regression-progression principle, if there is regression (e.g, the regression made by the three scenarios, with the Big Bang theory as the scientific basis, about the ultimate fate of the universe—as shown in the remaining problems which the critics are quick to point out), there is progression (e.g, the progress made by the three scenarios about the ultimate fate of the universe in enhancing the understanding of the likely possibility that the universe is accelerating in its expansion). And the reverse direction also holds true.

And in relation to the same-difference principle, if there is similarity in outcome (e.g., the contribution to the understanding of the ultimate fate of the universe, regardless of whether this is done by way of the three scenarios with the Big Bang theory as the basis, or of the multiverse theory), there is difference in outcome (e.g., the contribution to the understanding of the ultimate fate of the universe by way of the three scenarios with the Big Bang theory as the basis—but the contribution to the understanding of the ultimate fate of the universe by way of the multiverse theory, such that the universe will not end completely, strictly speaking). And the reverse direction also holds true.

Its Speculative Ends—and Society

The evolution of the universe and its speculative ends can also be analyzed from the perspective of society, as it can illuminate some shocking aspects of the subject matter .

Two case studies are arranged here, namely, (3.4.1) Heaven's Gate, the end of the Earth, and cult suicide, and (3.4.2) MRTCG, apocalypticism, and doomsday deaths—to be addressed hereafter, respectively.

Heaven's Gate, the End of the Earth, and Cult Suicide

An excellent illustration of the evolution of the universe and its speculative ends from the perspective of society has to do with Heaven's Gate, the end of the Earth, and cult suicide.

The Belief System about the End of the Earth

The term "Heaven's Gate" refers to "an American UFO Religion based in San Diego, California, founded and led by Marshall Applewhite (1931–1997) and Bonnie Nettles (1928-1985)." (WK 2010ss; I. Hexham 1997)

It gained infamy when "on March 26, 1997, in a period that Comet Hale-Bopp was at its brightest, police discovered the bodies of 39 members of the group, who had apparently died by suicide." (WK 2010ss; NYT 1997) But why?

The answer has to do with the belief system of the Heaven's Gate about the speculative end of the Earth, in that its members "believed that the planet Earth was about to be recycled (wiped clean, renewed, refurbished and rejuvenated), and that the only chance to survive was to leave it immediately. While the group was formally against suicide, they defined 'suicide' in their own context to mean 'to turn against the Next Level when it is being offered,' and believed that their 'human' bodies were only vessels meant to help them on their journey." (WK 2010ss)

With this belief in mind, its members suggested "several paths for a person to leave the Earth and survive before the 'recycling,' one of which was hating this world strongly enough: 'It is also possible that part of our test of faith is our hating this world, even our flesh body, to the extent to be willing to leave it without any proof of the Next Level's existence.'" (WK 2010ss)

In fact, "the members of the group added '-ody' to the first names they adopted in lieu of their original given names, which defines 'children of the Next Level.' This is mentioned in Applewhite's final video, 'Do's Final Exit' [with "Do" as one of Applewhite's aliases], that was filmed on March 19, 1997, just days prior to the suicides." (WK 2010ss)

The Origin of the Belief System

The origin of the belief system about the coming end of the Earth can be traced back to some Christian ideas of "salvation and apocalypse." (WK 2010ss)

It all started with Marshall Applewhite in the early 1970's who "was recovering from a heart attack during which he claimed to have had a near-death experience. He came to believe that he and his nurse, Bonnie Nettles, were 'the Two,' that is, the two witnesses spoken of in Book of Revelation 11:3 in the Holy Bible." (WK 2010ss; J. Vallee 1979)

So, "they began traveling around the country giving talks about their belief system. As with some other New Age faiths, they combined Christian doctrine (particularly the ideas of salvation and apocalypse) with the concept of evolutionary advancement and elements of science fiction, particularly travel to other worlds and dimensions." (WK 2010ss; C. Partridge 2005; D. Wojik 2003)

Group Structure and Lifestyle

This belief system about the coming end of the Earth (with some Christian ideas of "salvation and apocalypse") transformed the group into a comparable "medieval monastic order. Group members gave up their material possessions and lived a highly ascetic life devoid of many indulgences. The group was tightly knit and everything was shared communally. Seven of the male members of the group, including Applewhite, voluntarily underwent castration in Mexico as an extreme means of maintaining the ascetic lifestyle." (WK 2010ss; R. Ross 1999)

Randall Bell in *Strategy 360* wrote that Applewhite "wanted to create an environment where all the thinking was already done. At the same time, no cult member was allowed to be alone. Monitoring devices were everywhere. There was a bizarre amount of wiring throughout the house and even down the chimney. Even when cult members spoke on the phone, someone was there to monitor the conversation." (WK 2010ss)

The Days of the Suicide

A few days before the suicide, that is, "on March 19, 1997, Marshall Applewhite taped himself speaking of mass suicide and asserted 'it was the only way to evacuate this Earth.' The Heaven's Gate group was against suicide but they believed they had no choice but to leave Earth as quickly as possible. After claiming that a space craft was trailing the comet Hale-Bopp, Applewhite convinced 38 followers to commit suicide so that their souls could board the supposed craft. Applewhite believed that after their deaths, a UFO would take their souls to another 'level of existence above human,' which Applewhite described as being both physical and spiritual. This and other UFO-related beliefs held by the group have led some observers to characterize the group as a type of UFO religion." (WK 2010ss)

Then, on March 26, 1997, "thirty-eight Heaven's Gate members, plus group leader Applewhite, were found dead in a rented mansion in the upscale San Diego-area community of Rancho Santa Fe, California....In the heat of the California spring, many of the bodies had begun to decompose by the time they were discovered. The corpses underwent autopsies, where cyanide and arsenic were found. The bodies were later cremated." (WK 2010ss)

The suicide was committed with the "ingestion of phenobarbital mixed with applesauce or pudding, washed down with vodka. Additionally, plastic bags were secured around their heads after ingesting the mix to induce asphyxiation. Authorities found the dead lying neatly in their own bunk beds, faces and torsos covered by a square, purple cloth. Each member carried a five dollar bill and three quarters in their pockets. All 39 were dressed in identical black shirts and sweat pants, brand new black-and-white Nike Windrunner athletic shoes, and armband patches reading 'Heaven's Gate Away Team' (one of many instances of the group's use of the Star Trek fictional universe's nomenclature). Six males of the deceased, including Applewhite, were discovered to have undergone castration in the past. The adherents, between the ages of 26 and 72, are believed to have died in three groups over three successive days, with remaining participants cleaning up after each prior group's death. Fifteen members died on March 24, 15 more on March 25, and nine on March 26. Leader Applewhite was the third to last member to die; two women remained after him and were the only ones found without bags over their heads." (WK 2010ss; K. Ramsland 2006)

Later, "two former members of Heaven's Gate, Wayne Cooke and Charlie Humphreys,...committed suicide in a similar manner to the group. Humphreys had survived a suicide pact with Cooke in May 1997, but successfully committed suicide in February 1998." (WK 2010ss; SDHT 2007; NYT 1997a)

The Redemption of the Tragedy

A tragedy as this cult suicide is, the story can show us the other (dark) side of the possibility and desirability of the evolution of the universe (and its speculative ends) from the perspective of society (this time, in relation to the speculative end of the Earth and cult suicide as a case study), something that the respective spokespersons would not like us to know.

Furthermore, the analysis of Heaven's Gate, the speculative end of the Earth and cult suicide can teach us something valuable about the ontological principles in existential dialectics, and good examples include the absoluteness-relativeness principle, the predictability-unpredictability principle, the explicability-inexplicability principle, the preciseness-vagueness principle, the simpleness-complicatedness principle, the openness-hiddenness principle, the denseness-emptiness principle, the slowness-quickness principle, the expansion-contraction principle, the theory-praxis principle, the convention-novelty principle, the evolution-transformation principle, the symmetry-asymmetry principle, the softenss-hardness principle, the regression-progression principle, and the same-difference principle.

For instance, in relation to the absoluteness-relativeness principle, if there is absoluteness (e.g., the absolutely ascetic lifestyle of the members of Heaven's Gate comparable to "medieval monastic order," since "group members gave up their material possessions and lived a highly ascetic life devoid of many indulgences"), there is relativeness (e.g., what is good for the members of Heaven's Gate to live a strictly ascetic lifestyle is not so for those in the Playboy Club who enjoy living a life full of the pleasures of the flesh). And the reverse direction also holds true.

In relation to the predictability-unpredictability principle, if there is predictability (e.g., the predictable tendency of the members of Heaven's

Gate to indulge in apocalyptic thinking), there is unpredictability (e.g., no one outside Heaven's Gate had predicted that the members would commit suicide on March 24-26, 1997). And the reverse direction also holds true.

In relation to the explicability-inexplicability principle, if there is explicability (e.g., the explanation by Applewhite to his followers that, since "the planet Earth was about to be recycled [wiped clean, renewed, refurbished and rejuvenated],...the only chance to survive was to leave it immediately" by committing suicide), there is inexplicability (e.g., the lack of sufficient explanation by Applewhite, of why it was in fact true that "the planet Earth was about to be recycled [wiped clean, renewed, refurbished and rejuvenated]" and that "the only chance to survive was to leave it immediately" by committing suicide, without any other options). And the reverse direction also holds true.

In relation to the preciseness-vagueness principle, if there is preciseness (e.g., the precise identification of March 26, 1997 as a period when "Comet Hale-Bopp was at its brightest," by Applewhite to his followers), there is vagueness (e.g., the vagueness in the identfication of March 26, 1997 as a period when "Comet Hale-Bopp was at its brightest," by Applewhite to his followers, since it would mean just anything that he wanted to impose meaning on it, like the story of "a space craft...trailing the comet Hale" on that date, as the best time to commit suicide). And the reverse direction also holds true.

In relation to the simpleness-complicatedness principle, if there is simpleness (e.g., the relatively simplistic story about the end of the Earth by Applewhite to his followers), there is complicatedness (e.g., the relatively more complicated analysis of the story about Applewhite, as revealed by Jacques Vallée in his 1979 *Messengers of Deception*, in which Vallée critically questioned the "authoritarian political and religious outlook" of the group). (WK 2010ss) And the reverse direction also holds true.

In relation to the openness-hiddenness principle, if there is openness (e.g., the open exploration of the coming end of the Earth and the need to commit suicide, as told by Applewhite to his followers), there is hiddenness (e.g., the hidden bias in Applewhite's view, which could be traced back to his traumatic "near-death experience" in the early 1970's, which led him "to believe that he and his nurse, Bonnie Nettles, were 'the Two,' that is, the two witnesses spoken of in Book of Revelation 11:3 in the Holy Bible"). And the reverse direction also holds true.

In relation to the denseness-emptiness principle, if there is denseness (e.g., the relatively denser concentration of individuals living "a highly ascetic life devoid of many indulgences" in Heaven's Gate), there is emptiness (e.g., the relativley less dense, or more empty, concentration of individuals living "a highly ascetic life devoid of many indulgences" among members of the Playboy Club founded by Hugh Hefner). And the reverse direction also holds true.

In relation to the slowness-quickness principle, if there is slowness (e.g., the relatively slower readiness of those like Hugh Hefner who do not have any apocalyptic mindset to commit suicide), there is quickness (e.g., the relatively quicker readiness of those in Heaven 's Gate who have the apocalyptic mindset to commit suicide). And the reverse direction also holds true.

In relation to the expansion-contraction principle, if there is expansion (e.g., the relatively more developed ability of those in Heaven's Gate to confront life in an ascetic way), there is contraction (e.g., the relatively less developed ability of those in Heaven's Gate to confront life in a more positive way without having to kill oneself). And the reverse direction also holds true.

In relation to the theory-praxis principle, if there is theory (e.g., the theoretical construction of the belief system about the end of the Earth with some Christian ideas of "salvation and apocalypse"), there is praxis (e.g., the practical application of the belief system about the end of the Earth for cult suicide by members of Heaven's Gate). And the reverse direction also holds true.

In relation to the convention-novelty principle, if there is convention (e.g., the conventional wisdom about the end of the world based on some Christian ideas of "salvation and apocalypse"), there is novelty (e.g., the alternative novel revision of this conventional wisdom about the end of the world, by Applewhite, for his fabrication that he and his nurse were "the two witnesses spoken of in Book of Revelation 11:3 in the Holy Bible"). And the reverse direction also holds true.

In relation to the evolution-transformation principle, if there is evolution (e.g., the natural evolution of humans to live on earth in the state of nature), there is transformation (e.g., the technical transformation of the human view about living on Earth by way of the invention of the belief system about the end of the Earth and the need to leave it). And the reverse direction also holds true.

In relation to the symmetry-asymmetry principle, if there is symmetry (e.g., the co-existence of different beliefs about the end of the world through the ages), there is asymmetry (e.g., the acceptance of the beliefs about the end of the world by members of the traditional Christian ideas of "salvation and apocalypse"—but the acceptance of Applewhite's version of the end of the world by members of Heaven's Gate). And the reverse direction also holds true.

In relation to the softenss-hardness principle, if there is softness (e.g., the promise of "the Next Level's existence" after death by Applewhite to his followers), there is hardness (e.g., the call to kill oneself by Applewhite to his followers). And the reverse direction also holds true.

In relation to the regression-progression principle, if there is regression (e.g., the regression made by the belief system about the end of the Earth, as shown by the tragedy of Heaven's Gate), there is progression (e.g.,the progress made by the tragedy of Heaven's Gate, in that people are more aware of the "authoritarian political and religious outlook" of cults as described by Jacques Vallée in his 1979 *Messengers of Deception*). And the reverse direction also holds true.

And in relation to the same-difference principle, if there is similarity in outcome (e.g., the contribution to how life is to be lived, regardless of whether this is done by strict asceticism or sensual materialism), there is difference in outcome (e.g., the contribution to how life is to be lived, by way of strict asceticism like the one in Heaven's Gate—but the contribution to how life is to be lived, by way of sensual materialism like the one in the Playboy Club). And the reverse direction also holds true.

MRTCG, Apocalypticism, and Doomsday Deaths

Unfotunately, a few years after the tragedy of Heaven's Gate, another tragegy related to a comparable belief about the end of the world occurred in Uganda in 2000.

This second case study has to do with the Movement for the Restoration of the Ten Commandments of God (or MRTCG in short), apocalypticism, and doomsday deaths.

The Origins and Beliefs of MRTCG

The Movement for the Restoration of the Ten Commandments of God (or MRTCG) refers to "a breakaway cult from the Roman Catholic Church founded by Credonia Mwerinde and Joseph Kibweteere and Bee Tait in Uganda. It formed in the late 1980s after Mwerinde, a brewer of banana beer, and Kibweteere, a politician, claimed that they had visions of the Virgin Mary. The five primary leaders were Joseph Kibweteere, Joseph Kasapurari, John Kamagara, Dominic Kataribabo, and Credonia Mwerinde." (WK 2010yy)

In fact, "the earliest origins of the movement have been traced back to Credonia Mwerinde's father Paulo Kashaku. In 1960 he claimed to have had a vision of his deceased daughter Evangelista, who told him that he would have visions of heaven. This prediction passed in 1988, when he saw Jesus Christ, the Virgin Mary, and Saint Joseph." (WK 2010yy)

The mission of MRTCG was "to obey the Ten Commandments and preach the word of Jesus Christ. They taught that to avoid damnation in the apocalypse, one had to strictly follow the Commandments. The emphasis on the Commandments was so strong that the group discouraged talking, for fear of breaking the Ninth Commandment, 'Thou shalt not bear false witness against thy neighbor,' and on some days communication was only conducted in sign language. Fasts were conducted regularly, and only one meal was eaten on Fridays and Mondays. Sex was forbidden, as was soap." (WK 2010yy; BBC 2000)

At the beginning, the leaders of MRTCG "declared that the apocalypse would occur in the year 2000. The group had a strong emphasis on an apocalyptic end time, highlighted by their booklet *A Timely Message from Heaven: The End of the Present Time.* New members were required to study it and be trained in its text, reading it as many as six times. They also taught that the Virgin Mary had a special role in the end, and that she also communicated with their leadership. They held themselves akin to Noah's Ark, a ship of righteousness in a sea of depravity." (WK 2010yy; BBC 2000; H. Cauvin 2000)

The Unrealized Prophesy of Apocalypse in 2000

The prophesy of apocalypse in 2000 by the movement leaders put pressure on the group as "the new millennium approached"; consequently, "preparations for the end mounted. In 1999, the state owned *New Vision* newspaper ran an interview with a teenage member. He said, 'The world ends next year. There is no time to waste. Some of our leaders talk directly to god. Any minute from now, when the end comes, every believer who will be at an as yet undisclosed spot will be saved.'" (WK 2010yy; A. Borzello 2000)

Sometime near the new year of 2000, "activity by Movement members became frenzied," and "their leaders urged them to confess their sins in preparation for the end. Clothes and cattle were sold cheaply, past members were re-recruited, and all work in the fields ceased." (WK 2010yy)

However, things did not happen as promised, because "January 1, 2000 passed without the advent of the apocalypse, and the Movement began to unravel. Questions were asked of Mwerinde and Kibweteere, and payments to the Church decreased dramatically. Ugandan police believe that some members, who were required to sell their possessions and turn over the money to the Movement, rebelled and demanded the return of their money." (WK 2010yy; I. Fisher 2000 & 2000a)

For damage control, the movement leaders promised another date, which was set, this time, on March 17 as "the new end of the world, a doomsday that would come 'with ceremony, and finality' according to the *New York Times*. The Movement held a huge party at Kanungu, and roasted three bulls and drank 70 crates of soft drinks." (WK 2010yy; I. Fisher 2000a)

But on March 17, something tragic happened, because, when "group members arrived at their church in Kanangu to pray and sing, minutes later nearby villagers heard an explosion, and the building was gutted in an intense fire that killed all 530 in attendance, including dozens of children. The windows and doors of the building had been boarded up. The five principal cult leaders, Joseph Kibweteere, Joseph Kasapurari, John Kamagara, Dominic Kataribabo, and Credonia Mwerinde, were assumed to have died in the fire. The fire alerted the Ugandan authorities to what had been occurring in the Movement. Several days before Movement leader Dominic Kataribabo was seen buying 50 liters of sulfuric acid, which may have started the fire. Another party was planned for the eighteenth, which officials believe sect leaders had announced in order to mislead authorities as to their plans." (WK 2010yy; I. Fisher 2000; NYT 2000; A. Borzello 2000)

The Tragedy of Doomsday Deaths

Later, "police investigated Movement properties and discovered hundreds of bodies at sites across southern Uganda. Six bodies were discovered sealed in the latrine of the Kanungu compound, as well as 153 bodies at a compound in Buhunage, 155 bodies at Dominic Kataribabo's estate at Rugazi, where they had been poisoned and stabbed, and another 81 bodies at lay leader Joseph Nymurinda's farm. Forensics investigations indicated that they had been murdered weeks before the church inferno." (WK 2010yy; H. Cauvin 2000a; A. Borzello 2000a)

Also, "medical examiners determined that the majority of dead sect members had been poisoned. Early reports had suggested that they had been strangled based on the presence of twisted banana fibers around their necks. After searching all sites, the police concluded that earlier estimates of nearly a thousand dead had been exaggerated, and that the final death toll had settled at 778." (WK 2010yy; NYT 2000)

After an investigation was carried out, "the police ruled out a cult suicide, and instead consider it to be a mass murder conducted by Movement leadership. They believe that the failure of the doomsday prophecy led to a revolt in the ranks of the sect, and the leaders set a new date with a plan to eliminate their followers. The discovery of bodies at other sites, the fact the church had been boarded up, the presence of incendiaries, and the possible disappearance of sect leaders all point to this theory." (WK 2010yy; I. Fisher 2000; H. Cauvin 2000a)

Looking Back

This additional tragedy in Uganda, also in the context of the beliefs about the end of the world, shows us the other (dark) side of the possibility and desirability of the evolution of the universe (and its speculative ends) from the perspective of society (this time, in relation to MRTCG, apocalpyticism, and doomsday deaths as a case study), something that the respective spokespersons would not like us to know.

In addition, the analysis can reveal something refreshing about the ontological principles in existential dialectics, and good examples include the absoluteness-relativeness principle, the predictability-unpredictability principle, the explicability-inexplicability principle, the preciseness-vagueness principle, the simpleness-complicatedness principle, the openness-hiddenness principle, the denseness-emptiness principle, the slowness-quickness principle, the expansion-contraction principle, the theory-praxis principle, the convention-novelty principle, the evolution-transformation principle, the symmetry-asymmetry principle, the softenss-hardness principle, the regression-progression principle, and the same-difference principle.

For instance, in relation to the absoluteness-relativeness principle, if there is absoluteness (e.g., the absolute view of MRTCG that "one had to strictly follow the Commandments," so "[t]he emphasis on the Commandments was so strong that the group discouraged talking, for fear of breaking the Ninth Commandment"), there is relativeness (e.g., what is right for MRTCG to demand its members "to strictly follow the Commandments" is not necessarily so for those on the outside who do not accept the extreme Christian teachings in MRTCG). And the reverse direction also holds true.

In relation to the predictability-unpredictability principle, if there is predictability (e.g., the predictable tendency of MRTCG members to obey the Ten Commandments), there is unpredictability (e.g., no one had predicted that MRTCG members would have ended up in doomsday deaths, which the police finally concluded as "mass murder," contrary to one of the Christian commandments that "thou shalt not kill"). And the reverse direction also holds true.

In relation to the explicability-inexplicability principle, if there is explicability (e.g., the explanation by the Movement leaders that "the apocalypse would occur in the year 2000," on the basis of the Christian teachings of salvation and apocalypse, as revealed in the Movement booklet *A Timely Message from Heaven: The End of the Present Time*), there is inexplicability (e.g., the lack of convincing explanation, by the Movement leaders, of why the apocalypse had to be on January 01 2000, which was later revised to mean March 17, 2000 instead). And the reverse direction also holds true.

In relation to the preciseness-vagueness principle, if there is precision (e.g., the precise identification of January 01, 2000 as the date for the occurrence of apocalpyse, in accordance to the Movement leaders), there is vagueness (e.g., the vagueness in the precise identification of the date of January 01, 2000 as the date for the occurrence of apocalpyse, because it did not say whether it had to be taken literally or could be changed to a later date, which was precisely what the Movement leaders did when the first date turned out to be untrue). And the reverse direction also holds true.

In relation to the simpleness-complicatedness principle, if there is simpleness (e.g, the relatively simplistic prophesy of January 01, 2000 as the date for the occurrence of apocalpyse, by the Movement leaders to the followers), there is complicatedness (e.g., the relatively more complicated view of apocalpyse by the Movement members when the doomsday prophesy did not realize, which "led to a revolt in the ranks of the sect" and they came to question their leaders, before being mass-murdered). And the reverse direction also holds true.

In relation to the openness-hiddenness principle, if there is openness (e.g., the open exploration, by the Movement leaders to the followers, of another doomsday set for March 17, 2000, after the first doomsday of January 01, 2000 failed), there is hiddenness (e.g., the hidden motive of the Movement leader behind the new doomsday of March 17, 2000, in that "the leaders set a new date with a plan to eliminate [that is, to murder] their followers"). And the reverse direction also holds true.

In relation to the denseness-emptiness principle, if there is denseness (e.g., the relatively denser concentration of believers in doomsday among MRTCG members), there is emptiness (e.g., the relatively less dense, or more empty, concentration of believers in doomsday among the police who led the investigation of doomsday deaths). And the reverse direction also holds true.

In relation to the slowness-quickness principle, if there is slowness (e.g., the relativley slower readiness of the Movement members to have sex), there is quickness (e.g., the relatively quicker readiness of a prostitue to have sex). And the reverse direction also holds true.

In relation to the expansion-contraction principle, if there is expansion (e.g., the relatively more developed ability of the Movement members to live a strictly ascetic lifestyle, in that "fasts were conducted regularly, and only one meal was eaten on Fridays and Mondays," etc.), there is contraction (e.g., the relatively less developed ability of the Movement members to be sensual and joyful in life, without worrying about damnation and the Ten Commandments). And the reverse direction also holds true.

In relation to the theory-praxis principle, if there is theory (e.g., the theoretical construction of salvation and apocalypse in Christian teachings), there is praxis (e.g., the practical application of the Christian ideas of apocalyse and salvation by MRTCG members to prepare the doomsday of January 01, 2000, as "activity by Movement members became frenzied" shortly before that date). And the reverse direction also holds true.

In relation to the convention-novelty principle, if there is convention (e.g.,the conventional wisdom about the doomsday in Christian teachings of salvation and apocalypse), there is novelty (e.g, the alternative novel challenge to the conventional wisdom about the doomsday by the Movement leaders, who first set the date of January 01, 2000 for the doomsday and later changed it into March 17, 2000 instead). And the reverse direction also holds true.

In relation to the evolution-transformation principle, if there is evolution (e.g., the natural evolution of humans to die naturally in the state of nature, for instance, by reason of old age, hunger, disease, fatal injury, fighting, etc.), there is transformation (e.g., the technical transformation of human death by way of the invention of doomsday deaths, like those commited by MRTCG). And the reverse direction also holds true.

In relation to the symmetry-asymmetry principle, if there is symmetry (e.g., the co-existence of different versions of Christianity over time), there is asymmetry (e.g., the acceptance of Catholicism by the Roman Catholic Church—but the worship of MRTCG version of Christianity as written in its official booklet *A Timely Message from Heaven*, by the Movement members). And the reverse direction also holds true.

In relation to the softenss-hardness principle, if there is softness (e.g., the reward of salvation to those Christians who follow the Ten Commandments), there is hardness (e.g., the punishment by damnation against those Christians who violate the Ten Commandments). And the reverse direction also holds true.

In relation to the regression-progression principle, if there is regression (e.g., the regression made by the Christian ideas of salvation and apocalypse, as shown in the tragedy of doomsday deaths in MRTCG), there is progression (e.g., the progress made by the tragedy of doomsday deaths in MRTCG, in that it makes others more aware of the danger of blind worship in doomsday cults). And the reverse direction also holds true.

And in relation to the the same-difference principle, if there is similarity in outcome (e.g., the contribution to the regulation and control of human lifestyle, regardless of whether this is achieved by an ascetic form of existence or by a sensual form of existence), there is difference in outcome (e.g., the contribution to the regulation and control of human lifestyle, by way of an ascetic form of existence—but the contribution to the regulation and control of human lifestyle, by way of a sensual form of existence). And the reverse direction also holds true.

Its Speculative Ends—and Culture

The analysis of the beliefs about the end of the world with the traditional Christian ideas of "salvation and apocalypse" (in the previous section) paves the way for this section, since the evolution of the universe and its speculative ends can also be analyzed from the perspective of culture, as it can illuminate something apocalyptic about the subject matter

Eschatology, the End of the World, and Religion

Thus, an excellent case study of the evolution of the universe and its speculative ends from the perspective of culture has to do with eschatology, the end of the world, and religion.

The Meaning of a Technical Term

The word "eschatology" is "from the Greek ἔσχατος" ("eschatos"= "last") and "-logy" (to mean "the study of") and thus refers to "the study of the last." (WK 2010vv)

More specifically, "eschatology is a part of theology and philosophy concerned with what are believed to be the final events in history..., commonly referred to as the end of the world. The OED defines it as 'concerned with 'the four last things: death, judgement, heaven, and hell' (phrase from Book of Revelation). While in mysticism the phrase refers metaphorically to the end of ordinary reality and reunion with the Divine, in many traditional religions it is taught as an actual future event prophesied in sacred texts or folklore. More broadly, eschatology may encompass related concepts such as the Messiah or Messianic Age, the end time, and the end of days." (WK 2010vv)

Nowadays, "most modern eschatology and apocalypticism, both religious and secular, involve…the violent disruption or destruction of the world," although some other versions like "Christian and Jewish eschatologies view the end times as the consummation or perfection of God's creation of the world." (WK 2010vv)

Different Religious Views of Eschatology

For a more detailed illustration, let's consider several examples of eschatology from the standpoint of religion, in regard to the end of the world, namely, (a) Judeo-Christian eschatology, (b) Hindu eschatology, (c) Buddhist eschatology, and (d) Islamic eschatology—to be summarized in what follows, in that order (and also in *Table 3.4*).

(a) Judeo-Christian Eschatology

In the Judeo-Christian tradition, eschatology takes an apocalyptic form, which is from the "Greek Ἀποκάλυψις" ("apokálypsis"="lifting of the veil" or "revelation") and refers to "a disclosure of something hidden from the majority of mankind in an era dominated by falsehood and misconception, i.e., the veil to be lifted." (WK 2010ww)

However, in a more general sense, the term "apocalypse" is "associated with an eschatological final battle, the Armageddon, and the idea of an end of the world due to out of time. This perceptions may better be related to the phrase *apokalupsis eschaton*, literally 'revelation at [or of] the end of the *æon*, or age.'" (WK 2010ww)

In Christianity, for example, "*The Apocalypse of John* is the Book of Revelation, the last book of the Bible," where John "refers to the 'unveiling' or 'revelation' of Jesus Christ as Messiah. This term has come to mean, in common usage, the end of the world." (WK 2010ww)

In the Old Testament—the depiction of the end of the world often relies on "images of the judgment of the wicked, as well as the resurrection and glorification of those who were given righteousness before God. The dead are seen in the book of Job and in some of the Psalms as being in Sheol, awaiting the final judgment. The wicked will then be consigned to eternal torment in the fires of Gehinnom, or the Lake of Fire mentioned in Revelation." (WK 2010ww)

In the New Testament—"letters written by the Apostle Paul expand on this theme of the judgment of the wicked, and the glorification of those who belong to Christ or Messiah. In his letters to the Corinthians and the Thessalonians Paul expounds further on the destiny of the righteous. He speaks of the simultaneous resurrection and rapture of those who are in Christ, (or Messiah)." (WK 2010ww)

In general, "Christianity had a Millennial expectation for glorification of the righteous from the time it emerged from Judaism and spread out into the world in the first century. The poetic and prophetic literature of the Hebrew Bible, particularly in Isaiah, were rich in Millennial imagery. The New Testament Congregation after Pentecost carried on with this theme. During his imprisonment by the Romans on the Island of Patmos, John described the visions he experienced, writing the Book of Revelation. Revelation chapter 20 contains several reference to a thousand year reign of Christ/Messiah upon this earth." (WK 2010ww)

In the modern era—this Millennial expectation continued "in the 18th and 19th Centuries....Christian Apocalyptic eschatology was a continuation of the same two themes referred to throughout all of scripture as 'this age' and 'the age to come.' Evangelicals have been in the forefront popularizing the biblical prophecy of a major confrontation between good and evil at the end of this age, a coming Millennium to follow, and a final confrontation whereby the wicked are judged, the righteous are rewarded and the beginning of Eternity is viewed." (WK 2010ww)

More interestingly is the version of Millenialism "known as Dispensationalism, which arose in the 19th century. Dispensationalism sees separate destinies for the Christian Church and Israel. Its concept of a special Pre-Tribulation Rapture of the Church has become extremely popular....Dispensationalist interpretations find in biblical prophecy predictions of future events: the various periods of the church,...the throne of God in Heaven and his Glory; specific judgments that will occur on the earth; the final form of gentile power;...the second coming proper; a one-thousand year reign of Messiah; a last test of Mankind's sinful nature under ideal conditions by the loosing of Satan, with a judgment of fire coming down from Heaven that follows; the Great White Throne Judgment, and the destruction of the current heavens and the earth, to be recreated as a 'New Heaven and New Earth' ushering in the beginning of Eternity." (WK 2010ww)

(b) Hindu Eschatology

In the Hindu tradition, eschatology takes a cyclical form, which "is linked in the Vaishnavite tradition to the figure of Kalki, or the tenth and last avatar of Vishnu before the age draws to a close, and Shiva simultaneously dissolves and regenerates the universe." (WK 2010vv)

For the Hindus, in accordance to their eschatological cosmology, "we are living in the Kali Yuga (literally 'age of darkness'), the last of four periods (Yuga) that make up the current age. Each period has seen a successive degeneration in the moral order and character of human beings, to the point that in the Kali Yuga where quarrel and hypocrisy are prevalent. Often, the invocation of Kali Yuga denotes a certain helplessness in the face of the horrors and suffering of the human condition and a nostalgia for a golden past or a future salvation." (WK 2010vv)

But, unlike the Judeo-Christian tradition, "Hindu conceptions of time, like those found in other non-Western traditions, are cyclical in that one age may end but another will always begin. As such, the cycle of birth, growth, decay, death, and renewal at the individual level finds its echo in the cosmic order of all things, yet affected by the vagaries of the comings and goings of divine interventions in the Vaishnavite belief." (WK 2010vv)

In this cyclical sense, "most Hindus believe that Shiva will destroy the world at the end of the kalpa. Some Shaivites hold the view that he is incessantly destroying and creating the world." (WK 2010vv)

Therefore, unlike the Judeo-Christian final judgment for eternity—"in Hinduism, there is no eternal damnation of souls or end times." (WK 2010xx)

Even then, the world can be destroyed and recreated time and again, but one thing remains constant, namely, the transcendental level of reality, be it called "God," "the Supreme Cosmic Spirit," "the One," the Absolute," or any comparable transcendental term (as already discussed in *Sec. 2.2.1*).

(c) Buddhist Eschatology

In the Buddhist tradition, like the Hindu one, eschatology also takes a cyclical form.

The founder of Buddhism, Siddhārtha Gautama, "was a spiritual teacher from Nepal, and claimed that "his teachings would disappear after 5,000 years, when no one anymore practices Buddhism. According to the Sutta Pitaka [the scriptures of Theravada Buddhism], the 'ten moral courses of conduct' will disappear and people will follow the ten amoral concepts of theft, violence, murder, lying, evil speaking, adultery, abusive and idle talk, covetousness and ill will, wanton greed, and perverted lust resulting in skyrocketing poverty and the end of the worldly laws of true dharma." (WK 2010xx; MAI 2010)

In accordance to Buddhist eschatology, "it is believed that the era leading up to the coming of the next Buddha Maitreya will be characterized by impiety, physical weakness, sexual depravity and general societal disarray. Commentators like Buddhaghosa predicted a step-by-step disappearance of the Buddha's teachings. During the first stage, arahants would no longer appear in the world. Later, the content of the Buddha's true teachings would vanish, and only their form would be preserved. Finally, even the form of the Dharma would be forgotten. During the final stage, the memory of the Buddha himself would be forgotten, and the last of his relics would be gathered together in Bodh Gaya and cremated. Some time following this development a new Buddha named Maitreya will arise to renew the teachings of Buddhism and rediscover the path to Nirvana. Maitreya is believed to currently reside in the Tushita heaven, where he is awaiting his final rebirth in the world." (WK 2010xx)

In this cyclical way, "the decline of Buddhism in the world, and its eventual re-establishment by Maitreya, are in keeping with the general shape of Buddhist cosmology. Like Hindus, Buddhists generally believe in a cycle of creation and destruction, of which the current epoch represents only the latest step. The historical Buddha Shakyamuni is only the latest in a series of Buddhas that stretches back into the past." (WK 2010xx)

Again, like its Hindu counterpart, in Buddhist cosmology, the world can be destroyed and recreated time and again, but one thing remains constant, namely, the transcendental level of reality, be it called "God," "the Supreme Cosmic Spirit," "the One," the Absolute," or any comparable transcendental term (as already discussed in *Sec. 2.2.1*).

(d) Islamic Eschatology

In the Islamic tradition, like the Judeo-Christian one—eschatology takes an apocalyptic form.

For instance, in Islamic eschatology, "there are various signs (as many as up to 100) given in the Sunnah and Quran for the coming of Judgment Day. These signs can be divided into two parts, minor and major. The major signs include the following, for illustration. (WK 2010xx)

- "The coming of an Antichrist, Imam Mahdi and then Prophet Jesus (who will combine forces of good against evil). (Qur'an 43:61)"
- "The blowing of Trumpet and the minor signs will precede them."
- "The Sun will begin to die. (Qur'an 81: 1)"
- "The Earth will experience such big an earthquake that will cause mountains to crush down, the Earth's inner body will broke out and the Earth would be stretched out. (Qur'an 99:1, 69:13-14, 70:8, 84:3-4, 20:105-107, 99:1-6)"
- "The sun will swallow the moon. (Holy Qur'an 75:7-9)"

- "All the plants will be scattered."
- "The universe will stop expanding and contract on itself. (Holy Qur'an 51:47, 21:104)"

Surely, there are other signs, since there are "as many as up to 100" of them, so the ones as listed above are illustrative, not exhaustive. (WK 2010xx)

However, "like the other Abrahamic religions, Islam teaches the bodily resurrection of the dead, the fulfillment of a divine plan for creation, and the immortality of the human soul; the righteous are rewarded with the pleasures of *Jannah* (Heaven), while the unrighteous are punished in *Jahannam* (Hell). A significant fraction of the Quran deals with these beliefs, with many *hadith* elaborating on the themes and details. Islamic apocalyptic literature describing the Armageddon is often known as *fitna* (a test) and *malahim* (or *ghayba* in the shi'ite tradition)." (WK 2010xx)

Criticisms

These different religious views of eschatology are not without serious criticisms, however.

Consider, for illustration, a few of these criticisms.

Firstly, one serious criticism against religious eschatology is that it is easier to prophesize future events of the universe, but not everyone interpret them in the same way, even for those within the same religious tradition.

Two good examples of contrasting interpretations are shown below, with Christianity as a case in point.

The first contention about the prophesized future events in Judeo-Christian eschatology concerns the "one-thousand year reign of Messiah," because not every Christian interprets literally the meaning of the "one-thousand year reign of Messiah"; for instance, "the Amillennialists believe that the end times encompass the time from Christ's ascension to the Last day, and maintain that the mention of the 'thousand years' in the Book of Revelation is meant to be taken metaphorically (i.e., not literally, or 'spiritually'), a view which continues to cause divisions within evangelical Christianity." (WK 2010xx)

And the second contention about the prophesized future events in Judeo-Christian eschatology concerns the exact date for the "second coming of Christ." For instance, "William Miller and his followers fixed the time for the Second Coming by calendar calculations based on interpretations of the Biblical apocalypses; they originally set a date for the Second Coming in 1844"; but "Emanuel Swedenborg…considered the second coming to be symbolic, and to have occurred in 1757)." (WK 2010xx)

Secondly, another serious criticism against religious eschatology is that different religions offer different predictions about the future end of the world, such that this should caution us against accepting them uncritically, since they cannot be all equally valid.

For instance, the Abrahamic eschatology (as in Judaism, Christianity, and Islam) is relatively more apocalyptic in a linear way, whereas the Asian eschatology (as in Buddhism and Hinduism) is relatively more cyclical.

To privilege the Abrahamic version of eschatology so as to argue against the Asian version is just as biased as to favor the latter against the former.

And thirdly, still another serious criticism against religious eschatology is that the prophesized future events are based on metaphysical assumptions and claims which have been criticized as unscientific in the scientific community, like the debate between evolutionary theory and creation theory, or the dispute between religious cosmology and scientific cosmology.

While science of course does not have the last word on the ultimate fate of the world, its critical outlook against the empirical legitimacy of religious eschatology can make us think twice before accepting these different versions of religious eschatology dogmatically.

Remaining above the Fray

Once more, these criticisms do not imply that there is nothing to learn from religious eschatology about the end of the world.

On the contrary, they are useful to show us the other side of the possibility and desirability of the evolution of the universe (and its speculative ends) from the perspective of culture (this time, in relation to eschatology, the end of the world, and religion as a case study), something that the respective spokespersons would not like us to know.

In adddition, the analysis of eschatology, the end of the world, and religion can reveal valuable insights about the ontological principles in existential dialectics, and good examples include the absoluteness-relativeness principle, the predictability-unpredictability principle, the explicability-inexplicability principle, the preciseness-vagueness principle, the simpleness-complicatedness principle, the openness-hiddenness principle, the denseness-emptiness principle, the change-constancy principle, the order-chaos principle, the slowness-quickness principle, the expansion-contraction principle, the theory-praxis principle, the convention-novelty principle, the evolution-transformation principle, the symmetry-asymmetry principle, the softenss-hardness principle, the regression-progression principle, and the same-difference principle.

For instance, in relation to the absoluteness-relativeness principle, if there is absoluteness (e.g., the absolute view in Judeo-Christian eschatology about "the final judgment,...whereby the wicked are judged, the righteous are rewarded and the beginning of Eternity is viewed"), there is relativeness (e.g., what is final in Judeo-Christian eschatology in regard to the "final judgment" for "the beginning of Eternity" is not so for Buddhist eschatology in regard to the endless cycle of creation and destruction of the world). And the reverse direction also holds true.

In relation to the predictability-unpredictability principle, if there is predictability (e.g., the predictable tendency of a Christian to believe in the prophesized Second Coming of Christ), there is unpredictability (e.g., the more difficult task for a Christian to predict correctly when exactly Christ will return in the Second Coming). And the reverse direction also holds true.

In relation to the explicability-inexplicability principle, if there is explicability (e.g., the explanation, by those in Buddhist eschatology, of the end of Buddhism after 5000 years, paving the way for "the coming of the next Buddha Maitreya," on the basis of the endless cycle of creation and destruction of the world), there is inexplicability (e.g.., the lack of convincing explanation, by those in Buddhist eschatology, of why this endless cycle of destruction and creation of the world is necessary in the first place, especially to those who do not believe in Buddhist eschatology). And the reverse direction also holds true.

In relation to the preciseness-vagueness principle, if there is preciseness (e.g., the precise identification of the "one-thousand year reign of Messiah" as one of the prophesized future events in Judeo-Christian eschatology), there is vagueness (e.g., the vagueness in the identification of the "one-thousand year reign of Messiah," since this means literally for some but figuratively, or symbolically, for others). And the reverse direction also holds true.

In relation to the simpleness-complicatedness principle, if there is simpleness (e.g., the relatively simple delineation of the prophesized future events in religious eschatology), there is complicatedness (e.g., the relatively more complicated analysis of the prophesized future events by questioning its claims, as shown in the criticisms). And the reverse direction also holds true.

In relation to the openness-hiddenness principle, if there is openness (e.g., the open exploration, in Hindu eschatology, that "one age may end but another will always begin"), there is hiddenness (e.g., the hidden bias in Hindu eschatology, in terms of its cyclical assumption about the cosmic order of things, which other versions of eschatology, say, in the Abrahamic tradition, do not share).

In relation to the denseness-emptiness principle, if there is densenss (e.g., the relatively denser concentration of Christian eschatologists in Western Europe), there is emptiness (e.g., the relatively less dense, or more empty, concentration of Christian eschatologists in India). And the reverse direction also holds true.

In relation to the change-constancy principle, if there is change (e.g., the change of things, like birth, growth, decay, and death in life, in accordance to Hindu eschatology), there is constancy (e.g, the endless cycle that "one age may end but another will always begin," in accordance to Hindu eschatology). And the reverse direction also holds true.

In relation to the order-chaos principle, if there is order (e.g., "the cosmic order of all things" in regard to the cycle of birth, growth, decay, death, and renewal," in accordance to Hindu eschatology), there is chaos (e.g., the "degeneration" in the "age of darkness" in "the last of four periods, where quarrel, disorder, suffering, helplessness, and the like become common, in accordance to Hindu eschatology). And the reverse direction also holds true.

In relation to the slowness-quickness principle, if there is slowness (e.g., the relatively slower readiness of those in the Hindu tradition to think about the end of the world in a linear, apocalytpic way), there is quickness (e.g., the relatively quicker readiness of those in the Abrahmic tradition to think about the end of the world in a linear, apocalytpic way). And the reverse direction also holds true.

In relation to the expansion-contraction principle, if there is expansion (e.g., the recreation of "a 'New Heaven and New Earth' ushering in the beginning of Eternity," at the end of the world, in accordance to Christian eschatology), there is contraction (e.g., "the destruction of the current heavens and the earth" at the end of the world, in accordance to Christian eschatology). And the reverse direction also holds true.

In relation to the theory-praxis principle, if there is theory (e.g., the theoretical construction of Christian eschatology), there is praxis (e.g., the practical application of Christian eschatology by its followers for living a virtuous life in practice). And the reverse direction also holds true.

In relation to the convention-novelty principle, there is convention (e.g., the conventional wisdom about taking literally the "one-thousand year reign of Messiah" as one of the prophesized future events in Christian eschatology), there is novelty (e.g., the alternative novel challenge to this literal interpretation of the "one-thousand year reign of Messiah" by those known as "the Amillennialists [who]...maintain that the mention of the 'thousand years' in the Book of Revelation is meant to be taken metaphorically..., i.e., not literally, or 'spiritually,'...a view which continues to cause divisions within evangelical Christianity"). And the reverse direction also holds true.

In relation to the evolution-transformation principle, if there is evolution (e.g., the natural evolution of humans to live in the state of nature in its primitive way), there is transformation (e.g., the technical transformation of human living by the the invention of Buddhist eschatology for an ascetic way of life). And the reverse direction also holds true.

In relation to the symmetry-asymmetry principle, if there is symmetry (e.g., the co-existence of different religious versions of eschatology), there is asymmetry (e.g.., the acceptance of Christian eschatology in Western Europe—but the popularity of Hindu eschatology in India). And the reverse direction also holds true.

In relation to the softenss-hardness principle, if there is softness (e.g., "the righteous are rewarded with the pleasures of *Jannah* or Heaven, in accordance to Islamic eschatology), there is hardness (e.g., "the unrighteous are punished in *Jahannam* or Hell," in accordance to Islamic eschatology). And the reverse direction also holds true.

In relation to the regression-progression principle, if there is regression (e.g., the regression made by Christian eschatology, like the tragedy of Heaven's Gate), there is progression (e.g., the progress made by Christian eschatology for the living of a virtuous life in practice). And the reverse direction also holds true.

And in relation to the same-difference principle, if there is similarity in outcome (e.g., the contribution to the regulation and control of human life, regardless of whether this is done by the use of religious faith as in religious eschatology, or by the use of scientific knowledge as in the Big Bang theory and its prediction about the ultimate fate of the universe), there is difference in outcome (e.g., the contribution to the regulation and control of human life, by way of a scientific lifeworld lived by many scientists supporting the Big Bang theory and its prediction about the ultimate fate of the universe—but the contribution to the regulation and control of human life, by way of a lifeworld guided by religious faith as practiced by those in religious eschatology). And the reverse direction also holds true.

The Controversy of Its Speculative Ends

This comprehensive analysis of the possibility and desirability of the evolution of the universe (and its speculative ends), from the perspectives of the mind, nature, society, and culture, is important for the critical understanding that reality is not to the extent that the spokespersons from both sides of the debate would like us to believe.

Just as each view of a contested beginning of the universe and its aftermath (in Chapter Two) has its persuasion, each view of the evolution of the universe and its speculative end (in this chapter) has its wonder.

On the other hand, just as each view of a contested beginning of the universe and its aftermath (in Chapter Two) has its own share of questionability, each view of the evolution of the universe and its speculative end (in this chapter) has its own fair share of controversy.

For instance, just as those (in Chapter Two) who argue for initial cosmologies (that there was once a beginning of the universe) need to answer the follow-up question of why the universe had to begin like that or what really happened before that—those (in this chapter) who argue for the end of the world need to answer the follow-up question of why the universe will have to end like that or what will happen after that end.

Conversely, just as those (in Chapter Two) who argue for eternal cosmologies (that there was no beginning of the universe at all, if traced further back to space and time) also need to answer the follow-up question of why there is this infinite regress of causation or what exactly this infinity is supposed to refer to—those (in this chapter) who argue against the end of the world also need to answer the follow-up question of why there is this infinite projection of the world or what exactly this infinity is supposed to refer to.

In either way, the important point here (in this chapter)—just as this is the same for Chapter Two—is to not favor any one side of the debate at the expense of the other but to teach us the opposing sides of the debate, such that reality is not to the extent that each respective side would like us to believe.

Now that the intellectual journey about the universe, be it about its contested beginnings (and the aftermath) or the evolution of the universe (and its speculative ends), has come to an end here, what then can be concluded about the future of cosmology?

This is the question, to which the next chapter, that is, Chapter Four on the future of cosmology will answer—for which we now turn to the concluding chapter.

Table 3.1. The Universe and Its Speculative Ends

- **Its Speculative Ends—and the Mind**
  —Ex: involution, the mind, and the cosmic return to the Absolute
  —Ex: the esoteric seven, and the spiritual evolution of the universe

- **Its Speculative Ends—and Nature**
  —Ex: the scientific debate about the ultimate fate of the universe

- **Its Speculative Ends—and Society**
  —Ex: Heaven's Gate, the end of the Earth, and cult suicide
  —Ex: MRTCG, apocalypticism, and doomsday deaths

- **Its Speculative Ends—and Culture**
  —Ex: eschatology, the end of the world, and religion

Notes: The examples in the categories are solely illustrative (not exhaustive), and the comparison is relative (not absolute), nor are they necessarily mutually exclusive. And some can be easily re-classified elsewhere. As generalities, they allow exceptions.
Source: A summary of Ch.3 of *BCOS*

Table 3.2. Involution, and the Evolution of the Universe

- **The Development of Different Entities in the Evolution**
  —Ex: "inconscience" (e.g., "matter")
  —Ex: "life" (e.g., "movement," "sensation," "living physical beings")
  —Ex: "mind" (e.g., "conscious animals," "self-conscious" humans)
  —Ex: "spiritualized mind" (e.g., spiritual seekers)
  —Ex: "supermind" (e.g., "truth consciousness," "divine life")

- **Three Aspects of Involution**
  —Ex: Sat (for Existence)
  —Ex: Chit-Tapas (for Consciousness)
  —Ex: Ananda (for Bliss)

- **The Emanationist View of the Cosmic Return**
  —This Indian (Hindu) idea of involution in the context of the evolution of the universe points to a cosmic return to its original Absolute, which is "based on an emanationist view of reality." (WK 2010mm)
  —In accordance to emanation, "Creation proceeds as an outpouring or even a transformation in the original Absolute or Godhead. The Supreme Light or Consciousness descends through a series of stages, gradations, worlds or hypostases, becoming progressively more material and embodied, before finally turning around to return to the One, retracing its steps through spiritual knowledge, contemplation and ascent." (WK 2010mm)

Source: From *Sec. 3.2.1* of *BCOS*. See book for citations, like WK (2000kk), WK (2000ll), (WK 2010mm), etc.

Table 3.3. The Scientific Debate about the Ultimate Fate of the Universe

- **The Shape of the Universe**
  —Ex: "flat universe" (with "flat" geometry, for $\Omega=1$)
  —Ex: "open universe" (with "hyperbolic" geometry, for $\Omega < 1$)
  —Ex: "closed universe" (with "elliptic" geometry, for $\Omega > 1$)

- **The End of the Universe**
  —Ex: "the big freeze"
  —Ex: "the heat death"
  —Ex: "the big rip"
  —Ex: "multiverse" (with no complete end)

- **Remaining Problems**
  —Ex: it is not clear that the future can be so neatly outlined in these possibilities.
  —Ex: many of these possibilities assume the existence of "dark energy," which, however, is still not well understood.
  —Ex: it may well be motivated by religious beliefs.
  —Ex: it takes no account of the role of what I origianlly called "post-humans" in altering the universe (and, for that matter, the multiverse) in the distant future, as this is something that I went to great lengths to explain in many of my books, in regard to my different visions of the post-humans.

Source: From *Sec. 3.3.1* of *BCOS*. See book for citations, like WK (2000qq), etc.

Table 3.4. Different Religious Views of Eschatology

- **Different Religious Views**
  - —Ex: Judeo-Christian eschatology
  - —Ex: Hindu eschatology
  - —Ex: Buddhist eschatology
  - —Ex: Islamic eschatology

- **Main Criticisms**
  - —Ex: it is easier to prophesize future events of the universe, but not everyone interpret them in the same way, even for those within the same religious tradition.
  - —Ex: different religions offer different predictions about the future end of the world, such that this should caution us against accepting them uncritically, since they cannot be all equally valid.
  - —Ex: the prophesized future events are based on metaphysical assumptions and claims which have been criticized as unscientific in the scientific community, like the debate between evolutionary theory and creation theory, or the dispute between religious cosmology and scientific cosmology. While science of course does not have the last word on the ultimate fate of the world, its critical outlook against the empirical legitimacy of religious eschatology can make us think twice before accepting these different versions of religious eschatology dogmatically.

Source: From *Sec. 3.5.1* of *BCOS*. See book for citations, like WK (2010ww), WK (2010vv), WK (2000xx), etc.

• PART FOUR •

Conclusion

CHAPTER 4
CONCLUSION—THE FUTURE OF COSMOLOGY

> He who does not know what the world is does not know where he is, and he who does not know for what purpose the world exists, does not know who he is, nor what the world is.
> —Marcus Aurelius (WK 2010)

Beyond the Contested Beginnings and the Speculative Ends

The analyses of the universe and its contested beginnings (in Chapter Two) and of the universe and its speculative ends (in Chapter Three) allow us to rethink the very extent of their possibility and desirability, in that the understanding of the universe is neither possible nor desirable to the extent that its respective spokespersons would like us to believe.

Of course, this is not to suggest, as already indicated in Chapter One, that cosmology is a worthless field of study, or its other fields of study (like metaphysics, theology, astrology, and esotericism) should be dismissed. Needless to say, neither of these two extreme views is reasonable either.

The important point to remember is that this book provides an alternative (better) theory to understand the nature of the universe, especially in relation to its contested beginnings and speculative ends—while learning from different approaches in the literature but without favoring any one of them (nor integrating them, since they are not necessarily compatible with each other). Thus, this book offers a new theory to transcend the existing approaches in the literature on cosmology in a way not conceived before.

This new theory of mine is called *the theory of post-cosmology*, since it is to propose a new way to understand the universe.

With this clarification in mind—the theory of post-cosmology has four distinctive features to remember.

Firstly, my theory makes good use of all theoretical approaches in the literature on cosmology, be they about the scientific argument, the religious argument, the metaphysical argument, or the esoteric argument—especially from the most comprehensive combined perspectives of the mind, nature, society, and culture.

Secondly, just like many other theories of mine in my previous books, my theory here does not heavily favor any specific theory over others in the literature, nor trying to integrate them (as they are not necessarily compatible with each other).

Thirdly, my theory treats the issue of cosmology in the critical context of the so-called "contested" beginnings and "speculative" ends—especially in the larger dialectic context of transcending them unto post-cosmology.

And fourthly, it contains fifteen major theses, namely, (a) the first thesis: the absoluteness-relativeness principle, (b) the second thesis: the predictability-unpredictability principle, (c) the third thesis: the explicability-inexplicability principle, (d) the fourth thesis: the preciseness-vagueness principle, (e) the fifth thesis: the simpleness-complicatedness principle, (f) the sixth thesis: the openness-hiddenness principle, (g) the seventh thesis: the denseness-emptiness principle, (h) the eighth thesis: the slowness-quickness principle, (i) the nineth thesis: the expansion-contraction principle, (j) the tenth thesis: the convention-novelty principle, (k) the eleventh thesis: the evolution-transformation principle, (l) the twelveth thesis: the symmetry-asymmetry principle, (m) the thirteenth thesis: the regression-progression principle, (n) the fourteenth thesis: the sameness-difference principle, and (o) the fifteenth thesis: the post-human odyssey—to be elaborated in the rest of this concluding chapter.

If successful, this seminal project is to fundamentally change the way that we think about cosmology, from the combined perspectives of the mind, nature, society, and culture, with enormous implications for the human future and what I originally called its "post-human" fate.

Other Relevant Principles

Lest misunderstanding hastily occurs (as this is something that I regularly did in my previous books, *almost verbatim*), it should be clarified at the outset that, if other ontological principles in existential dialectics (other than the fourteen as listed in the fifteen theses) are not explicitly included in the listing of the theses in the Table of Contents, it is not because they are not important, but merely because they are not as equally relevant as the others in the current case study (in terms of the number of citations of each principle in the book), although different case studies will, of course, reveal different degrees of relevance of the principles.

Even then, in some cases (like this one), the difference between any two given principles, for instance, in terms of the number of citations in a book, is rather small, so the reason in those cases is more aesthetic (than otherwise), because it looks nicer to list only fourteen theses for fourteen principles (than 21 theses for all of the 21 principles) in the Table of Contents.

This is true, even if different studies of the same kind can yield different views about the degree of relevance for each ontological principle, depending on the specific nature of a research in question, needless to say. So, if a different author analyzes the subject matter in a different way, the relevance of the ontological principles will be different.

After all, the symmetry-asymmetry principle already implies that not all principles are equally relevant in any topic under examination, as they are not created equal for all cases under all circumstances (which would be an extreme view, of course).

In any event, the relevancy of other principles (besides the ones as already listed in the theory above) for the present study can be summarized below, in relation to the five categories of ontological principles in existential dialectics.

In the Category of Method

Firstly, in regard to the formalness-informalness principle (on the formal requirements of logical systems), if there is formalmess (e.g., the formal requirement of completeness in a system of ideas for provability, in *Sec. 2.2.2*; and the formal requirement of consistency in classical logic, in that there is either A or non-A, but not both, in *Sec. 2.5.1*), there is informalness too (e.g., the ambivalent nature of probability in regard to the Brahman in Advaita Vedanta, in that it cannot completely be known because it "cannot be correctly described," but the Atman, or self, and the Brahman can be converged into Oneness, under certain conditions for salvation as guided by a Guru, in *Sec. 2.2.2*; and the non-formal acceptance of both A and non-A, like the argument by Aquinas that the universe can be both contingent and necessary, since "Aquinas' argument is both *in fieri* and *in esse*...plus an additional argument from contingency," in *Sec. 2.5.1*). And the reverse direction also holds true.

And secondly, in regard to the partiality-totality principle (on the relationships between whole and parts), if there is partial analysis (e.g., the partial view of each individual self, in Advaita Vedanta, in *Sec. 2.2.1*; the partial viewpoint of a contingent entity that it is caused by something else, like "the light of the candle" which "is dependent on the candle's continued existence," in *Sec. 2.5.1*; and the partial view, from the perspective of each "verse," that it was born and it will end, in *Sec. 3.3.1*), there is also holistic analysis (e.g., the holistic view of the Brahman, in Advaita Vedanta, such that the whole is not simply the sum of its parts, in *Sec. 2.2.1*; the holistic view of all things in the universe, such that there can be the First Cause for all contingent entties, so that the whole is not the sum of its parts, in accordance to the cosmological argument, in *Sec. 2.5.1*; and the holistic view of the larger "multiverse," in that there can be a cosmic cycle, for it to not have a complete end, with uncertain consequences, so the view of the whole is not the sum of the views of the individual "verses," in *Sec. 3.3.1*). And the reverse direction also holds true.

In the Category of Structure

Firstly, in regard to the finiteness-transfiniteness principle (on the nature of numbers), on the one hand, there is finiteness (e.g., the finite number of the levels of truth or reality in Advaita Vedanta, that is, the three levels known as the transcendental, pragmatic, and apparent, in *Sec. 2.2.1*; the finte number of categories in the classification of universals, as shown in the three known as the "types," "properties," and "relations," in *Sec. 2.2.2*; the finite number of epochs which have been identified for the timeline of the Big Bang, in *Sec. 2.3.1*; the finite number of "correlations" in the universe as defined in Chinese correlative cosmology, such as those based on "pairs," "fours," "fives," "eights," and the like, in *Sec. 2.4.2*; the finite number of the First Cause, which cannot be further caused, in accordance to the cosmological argument, in *Sec. 2.5.1*; the finite number of kinds of form to be classified, like "substantial"forms vs. "accidental" forms, in *Sec. 2.5.2*; the finite number of the categories of entities in the evolution of the universe—that is, "matter," "life," "mind," "spiritualized mind," and "supermind," in accordance to Sri Aurobindo, in *Sec. 3.2.1*; the finite number of seven root races in Blavatsky's analysis of the spiritual evolution of human monads, in *Sec. 3.2.2*; and the finite number of the possibilities concerning the ultimate fate of the universe, as shown in the three scenarios, in *Sec. 3.3.1*).

On the other hand, there is transfiniteness (e.g., the transfinite nature of reality at the transcendental level, in accordance to Advaita Vedanta, such that it is "omnipresent, omnipotent, incorporeal, impersonal, transcendent," etc., in *Sec. 2.2.1*; the transfinite number of all the possible types, properties, and relations for universals in the world, in *Sec. 2.2.2*; the mind-boggling numbers of particles, stars, galaxies, quasars, etc., in the timeline of the Big Bang, in *Sec. 2.3.1*; the amazingly huge number of things that these "correlations" can possibly refer to in the entire universe, in *Sec. 2.4.2*; the transfinite number of causes in the universe after the First Cause, in accordance to the cosmological argument, in *Sec. 2.5.1*; the huge number of different particular things in the universe which could possibly be classified under "substantial" forms or under "accidental"forms, in *Sec. 2.5.2*; the amazingly huge number of particular things in the universe which can possibly be classified under each of the finite categories of "matter," "life," "mind," "spiritualized mind," and "supermind," in accordance to Sri Aurobindo, in *Sec. 3.2.1*; the amazingly huge number of particular things which could possibly be classified under the seven root races in Blavatksy's analysis, in *Sec. 3.2.2*; and the transfinite number of time for the universe to end, in accordance to any of the three scenarios, in *Sec. 3.3.1*).

And the reverse direction also holds true.

In the Category of Process

Firstly, in regard to the change-constancy principle (on the alteration of things), if there is change (e.g., the development of different entities in the evolution of the universe over time, in accordance to esoteric cosmology, in *Sec. 3.2.1*; and the change of things, like birth, growth, decay, and death in life, in accordance to Hindu eschatology, in *Sec. 3.5.1*), there is constancy (e.g., the recurrence of the cosmic cycle in the universe, in accordance to esoteric cosmology, in *Sec. 3.2.1*; and the endless cycle that "one age may end but another will always begin," in accordance to Hindu eschatology, in *Sec. 3.5.1*). And the reverse direction also holds true.

Secondly, in relation to the order-chaos principle (on the pattern of things), if there is order (e.g., "the universe began in its most orderly state," because of the second law of thermodynamics, in *Sec. 2.3.1*; the relatively orderly patterns of the universe, such that its ultimate fate can be scientifically discussed in terms of three major scenarios, in *Sec. 3.3.1*; and "the cosmic order of all things" in regard to the cycle of birth, growth, decay, death, and renewal," in accordance to Hindu eschatology, in *Sec. 3.5.1*), there is chaos (e.g., "the entropy of the universe—a measure of its disorder—increases with time," because of the second law of thermodynamics, in *Sec. 2.3.1*; the relatively chaotic nature of other aspects of the universe, like "the vaccum metastability event," with "the potential to fundamentally alter our universe" or that "all structures will be destroyed instantaneously, without any forewarning," in *Sec. 3.3.1*; and the "degeneration" in the "age of darkness" in "the last of four periods, where quarrel, disorder, suffering, helplessness, and the like become common, in accordance to Hindu eschatology, in *Sec. 3.5.1*). And the reverse direction also holds true.

In the Category of Agency

Firstly, in regard to the theory-praxis principle (on the duality of knowledge), on the one hand, there is theoretical construction (e.g., the theoretical construction of acosmism, in *Sec. 2.2.1*; the theoretical construction of the Scripture, in *Sec. 2.4.1*; the theoretical construction of correlative cosmology, in *Sec. 2.4.2*; the theoretical construction of the cosmological argument, in *Sec. 2.5.1*; the theoretical construction of hylomorphism by Aristotle, in *Sec. 2.5.2*; the theoretical construction of emanation in esoteric cosmology, in *Sec. 3.2.1*; the theoretical construction of the "seven root races" by Blavatsky in the Theosophical Society, in *Sec. 3.2.2*; the theoretical construction of different scenarios about the ultmate fate of the universe, in *Sec. 3.3.1*; the theoretical construction of the belief system about the end of the Earth with some Christian ideas of "salvation and apocalypse," in *Sec. 3.4.1*; the theoretical construction of salvation and apocalypse in Christian teachings, in *Sec. 3.4.2*; and the theoretical construction of Christian eschatology, in *Sec. 3.5.1*).

On the other hand, there is practical application (e.g., there is praxis (e.g., the practical application of acosmism, as in the use of Advaita Vedanta for spiritual salvation, in *Sec. 2.2.1*; the practical application of the Scripture to reject heliocentrism in the Galileo affair, in *Sec. 2.4.1*; the practical application of correlative cosmology, as in *Wuxing*, by those in power for "the politics of a unified, imperial order" in "about 400 B.C.," in *Sec. 2.4.2*; the practical application of the cosmological argument for the proof of the existence of God by Christian theologians like Aquinas, in *Sec. 2.5.1*; the practical application of hylomorphism for "the Christian doctrines such as the transubstantiation of the Eucharist's bread and wine into the body and blood of Jesus," in *Sec. 2.5.2*; the practical application of emanation in esoteric cosmology for the development of Christianity by Pierre Teilhard de Chardin, in *Sec. 3.2.1*; the practical application of the analysis of the "seven root races" by some Nazi groups for the ideology of white supramacy, in *Sec. 3.2.2*; the practical application of the theories concerning the ultimate fate of the universe for "countless sci-fi and fantasy works," which "use the threatened destruction of the universe as their plot device, usually with an evil supervillain and/or the incompetence of humanity as the cause, and usually with human ingenuity saving the day," in *Sec. 3.3.1*; the practical application of the belief system about the end of the Earth for cult suicide by members of Heaven's Gate, in *Sec. 3.4.1*; the practical application of the Christian ideas of apocalyse and salvation by MRTCG members to prepare the doomsday of January 01, 2000, as "activity by Movement members became frenzied" shortly before that date, in *Sec. 3.4.2*; and the practical application of Christian eschatology by its followers for living a virtuous life in practice, in *Sec. 3.5.1*).

And the reverse direction also holds true.

And secondly, in regard to the softness-hardness principle (on the force of change), on the one hand, there is softness (e.g., the concentration on meditation in Advaita Vedanta, in *Sec. 2.2.1*; "electrons and protons to combine into hydrogen atoms, p + e $\rightarrow$ H," in *Sec. 2.3.1*; the praise of those who conformed to the geocentric view of the Catholic Church at the time of Galileo, in *Sec. 2.4.1*; the refinement and acceptance of correlative cosmology by those in *Wuxing*, in *Sec. 2.4.2*; the promise of "the Next Level's existence" after death by Applewhite to his followers, in *Sec. 3.4.1*; the reward of salvation to those Christians who follow the Ten Commandments, in *Sec. 3.4.2*; and "the righteous are rewarded with the pleasures of *Jannah* or Heaven, in accordance to Islamic eschatology, in *Sec. 3.5.1*).

On the other hand, there is hardness (e.g., "the reunuciatio n of enjoyments of objects" in the material world, in Advaita Vedanta, in *Sec. 2.2.1*; when antimatter and matter, or antiparticles and partices, meet, they annihilate each other, instead of combining to form something new, in *Sec. 2.3.1*; the punishment of those who rejected the geocentric view of the Catholic Church at the time of Galileo, who was put under house arrest, for example, for his heliocentrism, in *Sec. 2.4.1*; the deconstruction and rejection of correlative cosmology by those critics like Liang Qichao and Gu Jiegang, in *Sec. 2.4.2*; the call to kill oneself by Applewhite to his followers, in *Sec. 3.4.1*; the punishment by damnation against those Christians who violate the Ten Commandments, in *Sec. 3.4.2*; and "the unrighteous are punished in *Jahannam* or Hell," in accordance to Islamic eschatology, in *Sec. 3.5.1*).

And the reverse direction also holds true.

1st Thesis: The Absoluteness-Relativeness Principle

With these other principles in mind—the first thesis refers to the absoluteness-relativeness principle (on the multiplicity of things) in the category of "method" in existential dialectics, which was first proposed in *FPHK*, in that there is the multiplicity of things in reality, be they about entities, qualities (or properties), and relationships. If there is something absolute, there is likewise something relative. And there is no absoluteness without relativeness—and vice versa.

Both absoluteness and relativeness here are also relevant to different modalities often cited in the literature on ontology, such as possibility (e.g., something "can" happen) and its opposite (e.g., impossibility), probability (e.g., something "will" happen) and its opposite (e.g., improbability), and necessity (e.g., something "should" happen) and its opposite (e.g., contingency).

For instance, on the one hand, there are absolute viewpoints (e.g., the absolute view in Advaita Vedanta that there is "the infinite unmanifest Absolute" at the transcendental level of truth, in *Sec. 2.2.1*; the absolute Realist viewpoint by Plato that universals exist in the metaphysical realm, in *Sec. 2.2.2*; the absolute view, wihtin the Big Bang theory, that the universe started in a singularity, in *Sec. 2.3.1*; the absolute view of the Catholic Church, at the time of Galileo, that "the earth was the center of the universe," on the basis of "a literalist interpretation of Scripture," in *Sec. 2.4.1*; the absolute view, in ancient Chinese correlative cosmology, of the universe as "an orderly system of correspondence among various domains of reality in the universe, correlating categories of the human world...with categories of the cosmos," without any transcendental entity like God outside it, in *Sec. 2.4.2*; the absolute view in the cosmological argument that the First Cause is exempt from any further causation, in *Sec. 2.5.1*; the absolute teleological view of the universe by Aristotle, in that the universe is "inherently purposeful," in the sense that form gives matter its actuality, in *Sec. 2.5.2*; the absolute view by Sri Aurobindo that there is "the Omnipresent Reality" as "the Absolute Brahman," which is a non-temporal precursor to the existence of the universe, in *Sec. 3.2.1*; the absolute view about the esoteric number seven in theosophy, which has been used to analyze the seven root races and the septenary, for instance, in *Sec. 3.2.2*; the absolute view of the three possibilities of the ultimate end of the "open" universe, with the Big Bang theory as its theoretical basis, in *Sec. 3.3.1*; the absolutely ascetic lifestyle of the members of

Heaven's Gate comparable to "medieval monastic order," since "group members gave up their material possessions and lived a highly ascetic life devoid of many indulgences," in *Sec. 3.4.1*; the absolute view of MRTCG that "one had to strictly follow the Commandments," so "[t]he emphasis on the Commandments was so strong that the group discouraged talking, for fear of breaking the Ninth Commandment," in *Sec. 3.4.2*; and the absolute view in Judeo-Christian eschatology about "the final judgment,...whereby the wicked are judged, the righteous are rewarded and the beginning of Eternity is viewed," in *Sec. 3.5.1*).

On the other hand, there are likewise relative counterparts (e.g., what is true for Advaita Vedanta in regard to "the infinite unmanifest Absolute" at the transcendental level is not necessarily so for astrophysics in regard to the beginning of the universe as the big bang, without the appeal to any transcendental level of reality, in *Sec. 2.2.1*; what is real for Plato in regard to the universals is not so for the idealists in regard to the illusion of the world, in *Sec. 2.2.2*; what is right for the Big Bang theory in regard to the need for the idea of singularity is not so for the "ekpyrotic" model in brane csomology, in that "the universe did not start in a singularity, but came about from the collision of two branes," in *Sec. 2.3.1*; what is correct for the Catholic Church in regard to geocentrism is not so for Galileo in regard to helocentrism, in *Sec. 2.4.1*; what is obvious for the Chinese correlative cosmology in regard to the absolute existence of the universe without the need of any transcendental entity like God is not so for Christian cosmology in regard to the need of a transcendental entity like God, who created the universe, in *Sec. 2.4.2*; what is reasonable for those in the cosmological argument in regard to the uncaused First Cause is not so for the critics who question this exemption, in *Sec. 2.5.1*; what is true for the Aristotelian teleological view of the universe in regard to the full realization of matter is not necessarily so for the contemporary neo-Darwinian evolutionary theory in regard to the sub-optimality of evolution in nature, in *Sec. 2.5.2*; what is true for Sri Aurobindo in regard to the Absolute Brahman is not so for those in the Big Bang theory, who do not accept the scientific legitimacy of any entity prior to the Big Bang, in *Sec. 3.2.1*; what is approriate for those in theosophy to appeal to the esoteric number seven is not necessarily so for the critics, who question its superstition, in *Sec. 3.2.2*; what is correct for the three possibilities of the ultimate theory of the "open" universe is not necessarily so for the theory of multiverse, in that it will not have a complete end, in *Sec. 3.3.1*; what is good for the members of Heaven's Gate to live a strictly ascetic lifestyle is not so for those in the Playboy Club who enjoy living a life full of the pleasures of the flesh, in *Sec. 3.4.1*; what is right for MRTCG to demand its members "to strictly follow the Commandments" is not necessarily so for those on the outside who do not accept the extreme Christian teachings in MRTCG, in *Sec. 3.4.2*; and what is final in Judeo-Christian eschatology in regard to the "final judgment" for "the beginning of Eternity" is not so for Buddhist eschatology in regard to the endless cycle of creation and destruction of the world, in *Sec. 3.5.1*).

However, it should be stressed (as this is something that I regularly did in my previous books, *almost verbatim*) that there are different shades of gray (or different degrees of truth) in the two opposites, and the classification is not necessarily mutually exclusive either. In fact, even when some combinations of the two occur, they only end up sharing the same dialectic relationship, but in a different degree.

Also, there are some other relationships (as a kind of family resemblance) which have something in common with the principle, but they are not exactly the same but only more or less comparable, which varies from case to case.

Good examples of family resemblance in relation to the principle are uniformity-diversity, internalness-externalness, immanence-transcendence, and so forth.

2nd Thesis: The Predictability-Unpredictability Principle

The second thesis is called the predictability-unpredictability principle (on the occurrence of events) in the category of "method" in existential dialectics (which was first proposed in *FC* and, later, other books of mine), in that both predictability and unpredictability have a major role to play in the occurrence of things, so that neither determinism nor indeterminism wins the centuries-old fight. There is no predictability without unpredictability—and vice versa.

There are events which are predictable, just as there are those which are not. Or what is regarded as unpredictable at one point in time may turn out to be predictable later, and, conversely, what is deemed as predictable may turn out to not be so predictable. Even in predictability, outcomes are subject to uncertainty, the degree of which varies from case to case.

For instance, on the one hand, there is predictability (e.g., the predictable tendency of a disciple of Advaita Vedanta to learn the Vedic scriptures, with the guidance of a Guru, in the quest for "moksha" as a "liberation from the cycle of births and deaths," with its recurrence of pain and suffering, in *Sec. 2.2.1*; the predictable tendency of an idealist to deny the existence of univerals in the external world, in *Sec. 2.2.2*; the predictable occurrence, in accordance to the chaotic inflation theory, of some universes which are "subject to 'selection' which determine each region's components based upon...the survivability of the quantum components within that region," in the larger context of the multiverse, in *Sec. 2.3.1*; the predictable tendency of the Catholic Church to interpret the Scripture in a way so as to defend it, in *Sec. 2.4.1*; the predictable tendency of a correlative cosmologist to correlate everything in the natural world with its corresponding equivalent in the human world, in *Sec. 2.4.2*; the predictable tendency of certain contingent entities to have their existence dependent on something else, like "a liquid"which "receives its shape from the vessel in which it is contained," in *Sec. 2.5.1*; the predictable tendency of those in the body-soul hylomorphism to argue that a body needs a soul to be alive, in *Sec. 2.5.2*; the predictable tendency of a cosmic cycle to go through the development of all entities from "inconscience" to "divine life," in accordance to the idea of involution, in *Sec. 3.2.1*; the predictable tendency of those in theosophy like Blavatsky to analyze the spiritual evolution of the universe in terms of "seven root races" and the "seven vestures of the soul," in *Sec. 3.2.2*; the predictable fate of the universe in that it will expan indefinitely, based on the current scientific evidences, in *Sec. 3.3.1*; the predictable tendency of the members of Heaven's Gate to indulge in apocalyptic thinking, in *Sec. 3.4.1*; the predictable tendency of MRTCG members to obey the Ten Commandments, in *Sec. 3.4.2*; and the predictable tendency of a Christian to believe in the prophesized Second Coming of Christ, in *Sec. 3.5.1*).

On the other hand, there is unpredictability (e.g., the more difficult task to predict which particular disciple of Advaita Vedanta will suceed in achieving salvation, since it requires some abilities which not everyone equally has, like "the reunuciation of enjoyments of objects" in the material world, concentration on "meditation," the discrimination between the "eternal" and the "transitory" in illusion, etc., in *Sec. 2.2.1*; the more difficult task to predict what a particular idealist can accept in relation to an understanding of what the mind really refers to, which can mean different things to different idealists, like the "divine cosmic" mind in Advaita Vedanta, in *Sec. 2.2.2*; the more difficult task to predict which particular universes will survive in the multiverse and end up with which particular physical laws, in *Sec. 2.3.1*; the more difficult task to predict what particular interpretation that the Catholic Church would creatively use to defend the Scripture in a given case is, since it varies from one era to another and from one authority to another, as shown by the shift of the church's position in regard to the Galileo affair over the centuries, in *Sec.*

2.4.1; the more difficult task to predict which particular correlation a correlative cosmologist would refer to in a particular case, since it varies from one version of correlative cosmology to another, from one historical era to another, from one expert to another, from one category of correlation to another, and so on, in *Sec. 2.4.2*; the more difficult task to predict something outside the realm of experience like the First Cause before the creation of the universe, because"it is unwise to draw conclusions from an extrapolation of causality beyond experience" and "it could be considered nonsensical to speak of pre-universal 'causes,' specifically a First Cause," in *Sec. 2.5.1*; the more difficult task to predict what would happen to the "intellect" after the death of the body, because Aristotle also said, in a contradictory way, that the intellect "can exist apart from the body" and that "the soul is a form or property of the body," in *Sec. 2.5.2*; the more difficult task to predict, if understood solely from the vantage point of involution, how many more rounds of the cosmic cycle there will be in future history, in *Sec. 3.2.1*; the more difficult task to predict correctly when exactly the seventh race, for instance, will actually emerge, if looked within Blavatsky's analysis alone, in *Sec. 3.2.2*; the more difficult task to predict correctly, based on the same current scientific evidences, how and when exactly the universe will end, and whether or not it will end completely, strictly speaking, in *Sec. 3.3.1*; no one outside Heaven's Gate had predicted that the members would commit suicide on March 24-26, 1997, in *Sec. 3.4.1*; no one had predicted that MRTCG members would have ended up in doomsday deaths, which the police finally concluded as "mass murder," contrary to one of the Christian commandments that "thou shalt not kill," in *Sec. 3.4.2*; and the more difficult task for a Christian to predict correctly when exactly Christ will return in the Second Coming, in *Sec. 3.5.1*).

Again, it should be stressed, however, that there are different shades of gray (or different degrees of truth) in the two opposites, and the classification is not necessarily mutually exclusive either. In fact, even when some combinations of the two occur, they only end up sharing the same dialectic relationship, but in a different degree.

Also, there are some other relationships (as a kind of family resemblance) which have something in common with the principle, but they are not exactly the same but only more or less comparable, which varies from case to case.

Good examples of family resemblance in relation to the principle are sureness-arbitrariness, and so forth.

3rd Thesis: The Explicability-Inexplicability Principle

The third thesis is called the explicability-inexplicability principle (on the underlying mechanisms of things) in the category of "method" in existential dialectics (which was proposed in *FPHU* and other books of mine), in that both explicability and inexplicability are part of the understanding of things. There is no explicability without inexplicability—and vice versa.

This principle tells us the dual nature of the research dilemma, in that, if reality can be explained in some ways, it also has its other ways which are not quite explainable, at a given point in time.

For instance, on the one hand, there is explicability (e.g., the explanation, in Advaita Vedanta, of the fact that "the effect has its self in the cause but not the cause in the effect," on the basis of the temporal sequence of cause and effect, in *Sec. 2.2.1*; the explanation, by those in Advaita Vedanta, that universals do not really describe the transcendental reality of the Brahman, because the Brahman "cannot be correctly described," in *Sec. 2.2.2*; the explanation, in accordance to the Big Bang theory, of the beginning of the universe on the basis of the Big Bang, with its different epochs, in *Sec. 2.3.1*; the explanation by Bellarmine "to treat…heliocentrism as a merely hypothetical phenomenon and not a physically real one," because Galileo "did not have the necessary conclusive proof," in *Sec. 2.4.1*; the explanation, by those in Chinese correlative cosmology, of the need of a "unified, imperial order," on the basis of the universe as "an orderly system of correspondence…correlating categories of the human world…with categories of the cosmos," in *Sec. 2.4.2*; the explanation, in accordance to the cosmological argument, that objects move because of some moving body as their cause, in *Sec. 2.5.1*; the explanation by Aristotle that "a soul is related to its body as form to matter," in *Sec. 2.5.2*; the explanation of the cosmic return to the Absolute for spiritual realization on the basis of "emanation" as "an outpouring or even a transformation in the original Absolute or Godhead," in conjunction with involution, in accordance to those in esoteric cosmology, in *Sec. 3.2.1*; the explanation by Blavatsky of the existence of the "inferior races" like "the Bushmen, the Veddhas of Ceylon, and some African tribes," on the basis of their "adjustment of nature which ever works in that direction," in *Sec. 3.2.2*; the explanation of the ultimate fate of the universe by scientists in terms of three possible scenarios, with the Big Bang theory as the scientific basis and the assumption of some entities like "dark energy," "dark matter," etc., in *Sec. 3.3.1*; the explanation by Applewhite to his followers that, since "the planet Earth was about to be recycled [wiped clean, renewed, refurbished and rejuvenated],...the only chance to survive was to leave it immediately" by committing suicide, in *Sec. 3.4.1*; the explanation by the Movement leaders that "the apocalypse would occur in the year 2000," on the basis of the Christian teachings of salvation and apocalypse, as revealed in the Movement booklet *A Timely Message from Heaven: The End of the Present Time*, in *Sec. 3.4.2*; and the explanation, by those in Buddhist eschatology, of the end of Buddhism after 5000 years, paving the way for "the coming of the next Buddha Maitreya," on the basis of the endless cycle of creation and destruction of the world, in *Sec. 3.5.1*).

On the other hand, there is inexplicability (e.g., the lack of convincing explanation, within Advaita Vedanta, of why the cause must therefore be "the infinite unmanifest Absolute," while the effect is the finite, manifest material world, in *Sec. 2.2.1*; the lack of sufficient explanation, within Advaita Vedanta, of why the Brahman is supposed to be real, if it cannot even be correctly described, in *Sec. 2.2.2*; the lack of sufficient explanation, if viewed within the Big Bang theory, of what happened prior to the Big Bang, as critically pointed out by those in pre-Big Bang theories, in *Sec. 2.3.1*; the lack of sufficient explanation by Bellarmine of why "it was not permissible to depart from the traditional interpretation of Scriptures," just because heliocentrism was not conclusively proved—or why geocentrism must be true, only because heliocentrism was not conclusively proved at the time, in *Sec. 2.4.1*; the lack of convincing explanation, by those in Chinese correlative cosmology, of why it must be true that the universe is "an orderly system of correspondence" like that, in *Sec. 2.4.2*; the lack of convincing explanation by those in the cosmological argument to rebuke the counter-argument by Kaku, in that "the law of conservation of energy and the laws governing molecular physics" show that "gas molecules may bounce against the walls of a container without requiring anything or anyone to get them moving," in *Sec. 2.5.1*; the lack of convincing explanation, by Aristotle, of how exactly the body-soul hylomorphism really explains much of anything, as the critics like Descartes are quick to point out that "no natural action at all can be explained by these substantial forms" and, if "their defenders...say that some action proceeds from a substantial form, it is as if they said it proceeds from something they do not understand, which explains nothing," in *Sec. 2.5.2*; the lack of convincing explanation, by those in esoteric cosmology, of why this emanation is necessary and, also, why it is true at all, in *Sec. 3.2.1*; the lack of convincing explanation, by Blavatsky, of why the superior Aryans are able to succeed, also by their "adjustment of nature," albeit in a different direction instead, in *Sec. 3.2.2*; the lack of sufficient explanation of what these assumed entities like "dark energy" and "dark matter" really are, since, "as the name is meant to imply, we know almost nothing of the real physics of the dark energy," in *Sec. 3.3.1*; the lack of sufficient explanation by Applewhite, of why it was in fact true that "the planet Earth was about to be recycled [wiped clean, renewed, refurbished and rejuvenated]" and that "the only chance to survive was to leave it immediately" by committing suicide, without any other options, in *Sec. 3.4.1*; the lack of convincing explanation, by the Movement leaders, of why the apocalypse had to be on January 01 2000, which was later revised to mean March 17, 2000 instead, in *Sec. 3.4.2*; and the lack of convincing explanation, by those in Buddhist eschatology, of why this endless cycle of destruction and creation of the world is necessary in the first place, especially to those who do not believe in Buddhist eschatology, in *Sec. 3.5.1*).

Once more, it should be stressed, however, that there are different shades of gray (or different degrees of truth) in the two opposites, and the classification is not necessarily mutually exclusive either. In fact, even when some combinations of the two occur, they only end up sharing the same dialectic relationship, but in a different degree.

Also, there are some other relationships (as a kind of family resemblance) which have something in common with the principle, but they are not exactly the same but only more or less comparable, which varies from case to case.

Good examples of family resemblance in relation to the principle are underlyingness-regularness, causation-regularity, causation-correlation and so forth.

4rd Thesis: The Preciseness-Vagueness Principle

The fourth thesis is called the preciseness-vagueness principle (on the refinement of things) in the category of "structure" in existential dialectics (which was first out worked out in *FIA*), in that both preciseness and vagueness are important, not that one is better than the other, but that both are used, in different degrees of preference, in accordance to the contextual application from the perspectives of nature, the mind, culture, and society. There is no preciseness without vagueness—and vice versa.

For instance, on the one hand, there is preciseness (e.g., the precise identification, in Advaita Vedanta, of Brahman as "the infinite unmanifest Absolute," in *Sec. 2.2.1*; the precise identification of "lacking any causal powers" as one criterion for abstract objects, in *Sec. 2.2.2*; the precise identification, within the Big Bang theory, that "the Universe today is 72% dark energy, 23% dark matter, 4.6% regular matter and less than 1% neutrinos," in *Sec. 2.3.1*; the precise identification of the Earth as the center of the universe in geocentrism, in *Sec. 2.4.1*; the precise identification of the historical origin of *Wuxing* in *Sifang*, which is "the archaic cosmology of the Four Quarters" during the Bronze Age of Chinese history, in *Sec. 2.4.2*; the precise identification of the First Cause as that which causes all contingent things in the universe, in accordance to the cosmological argument, in *Sec. 2.5.1*; the precise identification of the soul as the substantial form of the body, in *Sec. 2.5.2*; the precise identification of "the Omnipresent Reality" as "the Absolute Brahman," which is a non-temporal precursor to the existence of the universe, for Sri Aurobindo, in *Sec. 3.2.1*; the precise identification of the seven bodies in the literature on the septenary in theosophy, in *Sec. 3.2.2*; the precise identification of "dark energy" as crucial to understand the ultimate fate of the universe, in *Sec. 3.3.1*; the precise identification of March 26, 1997 as a period when "Comet Hale-Bopp was at its brightest," by Applewhite to his followers, in *Sec. 3.4.1*; the precise identification of January 01, 2000 as the date for the occurrence of apocalpyse, in accordance to the Movement leaders, in *Sec. 3.4.2*; and the precise identification of the "one-thousand year reign of Messiah" as one of the prophesized future events in Judeo-Christian eschatology, in *Sec. 3.5.1*).

On the other hand, there is vagueness (e.g., the unclarity in the conception of Brahman in Advaita Vedanta, because it "cannot be correctly described," in *Sec. 2.2.1*; the vagueness in the identification of "lacking any causal powers" as one criterion for abstract objects, because "it is not clear exactly what it is to have a causal power," for the critics, in *Sec. 2.2.2*; the vagueness in this identification, within the Big Bang theory, of "dark matter" and "dark energy," which, unfortunately, are not exactly clear as to what they really are, since they are still poorly understood in our time and their "exact nature…remains one of the great mysteries of the Big Bang," in *Sec. 2.3.1*; the vagueness in the geocentric view that the Earth is the center of the universe, as this can mean different things for different peoples, because one geocentric view is that "the stars, sun, and planets appear to revolve around the Earth each day, with the stars circling around the pole and those stars nearer the equator rising and setting each day and circling back to their rising point," and another geocentric view is that "the Earth is solid and stable it is not moving—but is at rest," in *Sec. 2.4.1*; the vagueness in understanding the historical origin itself, since it is subject to different historical interpretations, not only because of the "archaic" nature of *Sifang* at the time but also because of the "loosely defined and unstematically used" *Wuxing* at the time, in *Sec. 2.4.2*; the vagueness in the attribtutes of the First Cause, since, for the critics, any attempt to extract such attributes to the First Cause requires a separate proof, which is not in the cosmological argument, in *Sec. 2.5.1*; the vagueness in the identification of the soul as the substantial form of the body, since "[s]ome…thinkers argued that X's soul is X's only substantial form, responsible for all of the features of X's body," but "other…thinkers argued that a living being contains at least two substantial forms" like "the shape and structure of its body" and "its soul, which makes its body alive," in *Sec. 2.5.2*; the vagueness in this precise identification of "the Omnipresent Reality," since it is not clear what exactly it really refers to, as it is not "temporal" but "divine," which, unfortunately, cannot be totally grasped by humans, in *Sec. 3.2.1*; the vagueness in the analysis of each of the seven bodies, since it is not clear what such abstract terms like "pure consciousness" really refer to, since they can mean different things for different theosophical interpreters in different contexts, in *Sec. 3.2.2*; the vagueness in the idenfication of "dark energy," because, "as the name is meant to imply, we know almost nothing of the real physics of the dark energy," in *Sec. 3.3.1*; the vagueness in the identfication of March 26, 1997 as a period when "Comet Hale-Bopp was at its brightest," by Applewhite to his followers, since it would mean just anything that he wanted to impose meaning on it, like the story of "a space craft…trailing the comet Hale" on that date, as the best time to commit suicide, in *Sec. 3.4.1*; the vagueness in the precise identification of the date of January 01, 2000 as the date for the occurrence of apocalpyse, because it did not say whether it had to be taken literally or could be changed to a later date, which was precisely what the Movement leaders did when the first date turned out to be untrue, in *Sec. 3.4.2*; and the vagueness in the identification of the "one thousand year reign of Messiah," since this means literally for some but figuratively, or symbolically, for others, in *Sec. 3.5.1*).

Once more, it should be stressed, however, that there are different shades of gray (or different degrees of truth) in the two opposites, and the classification is not necessarily mutually exclusive either. In fact, even when some combinations of the two occur, they only end up sharing the same dialectic relationship, but in a different degree.

Also, there are some other relationships (as a kind of family resemblance) which have something in common with the principle, but they are not exactly the same but only more or less comparable, which varies from case to case.

Good examples of family resemblance in relation to the principle are clarity-ambiguity, directness-indirectness, quantitativeness-qualitativeness, specificity-obscurity, describability-nondescribability, specificity-generality, thickness-thinness, concretness-abstractness, and the like.

5th Thesis: The Simpleness-Complicatedness Principle

The fifth thesis refers to the simpleness-complicatedness principle (on the interconnection among things) in the category of "structure" in existential dialectics (which was first out worked out in *FIA*), in that both simpleness and complicatedness are vital, without favoring one over the other, and each is utilized, depending on the basis of the perspectives of nature, the mind, culture, and society. There is no simpleness without complicatedness—and vice versa.

For instance, on the one hand, there is simple analysis (e.g., the relatively simplistic analysis of the creation of the universe in a mystical way, in that the effect like the universe pre-existed in the cause like "the infinite unmanifest Absolute," in *Sec. 2.2.1*; the relatively simplistic claim by the idealists that universals do not exist, in *Sec. 2.2.2*; the relatively simple analysis of the beginning of the universe on the basis of the Big Bang, with its different epochs, especially in the context of the standard Big Bang theory, in *Sec. 2.3.1*; the relatively simplistic geocentric view of the Earth as the center of the universe, in *Sec. 2.4.1*; the relatively simplistic analysis of Chinese correlative cosmology as "a kind of anthropocosmology in which entities, processes, and classes of phenomena found in nature correspond to or 'go together with' various entities, processes, and classes of phenomena in the human world," in *Sec. 2.4.2*; the relatively simple analysis of the cosmological arguemnt in terms of its three variants, in *Sec. 2.5.1*; the relatively simple analysis by Aristotle of the body-soul hylomorphism, in *Sec. 2.5.2*; the relatively simple analysis of the cosmic cycle for spiritual realization, in accordance to esoteric cosmology, in *Sec. 3.2.1*; the relatively simple analysis of the races and the soul from the vantage point of the esoteric number seven, in theosophy, in *Sec. 3.2.2*; the relatively simple outline of the future fate of the universe on the basis of the three scenarios, in *Sec. 3.3.1*; the relatively simplistic story about the end of the Earth by Applewhite to his followers, in *Sec. 3.4.1*; the relatively simplistic prophesy of January 01, 2000 as the date for the occurrence of apocalpyse, by the Movement leaders to the followers, in *Sec. 3.4.2*; and the relatively simple delineation of the prophesized future events in religious eschatology, in *Sec. 3.5.1*).

On the other hand, there is also complicated counterpart (e.g., the relatively more complicated analysis of the Advaita Vedanta theory of creation by questioning its claims, as shown in the criticisms, in *Sec. 2.2.1*; there is complicatedness (e.g., the relativley more complicated analysis of the idealist claim by way of questioning its assumptions, as pointed out by the realists and the nominalists, in *Sec. 2.2.2*; the relatively more complicated analysis of the beginning of the universe by questioning the claims of the standard Big Bang theory and pointing out its major problems, as pointed out by the critics, in *Sec. 2.3.1*; the relatively complicated helocentric view that the Earth is not the center of the universe, but part of the solar system, with the Sun as the center and other objects revolving around it, in *Sec. 2.4.1*; the relatively more complicated analysis of Chinese correlative cosmology in terms of different conflicting interpretations as pointed out by the critics, who do not share a consensus, in *Sec. 2.4.2*; the relatively complicated analysis of the cosmological argument by pointing out various problems as shown by the critics, in *Sec. 2.5.1*; the relatively more complicated analysis of the body-soul hylomorphism as shown in the criticisms by the critics, in *Sec. 2.5.2*; the relatively more complicated analysis of the cosmic cycle for spiritual realization by questioning its assumptions and claims, as shown in the criticisms, in *Sec. 3.2.1*; the relatively more complicated analysis of the races and the soul by challenging its assumptions and claims, as pointed out by the critics, in *Sec. 3.2.2*; the relatively more complicated analysis of the ultimate fate of the universe by questioning the assumptions and claims in regard to the three scenarios, as shown in the problems as pointed out by the critics, in *Sec. 3.3.1*; the relatively more complicated analysis of the story about Applewhite, as revealed by Jacques Vallée in his 1979 *Messengers of Deception*, in which Vallée critically questioned the "authoritarian political and religious outlook" of the group, in *Sec. 3.4.1*; the relatively more complicated view of apocalpyse by the Movement members when the doomsday prophesy did not realize, which "led to a revolt in the ranks of the sect" and they came to question their leaders, before being mass-murdered, in *Sec. 3.4.2*; and the relatively more complicated analysis of the prophesized future events by questioning its claims, as shown in the criticisms, in *Sec. 3.5.1*).

Again, it should be stressed, however, that there are different shades of gray (or different degrees of truth) in the two opposites, and the classification is not necessarily mutually exclusive either. In fact, even when some combinations of the two occur, they only end up sharing the same dialectic relationship, but in a different degree.

Also, there are some other relationships (as a kind of family resemblance) which have something in common with the principle, but they are not exactly the same but only more or less comparable, which varies from case to case.

Good examples of family resemblance in relation to the principle are inflexibility-flexibility, standardization-specialization, imperfectness-perfectness, superficiality-depth, shallowness-deepness, economicalness-elaboratedness, plainness-circumspection, onesidedness-multisidedness, and the like.

6th Thesis: The Openness-Hiddenness Principle

The sixth thesis refers to the openness-hiddenenss principle (on the detection of things) in the category of "structure" in existential dialectics (which was already worked out in my previous books, especially in *FPHU*), in that reality has its hidden face, just as it is open to outside view in some other ways. There is no openness without hiddenness—and vice versa.

For instance, on the one hand, there is openness (e.g., the open exploration of the illusion in the creation of the world, in accordance to Advaita Vedanta, in *Sec. 2.2.1*; the open exploration, in idealism, of the relatonship between universals and the unreality of the world, in *Sec. 2.2.2*; the open exploration of the beginning of the universe "up to 10^{-43} seconds after the Big Bang" in the Planck epoch, in accordance to the Big Bang theory, in *Sec. 2.3.1*; the open defense of geocetrism by the Catholic Church against the helocentric view by Galileo, in *Sec. 2.4.1*; the open exploraton of the correlations between things in the natural world and those in the human world, in accordance to Chinese correlative cosmology, in *Sec. 2.4.2*; the open exploration of the relationship between the First Cause and the contingency of things, in accordance to the cosmological argument, in *Sec. 2.5.1*; the open exploration by Aristotle of the relationship between the body and the soul, in that "a soul is related to its body as form to matter," in *Sec. 2.5.2*; the open exploration of the relationship between the evolution of the universe and its involution for spiritual realization, in *Sec. 3.2.1*; the relatively open exploration of the universal brotherhood of humanity in theosophy, in *Sec. 3.2.2*; the open exploration of the ultimate fate of the universe, on the basis of the Big Bang theory and the assumpton of some entities like "dark energy" and "dark matter" in physics, in *Sec. 3.3.1*; the open exploration of the coming end of the Earth and the need to commit suicide, as told by Applewhite to his followers, in *Sec. 3.4.1*; the open exploration, by the Movement leaders to the followers, of another doomsday set for March 17, 2000, after the first doomsday of January 01, 2000 failed, in *Sec. 3.4.2*; and the open exploration, in Hindu eschatology, that "one age may end but another will always begin," in *Sec. 3.5.1*).

On the other hand, there is also hiddenness (e.g., the hidden bias in Advaita Vedanta, in its obsession with the spiritual quest for salvation in relation to "the infinite unmanifest Absolute," in *Sec. 2.2.1*; the hidden bias in the idealist view, as shown in Advaita Vedanta, because of its spiritual preference for the alleged "divine cosmic source" which is prior to the creation of the universe, in *Sec. 2.2.2*, the hiddenness of the pre-Big Bang period, such that the Big Bang theory cannot account for, in *Sec. 2.3.1*; the hidden bias in the Catholic Church because of its unquestioned faith in the Scripture, in *Sec. 2.4.1*; the hidden bias in interpreting the validity of correlative cosmology, because of the different political motives closely linked with the interpretations, in *Sec. 2.4.2*; the hidden bias in the cosmological argument, because of its assumption of a suprasensible being, a Prime Mover, a God, or something like that in the First Cause, in *Sec. 2.5.1*; the hidden bias in the Aristotelian body-soul hylomorphism, because of his teleological view of the universe as originally organized by the Prime Mover and subsequently set in motion, in *Sec. 2.5.2*; the hidden bias in the idea of the cosmic return to the Absolute in involution, because of its esoteric ideology of emanation, in *Sec. 3.2.1*; the hidden bias in the claims by Blavatsky, one of the founders of the Theosophical Society, on different races, with the Aryans as the superior one and many of the non-white folks as the "inferior" ones, in *Sec. 3.2.2*; the hidden nature of "dark matter" and "dark energy," such that the ultimate fate of the universe remains uncertain, in *Sec. 3.3.1*; the hidden bias in Applewhite's view, which could be traced back to his traumatic "near-death experience" in the early 1970's, which led him "to believe that he and his nurse, Bonnie Nettles, were 'the Two,' that is, the two witnesses spoken of in Book of Revelation 11:3 in the Holy Bible," in *Sec. 3.4.1*; the hidden motive of the Movement leader behind the new doomsday of March 17, 2000, in that "the leaders set a new date with a plan to eliminate [that is, to murder] their followers," in *Sec. 3.4.2*; and the hidden bias in Hindu eschatology, in terms of its cyclical assumption about the cosmic order of things, which other versions of eschatology, say, in the Abrahamic tradition, do not share, in *Sec. 3.5.1*).

Again, it should be stressed, however, that there are different shades of gray (or different degrees of truth) in the two opposites, and the classification is not necessarily mutually exclusive either. In fact, even when some combinations of the two occur, they only end up sharing the same dialectic relationship, but in a different degree.

Also, there are some other relationships (as a kind of family resemblance) which have something in common with the principle, but they are not exactly the same but only more or less comparable, which varies from case to case.

Good examples of family resemblance in relation to the principle are overtness-covertness, publicness-privateness, openness-closedness, transparency-secrecy, openness-biasedness, and so on.

7th Thesis: The Denseness-Emptiness Principle

The seventh thesis concerns the denseness-emptiness principle (on the distribution of entities in space) in the category of "structure" in existential dialectics, which was first proposed in *FPHUP*, in that both density and void are needed, in relation to the mind, nature, culture, and society, albeit in different ways. There is no denseness without emptiness—and vice versa.

Lest any misunderstanding occurs, the term "void" is used here only as an approximation of emptiness (depending on the degree of the lack of density), since, in physics, it is well known that "empty" space is not really empty all the way, because it can be full of energy (e.g., random quantum fluctuations at the sub-atomic level, and, for that matter, dark energy in the universe) and matter (e.g., different versions of sub-atomic particles, and, for that matter, dark matter in the universe). (F. Wilczek 2008)

For instance, on the one hand, there is denseness (e.g., the relatively denser concentration of spiritual concerns with "the reunuciation of enjoyments of objects" in the material world, in Advaita Vedanta, in *Sec. 2.2.1*; the relatively denser concentration of concerns, for a Platonic realist, with a metaphysical realm for universals which is independent of both the physical world and the mind, in *Sec. 2.2.2*; the extreme density of the universe when it started at the point of singularity, in accordance to the Big Bang theory, in *Sec. 2.3.1*; the relatively denser concentration of individuals with faith in geocentrism within the Catholic Church at the time of Galileo, in *Sec. 2.4.1*; the relatively denser concentration of those who accepted the "correlations" in the universe on the basis of "fours" as in *Sifang*, during the Bronze Age of ancient Chinese history, in *Sec. 2.4.2*;

the relatively denser concentration of concerns with the First Cause for those in the cosmological argument, in *Sec. 2.5.1*; the relatively denser concentration of the followers of the Aristotelian hylomorphism in the medieval era, especially when applied for "the Christian doctrines such as the transubstantiation of the Eucharist's bread and wine into the body and blood of Jesus," in *Sec. 2.5.2*; the relatively denser concentration of believers in involution in India, in *Sec. 3.2.1*; the relatively denser concentration of believers in the Theosophical Society with a relatively Eastern orientation in India, in *Sec. 3.2.2*; the reltively denser concentration of believers in the "open" universe as the ultimate fate of the universe nowadays, in *Sec. 3.3.1*; the relatively denser concentration of individuals living "a highly ascetic life devoid of many indulgences" in Heaven's Gate, in *Sec. 3.4.1*; the relatively denser concentration of believers in doomsday among MRTCG members, in *Sec. 3.4.2*; and the relatively denser concentration of Christian eschatologists in Western Europe, in *Sec. 3.5.1*).

On the other hand, there is emptiness (e.g., the relatively less dense, or more empty, concentration of mundane concerns with sensual life enjoyments, in Advaita Vedanta, in *Sec. 2.2.1*; the relatively more empty concentraiton of concerns, for a Platonic realist, with the cosmic mind as advocated by the idealists in Advaita Vedanta, in *Sec. 2.2.2*; the relatively lesser density of the universe in our present era, about "13.7 billion years since the big bang," because "the expansion of the universe appears to be accelerating," farther and farther away from the point of singularity, in accordance to the Big Bang theory, in *Sec. 2.3.1*; the relatively less dense, or more empty, concentration of individuals like Galileo and other astronomers working with good telescopes to have faith in geocentrism, in *Sec. 2.4.1*; the relatively less dense, or more empty, concentration of those who accepted the "correlations" in the universe on the basis of "fives," during the Bronze Age of ancient Chinese history, in *Sec. 2.4.2*; the relatively less dense, or more empty, concentration of concerns with the infinite regress of causation for those in the cosmological argument, in *Sec. 2.5.1*; the relatively less dense, or more empty, concentration of the followers of the Aristotelian hylomorphism in ancient China, in *Sec. 2.5.2*; the relatively less dense, or more empty, concentration of believers in involution among contemporary astrophysicists, especially those in the Big Bang theory, in *Sec. 3.2.1*; the relatively less dense, or more empty, concentration of believers in the Theosophical Society with a relatively Eastern orientation in Germany, especially among those in the Anthroposophical Society, in *Sec. 3.2.2*; the relatively less dense, or more empty, concentration of believers in the "flat" universe as the ultimate fate of the universe nowadays, in *Sec. 3.3.1*; the relativley less dense, or more empty, concentration of individuals living "a highly ascetic life devoid of many indulgences" among members of the Playboy Club founded by Hugh Hefner, *in Sec. 3.4.1*; the relatively less dense, or more empty, concentration of believers in doomsday among the police who led the investigation of doomsday deaths, in *Sec. 3.4.2*; and the relatively less dense, or more empty, concentration of Christian eschatologists in India, in *Sec. 3.5.1*).

Again, it should be stressed, however, that there are different shades of gray (or different degrees of truth) in the two opposites, and the classification is not necessarily mutually exclusive either. In fact, even when some combinations of the two occur, they only end up sharing the same dialectic relationship, but in a different degree.

Also, there are some other relationships (as a kind of family resemblance) which have something in common with the principle, but they are not exactly the same but only more or less comparable, which varies from case to case.

Good examples of family resemblance in relation to the principle are fullness-voidness, nearness-farness, concentration-dispersion, and the like.

8th Thesis: The Slowness-Quickness Principle

The eighth thesis concerns the slowness-quickness principle (on the speed of change) in the category of "process") in existential dialectics (which was first worked out in *FIA*), in that both slowness and quickness co-exist, with their own internal tension, to the extent that each fights for its own relevance with the other, in accordance to the perspectives of nature, the mind, culture, and society, without one being the victor and the other being the vanquished in the long haul. There is no slowness without quickness—and vice versa.

For instance, on the one hand, there is slowness (e.g., the relatively slower readiness of a follower of Advaita Vedanta to let himself preoccupied with actions for materialist life success, in *Sec. 2.2.1*; the relatively slower readiness of an idealist to accept universals as real, in *Sec. 2.2.2*; the relatively slower speed of the motion of particles through space, when compared with the expansion of space itself, in *Sec. 2.3.1*; the relatively slower readiness of those in the Catholic Church to accept heliocentrism unless it is conclusively proved, in *Sec. 2.4.1*; the relatively slower readiness of those "leading critics" like Liang Qichao and Gu Jiegang—who "deconstructed *Wuxing* cosmology" on the basis of revealing "the political motivations of the forgers of *Wuxing* texts"—to link correlative cosmology with empire building as proposed by A. Wang, in *Sec. 2.4.2*; the relatively slower readiness of those in the cosmological argument to accept that the universe has no beginning, in *Sec. 2.5.1*; the relatively slower readiness of the followers of the Aristotelian hylomorphism to understand the world on the basis of "different properties and forces" as understood in modern science like physics, without the messy body-soul hylomorphism, in *Sec. 2.5.2*; the relatively slower readiness of such entities as life in physical beings and animals to become self-conscious, in *Sec. 3.2.1*; the relatively slower readiness of someone like Steiner to accept a special boy like Jiddu Krishnamurti as "the incarnation of Maitreya...who was believed to have 'over-shadowed' Jesus Christ," in *Sec. 3.2.2*; the universe will take a long long time to end in the scenario of the "open" universe, if measured from the standpoint of average human lifespan, in *Sec. 3.3.1*; the relatively slower readiness of those like Hugh Hefner who do not have any apocalyptic mindset to commit suicide, in *Sec. 3.4.1*; the relativley slower readiness of the Movement members to have sex, in *Sec. 3.4.2*; and the relatively slower readiness of those in the Hindu tradition to think about the end of the world in a linear, apocalytpic way, in *Sec. 3.5.1*).

On the other hand, there is quickness (e.g., the relatively quicker readiness of a followre of Advaita Vedanta to concentrate on meditation for the "liberation from the cycle of births and deaths," in *Sec. 2.2.1*; the relatively quicker readiness of an idealist in Advaita Vednata to accept the existence of the divine cosmic mind, which cannot be described by words and is prior to the creation of the world, in *Sec. 2.2.2*; the relatively faster speed of the expansion of space itself, which can travel faster than the speed of light, and "[t]his does not violate the speed limit dictated by relativity, since the expansion of space itself is happening faster than light, rather than the motion of particles through that space," which is subject to the limit of the speed of light, in *Sec. 2.3.1*; the relatively quicker readiness of those like Galileo to accept heliocentrism, because "the available evidence strongly favoured heliocentrism," even though "he did not have the necessary conclusive proof," in *Sec. 2.4.1*; the relatively quicker readiness of those like A. Wang to interpret correlative cosmology as a cultural fabric in its constant interaction with the political realm, for the service of empire building, in *Sec. 2.4.2*; the relatively quicker readiness of those in the Big Bang theory to accept that "the question 'What was there before the Universe?' makes no sense," because "the concept of 'before' becomes meaningless when considering a situation without time," in *Sec. 2.5.1*; the relatively quicker readiness of the followers of the Aristotelian hylomorphism to understand the world on the basis of form and matter, in *Sec. 2.5.2*; the relatively quicker readiness of such entities as humans to become self-conscious, in *Sec. 3.2.1*; the relatively quicker readiness of those in The Order of the Star in the East, the Indian section of the Theosophical Society, to accept a special boy like Jiddu Krishnamurti as "the incarnation of Maitreya...who was believed to have 'over-shadowed' Jesus Christ," in *Sec. 3.2.2.*; the universe is accelerating very fast in its expansion, if measured from the standpoint of average human running speed, in *Sec. 3.3.1*; the relatively quicker readiness of those in Heaven 's Gate who have the apocalyptic mindset to commit suicide, in *Sec. 3.4.1*; the relatively quicker readiness of a prostitue to have sex, in *Sec. 3.4.2*; and the relatively quicker readiness of those in the Abrahmic tradition to think about the end of the world in a linear, apocalytpic way, in *Sec. 3.5.1*).

Again, it should be stressed, however, that there are different shades of gray (or different degrees of truth) in the two opposites, and the classification is not necessarily mutually exclusive either. In fact, even when some combinations of the two occur, they only end up sharing the same dialectic relationship, but in a different degree.

Also, there are some other relationships (as a kind of family resemblance) which have something in common with the principle, but they are not exactly the same but only more or less comparable, which varies from case to case.

Good examples of family resemblance in relation to the principle are inconvenience-convenience, passiveness-activenessness, gradualness-abruptness, deceleration-acceleration, and the like.

9th Thesis: The Expansion-Contraction Principle

The nineth thesis is called the expansion-contraction principle (on the growth of things) in the category of "process" in existential dialectics, in that entities in the world can both expand in some ways and contract in other ones, as part of their nature. There is no expansion without contraction—and vice versa.

This principle, although not so explicitly called, was already used in my previous works on different topics (e.g., the theory of floating consciousness in *FCD* and *FPHC*, the union of the unions in *BWT*, and the cyclical progression of system fragmentation and integration in *FCD*).

For instance, on the one hand, there is expansion (e.g., the relatively more developed ability of a follower of Advaita Vedanta to understand "the canonical texts consisting of [a] the Upanishads, [b] the Bhagavad

Gita and [c] the Brahma Sutras, in *Sec. 2.2.1*; the relatively more developed ability of a conceptualist to be frugal "against assuming any entities that were not necessary for explanations," also kwown as Ockham's razor, in *Sec. 2.2.2*; some universes within the multiverse can "rapidly expand forever," in accordance to the chaotic inflation theory, in *Sec. 2.3.1*; the relatively more developed ability of those in the Catholic Church to be good in biblical hermeneutics, in *Sec. 2.4.1*; the relatively more developed ability of those who specialize in correlative cosmology to think about reality in terms of "correlations" between the natural world and the human world, in *Sec. 2.4.2*; the relatively more developed ability of those in the cosmological argument to manipulate deduction for the search of necessity, even though the premise of causality in the world of experience is inductive and uncertain, in *Sec. 2.5.1*; the relatively more developed ability of the followers of the Aristotelian hylomorphism to deal with philosophical, especially metaphysical, problems, in *Sec. 2.5.2*; the relatively more developed ability of spiritual seekers in esoteric cosmology to develop their mind for spiritual realization, in *Sec. 3.2.1*; "in the last several decades" of the 20th century and the early 21st, "there was a steady increase in membership in India," in regard to the Theosophical Society, in *Sec. 3.2.2*; "new 'verses' would be formed" somewhere, in accordance to the multiverse theory, in *Sec. 3.3.1*; the relatively more developed ability of those in Heaven's Gate to confront life in an ascetic way, in *Sec. 3.4.1*; the relatively more developed ability of the Movement members to live a strictly ascetic lifestyle, in that "fasts were conducted regularly, and only one meal was eaten on Fridays and Mondays," etc., in *Sec. 3.4.2*; and the recreation of "a 'New Heaven and New Earth' ushering in the beginning of Eternity," at the end of the world, in accordance to Christian eschatology, in *Sec. 3.5.1*).

On the other hand, there is contraction (e.g., the relatively less developed ability of a follower of Advaita Vedanta to understand non-Hindu schools of thought, like the theory of relativity in physical cosmology, in *Sec. 2.2.1*; the relatively less developed ability of a conceptualist to be complicated in assuming many entities and their properties like those about the Brahman by the idealists, in *Sec. 2.2.2*; some universes withiin the multiverse can even "decay" over time and do not survive the "selection" process, in accordance to the chaotic inflation theory, in *Sec. 2.3.1*; the relatively less developed ability of those in the Catholic Church to be good in working with telescopes for scientific discovery, in *Sec. 2.4.1*; the relatively less developed ability of those who specialize in correlative cosmology to think logically, not correlatively, in a more systematic way as in Western symbolic logic, in *Sec. 2.4.2*; the relatively less developed ability of those in the cosmological argument to use more induction for the understanding of the world with a high dose of uncertainty, in *Sec. 2.5.1*; the relatively less developed ability of the followers of the Aristotelian hylomorphism to deal with technical and scientific problems like those in modern physics in understanding the world, in *Sec. 2.5.2*; the relatively less developed ability of spiritual seekers in eosteric cosmology to cope with materialist development, in *Sec. 3.2.1*; "outside India, the membership has been dropping," in regard to the Theosophical Society, in the same period of the the last several decades of the 20th century and the early 21st, in *Sec. 3.2.2*; "old 'verses'" would be "annihilated" somewhere, in accordance to the multiverse theory, in *Sec. 3.3.1*; the relatively less developed ability of those in Heaven's Gate to confront life in a more positive way without having to kill oneself, in *Sec. 3.4.1*; the relatively less developed ability of the Movement members to be sensual and joyful in life, without worrying about damnation and the Ten Commandments, in *Sec. 3.4.2*; and "the destruction of the current heavens and the earth" at the end of the world, in accordance to Christian eschatology, in *Sec. 3.5.1*).

Again, it should be stressed, however, that there are different shades of gray (or different degrees of truth) in the two opposites, and the classification is not necessarily mutually exclusive either. In fact, even when some combinations of the two occur, they only end up sharing the same dialectic relationship, but in a different degree.

Also, there are some other relationships (as a kind of family resemblance) which have something in common with the principle, but they are not exactly the same but only more or less comparable, which varies from case to case.

Good examples of family resemblance in relation to the principle are conquest-autarky, rise-fall (or up-down), spread-shrink, extendingness-shorteningness, and so forth.

10th Thesis: The Convention-Novelty Principle

The tenth thesis is the convention-novelty principle (on the nature of creative thinking) in the category of "agency" in existential dialectics (which was first worked out in *FPHCT*), in that, if there is conventional wisdom, there is likewise novel challenge, to the extent that both convergent and divergent thinking are part of life. There is no convention without novelty—and vice versa.

As srummarized from *FPHCT* (almost verbatim here), there are (a) "creative techniques" and (b) "creative traits," which, when satisfied—in relation to the larger context of the mind, nature, society, and culture—can be used to enhance creative works.

In addition, creative thinking has its own possibilities and limits (in relation to invention), just as it has its own promises and pitfalls (in relation to innovation)— as already analyzed in *FPHCT*.

In the end, creative thinking has its own desirability and dark sides (as also already analyzed in *FPHCT*).

With these clarifications in mind—there are good empirical examples for the convention-novelty princple.

For instance, on the one hand, there is convention (e.g., the conventional wisdom about non-duality in Hindu philosophy, like the one in Advaita Vedanta, in *Sec. 2.2.1*; the conventional wisdom, as argued by Plato, about the reality of universals, in *Sec. 2.2.2*; the conventional wisdom about the beginning of the universe in terms of the Big Bang, in *Sec. 2.3.1*; the conventional wisdom about geocentrism at the time of Galileo, in *Sec. 2.4.1*; the conventional wisdom about *Sifang* cosmology during the Bronze Age of ancient Chinese history, in *Sec. 2.4.2*; the conventional wisdom about the cosmological argument as defended by Aristotle in antiquity and by Aquinas in the medieval era, in *Sec. 2.5.1*; the conventional wisdom about hylomorphism and the teleology of the universe in Greek antiquity, in *Sec. 2.5.2*; the conventional wisdom about involution in the Indian, particularly Hindu, spiriutal tradition, in *Sec. 3.2.1*; the conventional wisdom about the relatively Eastern orientation in the original Theosophical Society, in *Sec. 3.2.2*; the conventional wisdom about the "static universe" which Einstein initially accepted, in *Sec. 3.3.1*; the conventional wisdom about the end of the world based on some Christian ideas of "salvation and apocalypse," in *Sec. 3.4.1*; the conventional wisdom about the doomsday in Christian teachings of salvation and apocalypse, in *Sec. 3.4.2*; and the conventional wisdom about taking literally the "one-thousand year reign of Messiah" as one of the prophesized future events in Christian eschatology, in *Sec. 3.5.1*).

On the other hand, there is novelty (e.g., the alternative novel challenge to this conventional wisdom by Dvaita Vedanta, which allows duality, in *Sec. 2.2.1*; the alternative novel challenge to the conventional wisdom about the Platonic reality of universals by the nominalists like Abelard and Ockham, who argued instead their existence as concepts in the human mind only, in *Sec. 2.2.2*; the alternative novel challenge to the conventional wisdom about the Big Bang, by those scholars favoring pre-Big Bang theories, like Steinhardt and Turok, for "a universe exploding into existence not just once, but repeatedly over time," in *Sec. 2.3.1*; the alternative novel challenge to geocentrism by Galileo, by way of his original heliocentric observations, in *Sec. 2.4.1*; the alternative novel challenge to the conventional wisdom about *Sifang* cosmology by its successor known as *Wuxing* cosmology for a different system of correlations based on fives, in *Sec. 2.4.2*; the alternative novel challenge to this conventional wisdom about the cosmology argument, by those in the Big Bang theory in the modern times, who instead hold that "the question 'What was there before the Universe?' makes no sense," in *Sec. 2.5.1*; the altenative novel challenge to this conventional wisdom about hylomorphism, by contemporary neo-Darwinian evolutionary theory that things can come into existence without implying any evolutionary purpose for optimal functionality, in *Sec. 2.5.2*; the alternative novel challenge to this conventional wisdom about involution, by Sri Aurobindo, who revised and expanded the Indian, particularly Hindu, tradition for his vision of the "supermind" for "divine life," in *Sec. 3.2.1*; the alternative novel challenge to the conventional wisdom about the relativley Eastern orientation in the original Theosophical Society, by Steiner for his Anthroposophical Society with a relatively more Western orientation, in *Sec. 3.2.2*; the alternative novel challenge to the conventional wisdom about the "static universe," by Hubble who "announced his conclusion that the universe was expanding," in *Sec. 3.3.1*; the alternative novel revision of this conventional wisdom about the end of the world, by Applewhite, for his fabrication that he and his nurse were "the two witnesses spoken of in Book of Revelation 11:3 in the Holy Bible," in *Sec. 3.4.1*; the alternative novel challenge to the conventional wisdom about the doomsday by the Movement leaders, who first set the date of January 01, 2000 for the doomsday and later changed it into March 17, 2000 instead, in *Sec. 3.4.2*; and the alternative novel challenge to this literal interpretation of the "one-thousand year reign of Messiah" by those known as "the Amillennialists [who]...maintain that the mention of the 'thousand years' in the Book of Revelation is meant to be taken metaphorically..., i.e., not literally, or 'spiritually,'...a view which continues to cause divisions within evangelical Christianity," in *Sec. 3.5.1*).

Again, it should be stressed, however, that there are different shades of gray (or different degrees of truth) in the two opposites, and the classification is not necessarily mutually exclusive either. In fact, even when some combinations of the two occur, they only end up sharing the same dialectic relationship, but in a different degree.

Also, there are some other relationships (as a kind of family resemblance) which have something in common with the principle, but they are not exactly the same but only more or less comparable, which varies from case to case.

Good examples of family resemblance in relation to the principle are convergence-divergence (or convergent thinking vs. divergent thinking), normalness-nonnormalness, conformity-nonconformity, and so on.

11th Thesis: The Evolution-Transformation Principle

The eleventh thesis refers to the evolution-transformation principle (on the multiple kinds of agency). in the category of "agency" in existential dialectics (which was first worked out in *FAE* and then in other books of mine).

This principle—and the symmetry-asymmetry principle, for instance—are both about agency. More precisely, the word "agency," in a formal definition, refers to "a person or thing through which power is exerted or an end is achieved." (MWD 2007) It therefore does not have to necessarily involve an intelligent lifeform.

Because of this dual meaning in agency, the evolution-transformation principle is more concerned with *the kind of agency*, that is, both about the *evolution* in the state of nature (e.g., an object of natural beauty) and the *transformation* in the world of intelligent lifeforms (e.g., a work of art). There is no evolution without transformation—and vice versa.

In classical Darwinian evolutionary theory (as more detailedly analyzed in *BNN*), evolution is "blind." But in the human world, change often takes place because of the conscious intervention of humans in transforming society and culture, just to cite two instances.

And the transformative part of the principle precisely refers to the other dimension in the dual meaning of agency, in giving technology (as invented by intelligent lifeforms like humans and, soon, post-humans) a major role to play in the change of the world, which is something that I extensively analyzed in *FHC* in the context of the technophilic lifeworld, espeically though not exclusively since modern times.

For instance, on the one hand, there is evolution (e.g., the natural evolution of humans to live in awe of the universe or nature, in *Sec. 2.2.1*; the natural evolution of humans to communicate with each other in a primitive way in the state of nature, without the existence of elaborated systems of language yet, in *Sec. 2.2.2*; the natural evolution of humans to observe with their naked eyes in the state of nature, in *Sec. 2.3.1*; the natural evolution of humans to observe things with their naked eyes in the state of nature, in *Sec. 2.4.1*; the natural evolution of humans to observe things in the world in the state of nature, in *Sec. 2.4.2*; the natural evolution of humans in the state of nature to understand the world from the realm of experience in a primitive way, in *Sec. 2.5.1*; the natural evolution of humans in the state of nature to understand the world by observations in their relatively primitive life, in *Sec. 2.5.2*; the natural evolution of different entities in the history of the universe, in accordance to Sri Aurobindo, in *Sec. 3.2.1*; the natural evolution of humans to interact with nature in a primitive way in the state of nature, in *Sec. 3.2.2*; the natural evolution of humans to observe the universe with awe and incomprehension in the state of nature, in *Sec. 3.3.1*; the natural evolution of humans to live on earth in the state of nature, in *Sec. 3.4.1*; the natural evolution of humans to die naturally in the state of nature, for instance, by reason of old age, hunger, disease, fatal injury, fighting, etc., in *Sec. 3.4.2*; and the natural evolution of humans to live in the state of nature in its primitive way, in *Sec. 3.5.1*).

On the other hand, there is transformation (e.g., the technical transformation of human relation with the universe by way of the invention of systems of thought like Advaita Vedanta, which searches for that blissful state where "there is no difference between Self and God," in *Sec. 2.2.1*; the technical transformation of human communication by way of the invention of metaphysics, like elaborated systems of universals, in *Sec. 2.2.2*; the technical transformation of human ability to observe the world by way of the invention of telescopes, which Hubble in 1929 used to "discover…a correlation between distance and recession velocity—now known as Hubble's law," in *Sec. 2.3.1*; the technical transformation of human observation by the invention of telescopes, in *Sec. 2.4.1*; the technical transformation of human observations about the world by way of the invention of correlative cosmology for "an immense system of correlation-building," in *Sec. 2.4.2*; the technical transformation of the human mode of thinking about the world by way of the systematic invention of different variants of the cosmological argument for the First Cause, in *Sec. 2.5.1*; the technical transformtion of human understanding of the world by way of the invention of metaphysics like body-soul hylomorphism, in *Sec. 2.5.2*; the technical transformation of the existence of entities in the world, like that of self-conscious humans, by way of the invention of esoteric cosmology for a more systematic spritual realization, in *Sec. 3.2.1*; the techical transformation of human understanding of nature by way of the invention of a systematic philosophy and religion known as theosophy, with its analyses of the "seven root races" and the "seven vestures of the soul," in *Sec. 3.2.2*; the technical transformation of human observation at the universe by way of the invention of telescopes and the scientific construction of the Big Bang theory, in *Sec. 3.3.1*; the technical transformation of the human view about living on Earth by way of the invention of the belief system about the end of the Earth and the need to leave it, in *Sec. 3.4.1*; the technical transformation of human death by way of the invention of doomsday deaths, like those commited by MRTCG, in *Sec. 3.4.2*; and the technical transformation of human living by the the invention of Buddhist eschatology for an ascetic way of life, in *Sec. 3.5.1*).

Again, it should be stressed, however, that there are different shades of gray (or different degrees of truth) in the two opposites, and the classification is not necessarily mutually exclusive either. In fact, even when some combinations of the two occur, they only end up sharing the same dialectic relationship, but in a different degree.

Also, there are some other relationships (as a kind of family resemblance) which have something in common with the principle, but they are not exactly the same but only more or less comparable, which varies from case to case.

Good examples of family resemblance in relation to the principle are inorganicness-volition, nonwillingness-willingness, naturalness-technologicalness, naturalness-nonnaturalness, and so on.

12th Thesis: The Symmetry-Asymmetry Principle

The twelveth thesis is labeled as the symmetry-asymmetry principle (on the relationships among existents) in the category of "agency" in existential dialectics (which was also already worked out in my previous books), in that there is no asymmetry without symmetry—and vice versa.

For instance, the Same can be symmetric and asymmetric towards the Others. But in case of asymmetry (as analyzed in *BDPD*), oppression and self-oppression can occur. So, when the Same is asymmetric towards the Others, the Same can also be relatively asymmetric towards itself in self-oppression, just as the Others can be likewise towards themselves. The subsequent oppressiveness is dualistic, as much by the Same against the Others and itself, as by the Others against the Same and themselves. Both oppression and self-oppression can be achieved by way of downgrading differences between the Same and the Others and of accentuating them.

This is true, even though not all forms of asymmetry have to be about oppression and self-oppression.

In addition, from Chapter Three of *FPHG*, symmetry is not perfect, to be understood in an approximate sense under many life circumstances. With this clarification in mind, hereafter are some empirical examples.

For instance, on the one hand, there is symmetry (e.g., the co-existence of different Vedanta schools in the history of Hindu philosophy, in *Sec. 2.2.1*; the co-existence of different schools of thought about the problem of universals through the ages, in *Sec. 2.2.2*; the co-existence of different cosmological models in the history of cosmology to understand the beginning of the universe, be they about the Big Bang theory and its pre-Big Bang alternatives, in *Sec. 2.3.1*; the co-existence of different views of the universe in human history, like geocentrism and heliocentrism, in *Sec. 2.4.1*; the co-existence of different versions of correlative cosmology in Chinese history, be they about *Wuxing*, *Sifang*, and the like, in *Sec. 2.4.2*; the co-existence of different views about the existence of the universe over the ages, in *Sec. 2.5.1*; the co-existence of different views about the existence of the world in human history, in *Sec. 2.5.2*; the co-existence of different spiritual schools of thought in India, regardless of whether it is about the original Vedas in antiquity or else, in *Sec. 3.2.1*; the co-existence of different groups in the theosophical movement over time, regardless of whether they are located in India, Germany, or in the U.S., in *Sec. 3.2.2*; the co-existence of different scenarios about the ultimate fate of the universe in the history of cosmology, in *Sec. 3.3.1*; the co-existence of different beliefs about the end of the world through the ages, in *Sec. 3.4.1*; the co-existence of different versions of Christianity over time, in *Sec. 3.4.2*; and the co-existence of different religious versions of eschatology, in *Sec. 3.5.1*).

On the other hand, there is also asymmetry (e.g., the popularity of Advaita Vedanta among those who accept the teachings of Adi Shankara—but the acceptance of other sub-schools of Vedanta by those who follow different teachings of Dvaita Vedanta and others, in *Sec. 2.2.1*; the popularity of Platonic realism in ancient Greece—but the acceptance of conceptualism in the Middle Ages, especially with the works of Abelard and Ockham, in *Sec. 2.2.2*; the domiannce of the Big Bang theory, especially since "the discovery and confirmation of the cosmic microwave background radiation in 1964"—but the rise of pre-Big Bang theories in our time, especially in the last two decades, in *Sec. 2.3.1*; the popularity of geocentrism in antiquity until the 16th century or so—but the increasing acceptance of heliocentrism from the late 16th century onward, in *Sec. 2.4.1*; the popularity of *Sifang* durng the Bronze Age—but the acceptance of *Wuxing* "from the fourth to third centuries B.C., during the transition from the Bronze Age to the imperial era," in *Sec. 2.4.2*; the popularity of the cosmological argument by Aquinas in the medieval era—but the acceptance of the Big Bang theory in the modern era by scientists, in *Sec. 2.5.1*; the popularity of hylomorphism in Greek antiquity and medieval Christendom—but the rise of modern science like physics to explain the working of the universe, without the old idea of "substantial forms," etc., in *Sec. 2.5.2*; the popularity of the original Vedas in ancient India and also later—but the more acceptance of the version by Sri Aurobindo among his followers in modern time, in *Sec. 3.2.1*; the popularity of the Order of the Star in the East, the eastern section of the Theosophical Society, in India—but the more acceptance of the Anthroposophical Society in Germany, in *Sec. 3.2.2*; the acceptance of the static universe by Einstein and his contemporaries in the old days—but the acceptance of the expanding universe by many astrophysicists nowadays, in *Sec. 3.3.1*; the acceptance of the beliefs about the end of the world by members of the traditional Christian ideas of "salvation and apocalypse"—but the acceptance of Applewhite's version of the end of the world by members of Heaven's Gate, in *Sec. 3.4.1*; the acceptance of Catholicism by the Roman Catholic Church—but the worship of MRTCG version of Christianity as written in its official booklet *A Timely Message from Heaven*, by the Movement members, in *Sec. 3.4.2*; and the acceptance of Christian eschatology in Western Europe—but the popularity of Hindu eschatology in India, in *Sec. 3.5.1*).

Again, for the last time, it should be stressed, however, that there are different shades of gray (or different degrees of truth) in the two opposites, and the classification is not necessarily mutually exclusive either. In fact, even when some combinations of the two occur, they only end up sharing the same dialectic relationship, but in a different degree.

Also, there are some other relationships (as a kind of family resemblance) which have something in common with the principle, but they are not exactly the same but only more or less comparable, which varies from case to case.

Good examples of family resemblance in relation to the principle are strength-weakness, balance-extremity, harmoniousness-unharrmoniousness, potence-impotence, reflexiveness-unreflexiveness, equality-inequality, and something like that.

13th Thesis: The Regression-Progression Principle

The thirteenth thesis refers to the regression-progression princple (on the direction of history) in the category of "outcome" in existential dialectics (which was also already worked out in my previous books), in that neither the cyclical nor the linear views are adequate for explaining many phenomena at all levels. There is no regression without progression—and vice versa.

History progresses to more advanced forms, but with a regressive touch. Examples include no freedom without unfreedom, no equality without inequality, and no civilization without barbarity. This is not an inevitable law, but merely a highly likely empirical trend.

For instance, on the one hand, there is regression (e.g., the regression made in Advaita Vedanta, as shown by the criticisms against its claims, in *Sec. 2.2.1*; the regression made in idealism, as shown in the criticisms as made by its criticcs, like the realists and the nominalists, in *Sec. 2.2.2*; the regression made by the Big Bang theory in regard to some serious problems, as pointed out by the critics, in *Sec. 2.3.1*; the regression made by heliocentrism at the time of Galileo, which "not only…irritate[d]…scholastic philosophers and theologians, but also…harm[ed] the Holy Faith by rendering Holy Scripture false" and was not even backed up with "necessary conclusive proof," as pointed out by Bellarmine, in *Sec. 2.4.1*; the problems in *Wuxing* cosmology, as pointed out by the critics, with their different interpretations, in *Sec. 2.4.2*; the regression made by the cosmological argument, as shown by the critics in terms of its major problems, in *Sec. 2.5.1*; the regression made by the Aristotelian teleological view of the universe, as shown in the criticisms by the critics, in *Sec. 2.5.2*; the serious problems in the idea of the cosmic return to the Absolute for spiritual realization, as shown in the criticisms which question its assumptions and claims, in *Sec. 3.2.1*; the regression made by the Theosophical Society, like the problems about its claims concerning the "inferiority" of non-Western races and ethnic groups, as pointed out by the critics, in *Sec. 3.2.2*; the regression made by the three scenarios, with the Big Bang theory as the scientific basis, about the ultimate fate of the universe—as shown in the remaining problems which the critics are quick to point out, in *Sec. 3.3.1*; the regression made by the belief system about the end of the Earth, as shown by the tragedy of Heaven's Gate, in *Sec. 3.4.1*; the regression made by the Christian ideas of salvation and apocalypse, as shown in the tragedy of doomsday deaths in MRTCG, in *Sec. 3.4.2*; and the regression made by Christian eschatology, like the tragedy of Heaven's Gate, in *Sec. 3.5.1*).

On the other hand, there is progress (e.g., the progress made in Advaita Vedanta for the better understanding of the nature of suffering inherent in the obsession with the "I," in *Sec. 2.2.1*; the progress made in idealism for the better understanding of why universals do not exist in the external world, in *Sec. 2.2.2*; the progress made by the Big Bang theory to help us understand better the beginning of the universe, at least "up to 10^{-43} seconds after the Big Bang" in the Planck epoch, in *Sec. 2.3.1*; the progress made by heliocentrism, like the one by Galileo, to help us understand better why the Earth is not the center of the universe, in *Sec. 2.4.1*; the progress made in correlative cosmology to help us understand some correlations in category between the natural world and the human world, in *Sec. 2.4.2*; the progress made by the cosmological argument in raising anew the unresolved question concernng the First Cause of the universe, in *Sec. 2.5.1*; the progress made by the Aristotelian teleological view of the universe to help us understand the virtue of realizing the potential of things in the world, in *Sec. 2.5.2*; the progress made by the idea of the cosmic return to the Absolute for a better understanding of the virtue of seeking spiritualtiy for the human mind in the evolution of the universe, in *Sec. 3.2.1*; the progress made by the Theosophical Society for the possibility of the Universal Brotherhood of Humanity, in *Sec. 3.2.2*; the progress made by the three scenarios about the ultimate fate of the universe in enhancing the understanding of the likely possibility that the universe is accelerating in its expansion, in *Sec. 3.3.1*; the progress made by the tragedy of Heaven's Gate, in that people are more aware of the "authoritarian political and religious outlook" of cults as described by Jacques Vallée in his 1979 *Messengers of Deception*, in *Sec. 3.4.1*; the progress made by the tragedy of doomsday deaths in MRTCG, in that it makes others more aware of the danger of blind worship in doomsday cults, in *Sec. 3.4.2*; and the progress made by Christian eschatology for the living of a virtuous life in practice, in *Sec. 3.5.1*).

Again, it should be stressed, however, that there are different shades of gray (or different degrees of truth) in the two opposites, and the classification is not necessarily mutually exclusive either. In fact, even when some combinations of the two occur, they only end up sharing the same dialectic relationship, but in a different degree.

Also, there are some other relationships (as a kind of family resemblance) which have something in common with the principle, but they are not exactly the same but only more or less comparable, which varies from case to case.

Good examples of family resemblance in relation to the principle are cost-benefit, undesirability-desirability, badness-goodness, risk-opportunity, and so forth.

14th Thesis: The Same-Difference Principle

The fourteenth thesis is referred to as the same-difference princple (on the metamorphosis of change) in the category of "outcome" in existential dialectics (which was worked out in *ALD* and other books of mine), in that an entity, as it evolves over time, can be both different from and similar to its opposing alternatives and does not have to be solely more different from them over time. There is no similarity without difference—and vice versa.

Opposites are not absolute in a black-or-white fashion; so, an entity can become relatively more similar to (or more different from) its opposite over time.

This is further constrained by another principle, that is, the symmetry-asymmetry principle about the relationships among existents under the category about agency in existential dialectics, in that if there is symmetry (equality) between two entities, there is likewise asymmetry (inequality) emerging in a different way.

For instance, on the one hand, there is similarity in outcome (e.g., the contribution to a happy life, regardless of whether this is done by way of a spiritual quest or by way of a material pursuit, in *Sec. 2.2.1*; the contribution to the understanding of the world, be this done with the use of universals or the rejection of them, in *Sec. 2.2.2*; the contribution to the understanding of the beginning of the universe, be it by way of the Big Bang theory or the pre-Big Bang alternatives, in *Sec. 2.3.1*; the contribution to our understanding of the world, be this so by way of heliocentrism or geocentrism, in *Sec. 2.4.1*; the contribution to the better understanding of correlative cosmology, regardless of the existence of different interpretations, in *Sec. 2.4.2*; the contribution to the better understanding of the origins of the universe, be this so by way of the cosmological argument or by the Big Bang theory, in *Sec. 2.5.1*; the contribution to the understanding of the world, regardless of whether it is by way of a teleology of the universe or an impartial scientific viewpoint, in *Sec. 2.5.2*; the contribution to the enhancement of life, regardless of whether it is done by spirituality or by materiality, in *Sec. 3.2.1*; the contribution to the better understanding of the spiriutal evolution of the universe, regardless of whether this is so by way of the cosmic return to the Absolute or the second coming of Christ, in *Sec. 3.2.2*; the contribution to the understanding of the ultimate fate of the universe, regardless of whether this is done by way of the three scenarios with the Big Bang theory as the basis, or of the multiverse theory, in *Sec. 3.3.1*; the contribution to how life is to be lived, regardless of whether this is done by strict asceticism or sensual materialism, in *Sec. 3.4.1*; the contribution to the regulation and control of human lifestyle, regardless of whether this is achieved by an ascetic form of existence or by a sensual form of existence, in *Sec. 3.4.2*; and the contribution to the regulation and control of human life, regardless of whether this is done by the use of religious faith as in religious eschatology, or by the use of scientific knowledge as in the Big

Bang theory and its prediction about the ultimate fate of the universe, in *Sec. 3.5.1*).

On the other hand, there is difference in outcome (e.g., the contribution to a happy life by way of the spiritual quest for salvation—but the contribution to a happy life by way of the craving for enjoyments in the material world, in *Sec. 2.2.1*; the contribution to the understanding of the world by way of the use of universals as concepts, with implications for Ockham's razor—but the contribution to the understanding of the world by way of the rejection of universals as adequately desccribing the Brahman, which is prior to the creation of the universe, for those in Advaita Vedanta, in *Sec. 2.2.2*; the contribution to the understanding of the beginning of the universe "up to 10^{-43} seconds after the Big Bang" by way of the Big Bang theory—but the contribution to the understanding of the beginning of the universe prior to the Big Bang by way of the pre-Big Bang cosmological models, in *Sec. 2.3.1*; the contribution to our understandiing of the world by geocentrism from a religious standpoint in accordance to the Scripture—but the contribution to our understanding of the world by heliocentrism from a scientific standpoint in accordance to observations with telescopes, in *Sec. 2.4.1*; the contribution to the better understanding of correlative cosmology by way of identifying its forgers—but the contribution to the better understanding of correlative cosmology by way of linkage its role to empire building, in *Sec. 2.4.2*; the contribution to the better understanding of the origins of the universe, by way of the cosmological argument in terms of the First Cause—but the contribution to the better understanding of the origins of the universe, by way of the Big Bang theory in terms of a Big Bang, and nothing before that, in *Sec. 2.5.1*; the contribution to the understanding of the world by way of the Aristotelian emphasis on the importance of full actualization of potential—but the contribution to the understanding of the world by way of an impartial scientific analysis of "different properties and forces" as in modern science, for instance, without any normative appeal to a universal telos, in *Sec. 2.5.2*; the contribution to the enhancement of life, by way of spirituality as in esoteric cosmology—but the contribution to the enhancement of life, by way of materiality as in capitalism, in *Sec. 3.2.1*; the contribution to the better understanding of the spiriutal evolution of the universe, by way of the cosmic return to the Absolute as in the eastern section of the Theosophical Society—but the contribution to the better understanding of the spiriutal evolution of the universe, by way of the second coming of Christ, as in the Anthroposophical Society, in *Sec. 3.2.2*; the contribution to the understanding of the ultimate fate of the universe by way of the three scenarios with the Big Bang theory as the basis—but the contribution to the understanding of the ultimate fate of the universe by way of the multiverse theory, such that the universe will not end completely, strictly speaking, in *Sec. 3.3.1*; the contribution to how life is to be lived, by way of strict asceticism like the one in Heaven's Gate—but the contribution to how life is to be lived, by way of sensual materialism like the one in the Playboy Club, in *Sec. 3.4.1*; the contribution to the regulation and control of human lifestyle, by way of an ascetic form of existence—but the contribution to the regulation and control of human lifestyle, by way of a sensual form of existence, in *Sec. 3.4.2*; and the contribution to the regulation and control of human life, by way of a scientific lifeworld lived by many scientists supporting the Big Bang theory and its prediction about the ultimate fate of the universe—but the contribution to the regulation and control of human life, by way of a lifeworld guided by religious faith as practiced by those in religious eschatology, in *Sec. 3.5.1*).

Again, it should be stressed, however, that there are different shades of gray (or different degrees of truth) in the two opposites, and the classification is not necessarily mutually exclusive either. In fact, even when some combinations of the two occur, they only end up sharing the same dialectic relationship, but in a different degree.

Also, there are some other relationships (as a kind of family resemblance) which have something in common with the principle, but they are not exactly the same but only more or less comparable, which varies from case to case.

Good examples of family resemblance in relation to the principle are homogeneity-heterogeneity, we-they, and so forth.

15th Thesis: The Post-Human Odyssey

The fifteenth thesis is about the role of "post-humans," which I originally proposed in my previous books, starting with the first book titled *The Future of Human Civilization* in 2000 and all others afterwards.

As already pointed out in *Sec. 1.8*, I need to emphasize, as this is something that I used to repeat (*almost verbatim*) from my previous books, two clarifications here about the term "post-human" as a neologism in my works.

Firstly, the word "post-human" here should *not* be confused with another term which looks similar but has a totally different meaning in the literature of postmodernism, namely, "post-humanism"—which constitutes a critique of "humanism" as tradionally understood (especially, though not exclusively, in relation to the idea of progress in science and reason in the Enlightenment project). (WK 2008)

My works reject the project of "postmodernism" and propose the future world of what I orignally called "after-postmodernity" in *FHC* and *FCD*, for instance.

And secondly, the word "post-human" here should also *not* be confused with a similar term which is used to champion the ideology of technology for the future co-existence between humans and cyborgs in "trans-humanism." (WK 2008a)

Instead, my term "post-human" in relation to "posthuman-ism" also rejects "transhumanism" (especially, though not exclusively, in relation to the promises of technology) and refers to something else altogether, that is, the future extinction of humans and its post-human successors in deep space and beyond unto multiverses.

Later, I also went to great lengths in *BEPE* to provide a critique of "trans-humanism."

With these two clarifications in mind (as summarized in *Table 1.31*)—the post-human odyssey (in the context of cosmology) can be analyzed in relation to five main directions, based on the theory of post-cosmology as analysed in this book (and also on the visions as already worked out in my previous books).

Firstly, the "post-human" odyssey (in the context of cosmology) will take guidance from the theory of post-cosmology in relation to the context of the contested beginnings of the universe (and the aftermath), as well as the evolution of the universe and its speculative ends.

Secondly, the "post-human" odyssey (in the context of cosmology) will also take guidance from the theory of post-cosmology in relation to the four perspectives of the mind, nature, society, and culture.

Thirdly, the "post-human" odyssey (in the context of cosmology) will also take guidance from the theory of post-cosmology in regard to not heavily favoring any specific theory over others in the literature, nor trying to integrate them (as they are not necessarily compatible with each other).

Fourthly, the "post-human" odyssey (in the context of cosmology) will have to confront the new challenges as posed by the long-term civilizational development of intelligent life in the distant future, both here on this planet Earth and elsewhere in deep space until multiverses.

As this is something that I regularly pointed out in my previous books for background information (and summarize here, *almost verbatim*), I already worked out, in my numerous books, what these new challenges will be and provide different original visions to meet them—especially, in regard to the future evolutions of the mind (e.g., "the contrarian personality," "the hyper-martial body," "the hyper-sexual body," "floating consciousness," "hyper-spatial consciousness," "unfolding unconsciousness," "thinking machines," "thinking robots," "genetically altered superior beings," "cyborgs," and others), of nature (e.g., "post-cosmology," "the alteration of space-time," "the creation of new matter-energy," "selective geometry," and the like), of society (e.g., "multifaceted war and peace," virtual organizations, different versions of "post-capitalism" and "post-democracy," the movement of "cyclical progression" at both structural and systemic levels, "ambivalent technology," and so forth), and of culture (e.g., the rise of "post-civilization," "transformative aesthetic experience," "contrastive mathematical-logic," and the whatnot) in history (e.g., the age of "after-postmodernity"), in the context of my approaches in relation to methodology (e.g., "sophisticated methodologial holism," "critical-dialectic formal science," etc,) and ontology (e.g., "existential dialectics," "contrastive rationality").

Of course, the examples (as listed above) are not exhaustive but illustrative, since my numerous books have worked out many other visions (as already roughly summarized in *Sec.1.8* and *Sec. 1.9*). Many, though not all, of my visions on the mind, nature, society, culture, history, methodology, and ontology (as cited above) are summarized in the tables as shown in Chapter One and Chapter Four (especially in *Table 4.43* about my original theories on numerous topics).

And finally, or fifthly, the "post-human" odyssey (in the context of cosmology) will also take hint from the wisdom of existential dialectics (as revealed by the 21 ontological principles) in regard to the need to go beyond both the contested beginnings and the speculative ends of the universe, in a small sense, and cosmology, in a broad one (as further elaborated in the next section).

Towards Post-Cosmology

This "post-human" odyssey (in the context of cosmology) to transcend both the contested beginnings and the speculative ends of the universe, in a small sense, and cosmology, in a broad one, has profound implications for more important questions which have troubled the best and the brightest minds through the ages—Does the universe have a beginning? What happened before that beginning? Does the universe have an end? And what will happen after that end?

If the conclusion of this book is that the contested beginnings and the speculative ends of the universe are neither possible nor desirable to the extent that the respective spokespersons in all human history hitherto existing would want us to believe—then this conclusion can be carried to its logical completion.

Two aspects of this logical completion can be summarized hereafter.

Firstly, the different contested beginnings and the speculative ends of the universe are as problematic as the opposite views that the universe has an infinite past (like an infinite temporal regress of events) and an infinite future (like the cosmic cycling).

For instance, those (in Chapter Two) who argue for initial cosmologies (that there was once a beginning of the universe) need to answer the follow-up question of why the universe had to begin like that or what really happened before that—just as those (in Chapter Three) who argue for the end of the world need to answer the follow-up question of why the universe will have to end like that or what will happen after that end.

Conversely, those (in Chapter Two) who argue for eternal cosmologies (that there was no beginning of the universe at all, if traced further back to space and time) also need to answer the follow-up question of why there is this infinite regress of causation or what exactly this infinity is supposed to refer to—just as those (in Chapter Three) who argue against the end of the world also need to answer the follow-up question of why there is this infinite projection of the world (or the cosmic cycling) or what exactly this infinity is supposed to refer to.

After all, in *The Future of Post-Human Geometry*, I already went to great lengths to explain the problematic idea of infinity itself (which may not exist, although it can be used for mathematical convenience) and instead proposed the alternative idea of "transfinity" (as shown in the finiteness-transfiniteness principle).

In the end, the important point to remember here is to not favor any one side of the debate at the expense of the other but to teach us the opposing sides of the debate, such that reality is not to the extent that each respective side would like us to believe.

And secondly, in the long run of civilizational development for intelligent life, both on earth and, later, even more importantly, in deep space unto multiverses—the very field of study known as "cosmology" need to be transcended, since it will no longer be suficient to just study the "cosmos, the world," or "the universe."

Instead, the post-human odyssey will also open up a broader horizon of gazing at reality, not merely in relation to "the universe" but, even more fundamentally, in relation to different worlds (in different multiverses)—as this is something that I had already gone to great lengths to address in the context of my different visions, especially in my first two-volume work titled *The Future of Human Civilization* in 2000 (and all other books of mine afterwards).

This will be the beginning of what I originally propose here as the rise of *post-cosmology* (for the study of different worlds in different multiverses) in the post-human future.

Post-cosmologists will try to answer questions which cosmologists in all human history hitherto existing are not able to, due to the relatively primitive technological level which has so long delimited the human understanding of other worlds.

Good examples of these questions include: How can intelligent life cross from one universe to another, with each having its own space and time, its own matter and energy, its own physical laws, and its own lifeforms? How can intelligent life move into different dimensions of space-time and still survive?

My numerous visions (in my numerous books)—regarding "floating consciousness," "hyper-spatial consciousness," "unfolding unconsciousness," the "alteration of space-time," the "creation of new matter-energy," "selective geometry," different versions of "post-capitalism" and "post-democracy," the movement of "cyclical progression," "ambivalent technology," the rise of "post-civilization," "contrastive mathematical-logic," the age of "after-postmodernity," and so on—complicate other questions which post-cosmologists have to confront, when contextualized with the perspectives of the mind, nature, society, and culture.

Welcome, therefore, to the different post-human worlds in post-cosmology.

Table 4.1. Sophisticated Methodological Holism (Part I)

• "My methodological holism implies the partiality-totality principle in the ontology of existential dialectics (see the table on the partiality-totality principle for summary), which is against the varieties of (a) reductionism and (b) reverse-reductionism, in relation to (i) concept, (ii) theory, (iii) methodology, and (iv) ontology." (*FC*)

• "[M]y methodological holism here is not opposed to methodological individualism but includes it (and, for that matter, other methodologies too)…." (*FPHC*) For this reason (and others too, as summarized hereafter), my version of methodological holism is sophisticated—not vulgar as sometimes used by inapt scholars using the same term. (*FC*)

• "[M]y methodological holism does not democratically presume that all levels are equally valid, as all levels are not created equal. In other words, in relation to issue X, level A may be more relevant than level B, but in relation to Y, level B can be more relevant than level A instead." (*FPHC*) One excellent example of this vulgar democratic presumption is what I called in *BNN* "the compromise fallacy." (*FC*)

• My methodological holism does not presume that a lower level of analysis is more important than a higher level, solely because the former serves as the foundation for the latter—and vice versa, for that matter. One excellent example of this reductionistic presumption is what I called in *FPHST* "the foundation fallacy." (*FPHST, FC*)

(continued on next page)

Table 4.1. Sophisticated Methodological Holism (Part II)

• "[M]y methodological holism does not make any a-priori postulation that there must be a definite (and, for that matter, indefinite) number of levels" in any analysis. (*FPHC*) Nor does it dogmatically require that there must be a certain combination of levels of analysis in a given inquiry. (*FC*)

• "[M]ethodological holism, in my usage, does not assume that all levels…can necessarily be integrated, since methodological holism is not aimed to search for the holy grail of 'an integral theory…' (as is the case for Wilber). In other words, it allows that sometimes some levels may experience irreducible gaps between them, to be understood, at best, as empirical correlations, not as causal relations…." (*FPHC*)

• "[D]ifferent levels may overlap and even interact with each other in a given context (but sometimes may not), and the fact that I even proposed different ways of re-classifying the levels (whenever needed) in *FDC* reinforces this point….The dual danger here is either forcefully making different levels interact when they are just different (or, metaphorically speaking, apples and oranges) or inappropriately ignoring their interactions when some situations instead require them." (*FPHC*)

• "[T]o understand different levels from their own (unique) perspectives (as required by my methodological holism) is not the same as trying to reduce them to a preferred level in the process of learning from other levels. This second kind of multidisciplinary work is not genuine and does no justice to the unique complexities and merits inherent at each level." (*FPHC*)

(continued on next page)

Table 4.1. Sophisticated Methodological Holism (Part III)

• "[My] methodological holism walks a fine line between the artificial classification (separation) of levels and the simultaneous incorporation of them, if only for the sake of human scholarly endeavor. It should be reminded that nature does not impose upon itself the academic classification of the levels of analysis as humans have. The enterprise of classification is therefore anthropocentric." (*FPHC*)

• "[M]y methodological holism advocates neither epistemic subjectivism nor epistemic non-subjectivism (e.g., realism, idealism, and historicism), neither epistemic relativism (e.g., subjectivism, historicism) nor epistemic absolutism (e.g., realism, positivism), neither epistemic reductionism nor epistemic emergencism, and neither epistemic objectivism (e.g., realism, idealism) nor epistemic historicism....Neither does methodological holism, in my usage, accept the false meta-conceptual dichotomy between nominalism and realism....These false dichotomies...are to be transcended. In other words, methodological holism does not fully accept epistemic realism, positivism (a form of epistemic idealism), historicism, subjectivism, and reductionism in epistemology and philosophy of science but learns from the strengths and weaknesses of all of the opposing approaches without siding with any of them...." (*FPHC*)

• "Sophisticated methodological holism is subject to the constraints as imposed by the syntax of existential dialectics (e.g., the partiality-totality principle and the predictability-unpredictability principle). Even in predictability, outcomes are subject to uncertainty, the degree of which varies from case to case." (*FC*)

(continued on next page)

Table 4.1. Sophisticated Methodological Holism (Part IV)

• "Sophisticated methodological holism—when applied, especially though not exclusively, as illustrated in my numerous works—can enrich the understanding of reality in some distinctive ways. Here are three examples (as revealed in each of my books)." (*FPHCT*; *FPHL*)

—"Firstly, it provides a comprehensive analysis of a subject matter, from which much can be learned about reality," "both in relation to the perspectives of the mind, nature, society, and culture and also in relation to a new classification of the subject matter."

—"Secondly, it suggests some visions of the future in relation to the subject matter in question."

—"And thirdly, it proposes some insights on meta-theory (e.g., methodology and ontology) in general—with the clear understanding, however, of the dilemma of specific vs. general ontology (as shown in the table on the syntax of existential dialectics in the context of the dilemma of ontology). For this very reason, all of these ways are important, without reducing one into the analysis of another."

• "In the end, my meta-theory (both sophisticated methodological holism and existential dialectics) serves as a foundation to unify all domains of knowledge into *an unified theory of everything* (by way of some ontological principles and the comprehensive perspectives of the mind, nature, society, and culture). This is so, without commiting the sins of reductionism and reverse-reductionism (as often seen in many holistic approaches, with the fad of sysems approach—be it about systems theory, chaos theory, complexity theory, or else—as a most recent notorious example, which I debunked in *The Future of Complexity* and also in *The Future of Post-Human Formal Science*). My distinctive approach makes good use of different schools of thought without favoring any of them nor trying to integrate them (as they are not necessarily compatible with each other), so as to adjust for subjectivity, diversity, conflict, and complexity, for example. In this sense, the word "unified" does not have to mean integrative approach, in a narrow sense, nor systems approach, in a broad one, both of which I reject while learning from them." (*FPHFS*; *FPHS*)

Sources: A summary of *Sec.1.2* in *FPHC*—and also from *BNN*, *FPHST*, *ALD*, *FC*, *FPHCT*, and the rest of all other books of mine. See the books for more detail.

Table 4.2. On Reductionism and Reverse-Reductionism (Part I)

• **The Partiality-Totality Principle**
—The partiality-totality principle in the ontology of existential dialectics targets against the varieties of reductionism and reverse-reductionism (as already worked out in my previous books).

• **Against the Varieties of Reductionism**
—*Conceptual Reductionism*
- Some illustrative instances involve myriad dualities like mind vs. body, self vs. world, democracy vs. non-democracy, and the like (as already addressed in *FHC*, *FPHC*, and *BDPD*, for instance).

—*Theoretical Reductionism*
- A fascinating case study concerns what I originally called "the foundation fallacy" in *FPHST*, in any attempt to naively understand space-time from the physical perspective as the foundation and, consequently, to dangerously dismiss other perspectives.
- In *FAE*, I elaborated further these versions of reductionism in the literature on aesthetics (e.g., form vs. content, representation vs. expression, critics vs. artists, and externalism vs. internalism).
- In *FIA*, I revealed other forms of reductionism in the literature on information architecture (e.g., the constructivist argument).
- In *FPHU*, I showed the persistent legacy of reductionism, this time, in the literature on anomalous experience (e.g., the obsession with physics, chemistry, and biology for explaining anomalous experience).
- In *FPHE*, I examined another case of reductionism in action, in the context of engineering (e.g., technical constraints vs. normative constraints).
- In *FPHMM*, I elaborated one more version of reductionism, in relation to the three domains of communication (e.g., the competing views on sending, connecting, and receiving).
- In *FPHCT*, I explored another version of reductionism, in relation to invention and innovation (e.g., the bio-psychological argument vs. the socio-cultural arguments).

(continued on next page)

Table 4.2. On Reductionism and Reverse-Reductionism (Part II)

- **Against the Varieties of Reductionism *(cont'd)***
  —*Theoretical Reductionism (cont'd)*
    - In *FPHG*, I identify another version of reductionism, in relation to infinity, symmetry, and dimensionality (e.g., the Euclidean argument vs. the non-Euclidean arguments).
    - In *FPHUP*, I analyzed another version of reductionism, in relation to density and void (e.g., the engineering argument and the ecology argument).
    - In *FPHL*, I examined another version of reductionism, in relation to structure and context (e.g., the structuralist argument and the contextualist argument).
    - In *PFHO*, I explored another version of reductionism, in relation to communcation, decision-making, and leadership (e.g., the rational-system argument vs. the natural-system argument).
    - In *PFHMA*, I revealed another version of reductionism, in relation to the martial body and spirit (e.g., the spiritual argument vs. the materialist argument vs. the defensive argument).
    - In *PFHS*, I analyzed another version of reductionism, in relation to the sexual body and spirit (e.g., the naturalist argument vs. the constructivist argument).
    - In *PFHLAW*, I examined another version of reductionism, in relation to law (e.g., the necessity argument vs. the contengency argument).
    - In *FPHWP*, I showed another version of reductionism, in relation to war and peace (e.g., the aggressivist argument vs. the pacifist argument).
    - In *BEPE*, I analyzed another version of reductionism, in relation to morality and immorality (e.g., the objectivist argument vs. the non-objectivist argument vs. the skeptical argument).
    - And in *BCOS*, I examine another version of reductionism, in relation to the contested beginnings and speculative ends of the universe (e.g., the scientific argument vs. the religious argument vs. the esoteric argument vs. the metaphysical argument).

(continued on next page)

Table 4.2. On Reductionism and Reverse-Reductionism (Part III)

- **Against the Varieties of Reductionism** ***(cont'd)***
  —*Methodological Reductionism*
  - A good instance concerns the debate between different versions of qualitative and quantitative methods (as already analyzed in *FC* and also *FHC*). In *FPHML*, I examined similar reductionism, this time, in the literature on mathematical logic (e.g., the obsession with consistency, soundness, and completeness). And in *FPHFS*, I also explored the problems of reductionism in the context of formal science (e.g., the analytical argument).

  —*Ontological Reductionism*
  - An excellent example is the debate between emergentism and reductionism in complexity theory and also in psychology (as elaborated in *FPHC*, in the context of Being and Becoming).

- **Against the Varieties of Reverse-Reductionism**
  —*Conceptual Reverse-Reductionism*
  - Any concept of "art" (e.g., fine art, cave art, outsider art, junk art) is deemed acceptable in postmodernism (as already addressed in Ch.4 of *FHC*).

  —*Theoretical Reverse-Reductionism*
  - There are numerous art and literary theories co-exist. Take the case of literary studies, as there are now Literary Structuralism, Marxist Literary Criticism, New Criticism, Phenomenology, Hermeneutics, Language-Game Literary Criticism, Feminist Literary Criticism, Reception Theory, Reader Response Criticism, Poststructuralism, Semiotics, Pyschoanalytic Literary Criticism, just to cite some well-known ones, with no one being said to be better than any others (as detailedly analyzed in Ch.4 of *FHC*). (S. Raman 1997) In *BNN*, I even introduced "the compromise fallacy" as another good illustration of theoretical reverse-reductionism, in misleadingly treating both genetic and environmental approaches as equally valid.

(continued on next page)

Table 4.2. On Reductionism and Reverse-Reductionism (Part IV)

- **Against the Varieties of Reverse-Reductionism *(cont'd)***
  —*Methodological Reverse-Reductionism*
  - There is the "anything-goes" mentality in postmodernism (e.g., doing art without praxis, doing art with praxis, and doing art by sublation), as analyzed in Ch.4 of *FHC*. And in *FPHFS*, I also exposed the problems of reverse-reductionism in the context of formal science in relation to systems theory.

  —*Ontological Reverse-Reductionism*
  - There are likewise no privileged ontology, and the door is open for anything in postmodernism (e.g., the equal status of the ontology of Being vs. that of Becoming, as already addressed in Ch.4 of *FHC*—and also in *FPHC)*. In *FAE*, I also introduced another version of reverse-reductionism, that is, "the pluralist fallacy," in the context of understanding aesthetic experience, for instance—although this fallacy has been committed not exclusively in relation to the ontological level (but also at the conceptual, theoretical, and methodological ones).

Sources: From my previous books.

Table 4.3. The Concepton of Existential Dialectics (Part I)

- **Sets and Elements**
  - —Sets
    - Ex: the Same
    - Ex: the Others
  - —Elements
    - Ex: whites in 20th century America (in the set of "the Same")
    - Ex: Iraq during the U.S. invasion in 2003 (in the set of "the Others")

- **Relations, Operations, Functions**
  - —Relations (e.g., "belongs," "equals to," "is greater than")
    - Ex: symmetric interactions within the Same (or the Others)
    - Ex: asymmetric interactions between the Same and the Others
  - —Operations (e.g., "and," "or," "not," "if…then")
    - Ex: if the Same oppresses the Others, it will also oppress itself.
    - Ex: the Same is not the Others.
  - —Functions (e.g., goals)
    - Ex: the Same is hegemonic in relation to the Others.

- **Truth Values**
  - —"1" if True (in Symbolic Logic)
    - Ex: the proposition that imperial Japan was hegemonic to China during WWII
  - —"0" if False (in Symbolic Logic)
    - Ex: the proposition that Grenada invaded France in 2003
  - —"1" & "0" if Both True and False (in Dialectic Logic)
    - Ex: the proposition that the rabbit-duck picture refers to a duck
  - —"~1" & "~0" if Neither True Nor False (or N/A, in Dialectic Logic)
    - Ex: the proposition that God really exists

(continued on next page)

Table 4.3. The Conception of Existential Dialectics (Part II)

- **Axioms, Postulates, Theorems, Principles**
  - —Axioms
    - Ex: the reflexive axiom—"any quantity is equal to itself"
  - —Postulates
    - Ex: the SSS postulate—"if the three sides of a triangle are congruent to their corresponding parts, then the triangles are congruent"
  - —Theorems (and Principles) in Existential Dialectics
    - In Relation to Method
      - –#1: The formalness-informalness principle
      - –#2: The absoluteness-relativeness principle
      - –#3: The partiality-totality principle
      - –#4: The predictability-unpredictability principle
      - –#5: The explicability-inexplicability principle
    - In Relation to Structure
      - –#6: The finiteness-transfiniteness principle
      - –#7: The preciseness-vagueness principle
      - –#8: The simpleness-complicatedness principle
      - –#9: The openness-hiddenness principle
      - –#10: The denseness-emptiness principle
    - In Relation to Process
      - –#11: The change-constancy principle
      - –#12: The order-chaos principle
      - –#13: The slowness-quickness principle
      - –#14: The expansion-contraction principle
    - In Relation to Agency
      - –#15: The theory-praxis principle
      - –#16: The convention-novelty principle
      - –#17: The evolution-transformation principle
      - –#18: The symmetry-asymmetry principle
      - –#19: The softness-hardness principle
    - In Relation to Outcome
      - –#20: The regression-progression principle
      - –#21: The same-difference principle

(continued on next page)

Table 4.3. The Conception of Existential Dialectics (Part III)

Notes: The categories and examples in each are solely illustrative (not exhaustive). The comparison is also relative (not absolute), nor are they mutually exclusive. As generalities, they allow exceptions.
Sources: Starting from Ch.6 of *BCPC* and also from other books of mine

Table 4.4. The Syntax of Existential Dialectics I: The Principles (Part I)

- **In Relation to Method**

—*The Formalness-Informalness Principle*
(On the Formal Requirements of Logical Systems)

- The formal requirements of a logical system (e.g., consistency, soundness, and completeness) have both usefulness and non-usefulness, to the extent that, if there are formal systems requiring them, there are alternative ones which do not. Thus, it does *not* exclude classical logics but simply goes beyond both classical and non-classical logics, while learning something from each. There is no formalness without informalness—and vice versa.
- For instance, existential dialectics can make use of both classical logics under certain conditions (e.g., especially, though not exclusively, when they are clear-cut, etc.) and non-classical logics under alternative conditions (especially, though not exclusively, when they are "unknown," "irrelevant," "ambiguous," "possible," with "different degrees of truth," empirically inconsistent in a desirable way, etc.). (WK 2008u)
- Family resemblance: e.g., logicalness-nonlogicalness, rationality-nonrationality, etc.
- Sources: From *FPHML*. See also my later books.

—*The Explicability-Inexplicability Principle*
(On the Underlying Mechanisms of Things)

- Both explicability and inexplicability are part of the understanding of things. There is no explicability without inexplicability—and vice versa.
- This principle tells us the duality of the research dilemma, in that, if reality can be explained in some ways, it also has its other ways which are not quite explainable, at a given point in time.
- Family resemblance: e.g., underlyingness-regularness, causation-regularness, causation-correlation, etc.
- Sources: Especially from *FPHU*—and also *FPHC*. See also my later books.

(continued on next page)

Table 4.4. The Syntax of Existential Dialectics I: The Principles (Part II)

- **In Relation to Method** ***(cont'd)***

—The Predictability-Unpredictability Principle (On the Occurrence of Events)

- Both predictability and unpredictability have a major role to play in the occurrence of things, so that neither determinism nor indeterminism wins the centuries-old fight. There is no predictability without unpredictability—and vice versa.
- There are events which are predictable, just as there are those which are not. Or what is regarded as unpredictable at one point in time may turn out to be predictable later, and, conversely, what is deemed as predictable may turn out to not be so predictable. Even in predictability, outcomes are subject to uncertainty, the degree of which varies from case to case.
- Family resemblance: e.g., sureness-arbitrariness, etc.
- Sources: Especially from *FC*. See also my later books.

—The Partiality-Totality Principle (On the Relationships between Whole and Parts)

- The whole is not the sum of the parts. There is no partiality without totality—and vice versa.
- Any inquiry about a phenomenon in the work is to guard against the varieties of (a) reductionism and (b) reverse-reductionism.
- Reductionism and reverse-reductionism can be (i) conceptual, (ii) theoretical, (iii) methodological, and (iv) ontological.
- Family resemblance: e.g., individualisticness-holisticness, partness-wholeness, analysis-synthesis, etc.
- Sources: Especially from *FC*. See also my later books.

(continued on next page)

**Table 4.4. The Syntax of Existential Dialectics I:
The Principles
(Part III)**

• **In Relation to Method** ***(cont'd)***

—*The Absoluteness-Relativeness Principle*
(On the Multiplicity of Things)

- There is the multiplicity of things in reality, be they about entities, qualities (or properties), and relationships, such that what is acceptable from one standpoint may not be so from another. For instance, if there is something absolute, there is likewise something relative. There is no absoluteness without relativeness—and vice versa.
- Both absoluteness and relativeness here are also relevant to different modalities often cited in the literature on ontology, such as possibility (e.g., something "can" happen) and its opposite (e.g., impossibility), probability (e.g., something "will" happen) and its opposite (e.g., improbability), and necessity (e.g., something "should" happen) and its opposite (e.g., contingency).
- Family resemblance: e.g., uniformity-diversity, internalness-externalness, immanence-transcendence, etc.
- Sources: From *FPHK*. See also my later books.

(continued on next page)

Table 4.4. The Syntax of Existential Dialectics I: The Principles (Part IV)

- **In Relation to Structure**

  —*The Finiteness-Transfiniteness Principle*

  (On the Nature of Numbers)

  - If there are finite things, there are likewise transfinite ones. There is no finiteness without transfiniteness—and vice versa.
  - To avoid confusion, my usage of the word "transfinite" here differs radically from the one used by Cantor (and other mathematicians) for "relative" infinity—and is more limited, in light of the problems confronting any attempt to understand the idea of infinity, be it by intuition, imagination, and conception (as detailedly analyzed in *Sec. 2.2.3* of *FPHG*).
  - Instead, by "transfinity," I allow numbers which can be many times larger—or smaller, for that matter—than the finite things that we encounter in daily life, but they do not have to be related to the idea of infinity at all (which may not exist).
  - Of course, there may be some *borderline* cases, in which it is not clear whether the number in question is transfinite (in my usage) or simply a mathematical convenience. A good example of a borderline case is the Planck unit of length for "the smallest space possibly measured in nature," which is "less than billionths of trillionths of trillionths of an inch" (or something like 1.6×10^{-35} meters).
  - That said—my usage of "transfinity" can also resolve (or better, dissolve) an age-old problem in philosophy known as "Zeno's paradoxes" (as already explained in *Sec. 2.2.3* of *FPHG*).
  - Family resemblance: e.g., boundedness-quasiunboundedness, smallness-largeness, microness-macroness, etc.
  - Source: From *FPHG*. See also my later books.

(continued on next page)

Table 4.4. The Syntax of Existential Dialectics I: The Principles (Part V)

- **In Relation to Structure** ***(cont'd)***

  —*The Simpleness-Complicatedness Principle*
  (On the Interconnection among Things)

  - Both simpleness and complicatedness are vital, without favoring one over the other, in that each is utilized, depending on the basis of the perspectives of nature, the mind, culture, and society. And even when a combination of them is preferred, the dilemma is only shifted to a combinational degree of concern. There is no simpleness without complicatedness—and vice versa.
  - In relation to taxonomy, simpleness has its heuristic usefulness, just as complicatedness has its realistic representation, for instance. And in relation to network, simpleness has its economical attractiveness, just as complicatedness has its practical reliability, for instance.
  - Family resemblance: e.g., inflexibility-flexibility, standardization-specialization, imperfectness-perfectness, superficiality-depth, shallowness-deepness, economicalness-elaboratedness, plainness-circumspection, onesidedness-multisidedness, etc.
  - Sources: Especially from *FIA*. See also my later books.

  —*The Denseness-Emptiness Principle*
  (On the Distribution of Entities in Space)

  - Both density and void are needed, in relation to the mind, nature, culture, and society, albeit in different ways. There is no denseness without emptiness—and vice versa.
  - For clarity, the term "void" is used here only as an approximation of emptiness (depending on the degree of the lack of density), since, in physics, it is well known that "empty" space is not really empty all the way, because it can be full of energy (e.g., random quantum fluctuations at the sub-atomic level, and, for that matter, dark energy in the universe) and matter (e.g., different versions of sub-atomic particles, and, for that matter, dark matter in the universe).
  - Family resemblance: e.g., fullness-voidness, nearness-farness, concentration-dispersion, etc.
  - Sources: First worked out in *FPHUP*. See also my later books.

(continued on next page)

Table 4.4. The Syntax of Existential Dialectics I: The Principles (Part VI)

- **In Relation to Structure *(cont'd)***
  —The Preciseness-Vagueness Principle
  (On the Refinement of Things)
  - Both preciseness and vagueness are important, not that one is better than the other, but that both are used, in different degrees of preference, in accordance to the contextual application from the perspectives of nature, the mind, culture, and society. Even when both are used in a combination, the dilemma is shifted instead to one of combinational concern. There is no preciseness without vagueness—and vice versa.
  - In relation to taxonomy, preciseness has its taxonomic clarity, just as vagueness has its classificatory flexibility, for instance. And in relation to network, vagueness has its explorative liberty, just as preciseness has its conceptual definitiveness, for instance.
  - Family resemblance: e.g., clarity-ambiguity, directness-indirectness, quantitativeness-qualitativeness, describability-nondescribability, specificity-obscurity, specificity-generality, thickness-thinness, concretness-abstractness, etc.
  - Sources: Especially from *FIA*. See also my later books.

(continued on next page)

Table 4.4. The Syntax of Existential Dialectics I: The Principles (Part VII)

- **In Relation to Structure** ***(cont'd)***
  —*The Openness-Hiddenness Principle*
  (On the Detection of Things)
  - Reality has its hidden face, just as it is open to outside view in some other ways. There is no openness without hiddenness—and vice versa.
  - For instance, in the context of anomalous experience, certain aspects of reality can be open for examination (e.g., the experiment with SPECT images, the transmission of telepathemic bit, and the use of metonymies and metaphors). Other aspects, however, remain hidden, and examples include the elusive deeper nature of spiritual reality in the intangible realm, the undetectability of different branched universes, the underlying mechanisms of signals traveling between different folds of physical space-time, and the dependence on language and the bias for science and logic).
  - Family resemblance: e.g., overtness-covertness, publicness-privateness, openness-closedness, transparency-secrecy, openness-biasedness, etc.
  - Sources: Especially from *FPHU.* See also my later books.

- **In Relation to Process**
  —*The Change-Constancy Principle*
  (On the Alteration of Things)
  - Change occurs over time, although constancy is also allowed. There is no change without constancy—and vice versa.
  - Asymmetry undergoes changes over time, so does symmetry.
  - Old players fade away, and new ones emerges, with ever new causes and ever new forms.
  - Family resemblance: e.g., dynamicness-staticness, instability-stability, etc.
  - Sources: First named in *BCPC*. Especially from *FHC*, *FCD*, and *FPHC*. See also other books of mine (from that point on).

(continued on next page)

Table 4.4. The Syntax of Existential Dialectics I:
The Principles
(Part VIII)

• **In Relation to Process** ***(cont'd)***

—*The Expansion-Contraction Principle*
(On the Growth of Things)

- Entities in the world can both expand in some ways and contract in other ones, as part of their nature. There is no expansion without contraction—and vice versa.
- For instance, in the context of anomalous experience, one of its most fascinating expressions is none other than the analysis of different causes of the slow but steady expansion (and, furthermore, if put in a different way, unfolding) of unconsciousness.
- Family resemblance: e.g., conquest-autarky, rise-fall (or up-down), spread-shrink, extendingness-shorteningness, etc.
- Sources: Especially from *FPHU*. See also my later books.

—*The Slowness-Quickness Principle*
(On the Speed of Change)

- Both slowness and quickness co-exist, with their own internal tension, to the extent that each fights for its own relevance with the other, in accordance to the perspectives of nature, the mind, culture, and society, without one being the victor and the other being the vanquished in the long haul. Even when both are chosen in other cases, this dilemma is only transferred into something else with a combinational character. There is no slowness without quickness—and vice versa.
- In relation to taxonomy, quickness has its efficient usability, just as slowness has its aesthetic appeal, for instance. And in relation to network, quickness has its adventurous readiness, just as slowness has its risk-adverse convenience, for instance.
- Family resemblance: e.g., inconvenience-convenience, passiveness-activenessness, gradualness-abruptness, deceleration-acceleration, etc.
- Sources: Especially from *FIA*. See also my later books.

(continued on next page)

Table 4.4. The Syntax of Existential Dialectics I: The Principles (Part IX)

- **In Relation to Process *(cont'd)***
  —*The Order-Chaos Principle*
  (On the Pattern of Things)
  - Both order and chaos are vital in the process of change in the world. The preference for order is biased, since it does not give sufficient attention to the vital role of chaos in the transformation of the world (without somehow reducing it for the understanding of order). There is no order without chaos—and vice versa.
  - The scientific search for order in the world is often a hidden bias in its ontological obsession with order, since chaos is often treated as the "bad" guy, with order as the "good" guy (for the end goal of science).
  - Neither order nor chaos is the final end of the world, and one is not to be treated as the means for the other in the transformation of things. Both are fundamental in their recurrent dialectical interactions with each other over time, without reducing one for the other.
  - Family resemblance: e.g., lawfulness-disorder, order-disorder, etc.
  - Sources: Especially from *FC*. See also my later books.

- **In Relation to Agency**
  —*The Softness-Hardness Principle*
  (On the Force of Change)
  - This has to do with *the force of change* under the category about agency in existential dialectics, in that any change by an agent, be it organic (like humans) or non-organic (like natural objects), can occur in a forceful (aggressive) or gentle (pacific) way, which can come in all shapes and sizes, of course. There is no softness without hardness—and vice versa.
  - Family resemblance: e.g., peacefulness-violence, cooperation-competition, pacificity-aggression, dovishness-hawkishness, reward-punishment, peace-war, etc.
  - Sources: Especially from *ALD*. See also my later books.

(continued on next page)

Table 4.4. The Syntax of Existential Dialectics I: The Principles (Part X)

• **In Relation to Agency *(cont'd)***

—*The Symmetry-Asymmetry Principle*
(On the Relationships among Existents)

- There is no symmetry without asymmetry—and vice versa.
- For instance, the Same can be symmetric and asymmetric towards the Others. But in case of asymmetry, oppression and self-oppression can occur. So, when the Same is asymmetric towards the Others, the Same can also be relatively asymmetric towards itself in self-oppression, just as the Others can be likewise towards themselves.
- The subsequent oppressiveness is dualistic, as much by the Same against the Others and itself, as by the Others against the Same and themselves.
- Both oppression and self-oppression can be achieved by way of downgrading differences between the Same and the Others and of accentuating them.
- This is true, even though not all forms of asymmetry have to be about oppression and self-oppression.
- In addition, from Chapter Three of *FPHG*, symmetry is not perfect, to be understood in an approximate sense under many life circumstances.
- Family resemblance: e.g., strength-weakness, balance-extremity, harmoniousness-unharrmoniousness, potence-impotence, reflexiveness-unreflexiveness, equality-inequality, etc.
- Sources: From all my books, starting with *FHC*. First named in *BCPC*. See, for instance, *FCD* for more details on the Same and the Others.

(*continued on next page*)

Table 4.4. The Syntax of Existential Dialectics I: The Principles (Part XI)

- **In Relation to Agency** ***(cont'd)***
  —*The Theory-Praxis Principle*
  (On the Duality of Knowledge))
  - This has to do with the duality of knowledge in existential dialectics, in that, if there is theoretical construction, there is likewise its practical application, both technical and normative. There is no theory without praxis—and vice versa.
  - Family resemblance: e.g., discovery-application, knowledge-action, invention-innovation, etc.
  - Source: Especially from *FPHE*. See also my later books.

  —*The Evolution-Transformation Principle*
  (On the Multiple Kinds of Agency)
  - This principle (and the symmetry-asymmetry principle) are both about *the agency of change*. The word "agency," in a formal definition, refers to "a person or thing through which power is exerted or an end is achieved." (MWD 2007b) It therefore does not have to necessarily involve an intelligent lifeform.
  - Because of this dual meaning in agency, the evolution-transformation principle is more concerned with *the multiple kinds of agency*, that is, both about the *evolution* in the state of nature (e.g., an object of natural beauty) and the *transformation* in the world of intelligent lifeforms (e.g., a work of art, an air-conditioner, etc.). There is no evolution without transformation—and vice versa.
  - And the transformative part of the principle precisely refers to the other dimension in the dual meaning of agency, in giving technology a major role to play in the change of the world, which is something that I extensively analyzed in *FHC* in the context of the technophilic lifeworld, espeically though not exclusively since modern times.
  - Family resemblance: e.g., inorganicness-volition, nonwillingness-willingness, naturalness-technologicalness, naturalness-nonnaturalness, etc.
  - Sources: Especially from *FAE*. See also other books of mine.

(*continued on next page*)

Table 4.4. The Syntax of Existential Dialectics I:
The Principles
(Part XII)

- **In Relation to Agency** ***(cont'd)***
  —*The Convention-Novelty Principle*
  (On the Nature of Creative Thinking)
  - If there is conventional wisdom, there is likewise novel challenge, to the extent that both convergent and divergent thinking are part of life. There is no convention without novelty—and vice versa.
  - In addition, in this context of the convention-novelty principle, there are, in the absence of better words, what I want to call (a) *creative techniques* and (b) *creative traits*, which, when satisfied—in relation to the larger context of the mind, nature, society, and culture—can be used to enhance creative works. The list of creative techniques and traits (as summarized in *Table 4.1* of *FPHCT*) are not exhaustive, of course, but the examples here constitute a great beginning of understanding the structure of creative thinking.
  - Yet, lest the reader gets carried away by the euphoria about creative thinking in our time, it should be reminded that creative thinking has its own possibilities and limits (as shown in *Table 2.1* of *FPHCT* on invention), just as it has its own promises and pitfalls (as shown in *Table 3.1* of *FPHCT* on innovation). In the end, it should also be stressed that creative thinking has its own desirability and dark sides (as shown in *Table 4.2* of *FPHCT*).
  - This has important daily implications for an average individual, and the word "average" here allows different degrees of variation from one individual to another in a population, at any given point in history.
  - Family resemblance: e.g., convergence-divergence, normalness-nonnormalness, conformity-nonconformity, etc.
  - Sources: First from *FPHCT*. See also my later books.

(*continued on next page*)

Table 4.4. The Syntax of Existential Dialectics I: The Principles (Part XIII)

- **In Relation to Outcome**
  —*The Regression-Progression Principle*
  (On the Direction of History)
  - Neither the cyclical nor the linear views are adequate for explaining many phenomena at all levels. There is regression without progression—and vice versa.
  - History progresses to more advanced forms, but with a regressive touch. Examples include no freedom without unfreedom, no equality without inequality, and no civilization without barbarity.
  - This is not an inevitable law, but merely a highly likely empirical trend.
  - Family resemblance: e.g., cost-benefit, undesirability-desirability, badness-goodness, risk-opportunity, etc.
  - Sources: From all my books, starting with *FHC*. First named in *BCPC*.

  —*The Same-Difference Principle*
  (On the Metamorphosis of Change)
  - An entity, as it evolves over time, can be both different from and similar to its opposing alternatives and does not have to be solely more different from them over time. There is no similarity without difference—and vice versa.
  - Opposites are not absolute in a black-or-white fashion; so, an entity can become relatively more similar to (or more different from) its opposite over time.
  - Family resemblance: e.g., homogeneity-heterogeneity, we-they, etc.
  - Sources: Especially from ALD. See also my later books.

Notes: The features in each principle are solely illustrative (not exhaustive). The comparison is also relative (not absolute), nor are they mutually exclusive. As generalities, they allow exceptions.

Table 4.5. The Syntax of Existential Dialectics II: The Principles as Short Cuts (Part I)

• The principles should be treated with caution, lest misunderstanding occurs, since they do not constitute rigid dualities (or dichotomies).

• The reason is that each pair in an ontological principle consists of two opposites, which are, however, merely short cuts both for multiple variations and degrees, as well as for different interactions with multiple other entities unlike them. In this light, each pair can end up having hundred (or even thousand, if not more) different versions, which interact with hundred (or even thousand, if not more) other entities. There are two clarifications here.

• Firstly, this conception of shortcuts is not mutually exclusive nor absolute, in that the opposites can come in all shapes and sizes, with different degrees. For instance, by analogy, just as there are different degrees of the two colors "white" and "black"—there are likewise different degrees of the opposites in each ontological pair, to the extent that there can be multiple entities (not only two) interacting with each other in each pair.

• And secondly, this conception of shortcuts do not ignore other possible entities in interacting with the two opposites (with their different versions) in each pair. For instance, by analogy, there are not only the two colors "white" and "black" as opposites, since there are other colors too besides them like "yellow," "green," "purple," or else. The same logic can be applied to each ontological pair, in that they also interact with other entities, not with only two of them (with their different versions).

• Therefore, with these two clarifications in mind—each pair in an ontological principle serves only as an abbreviation for something more complicated and, therefore, although it contains two opposites, it should not be confused as a duality (dualism).

(continued on next page)

Table 4.5. The Syntax of Existential Dialectics II: The Principles as Short Cuts (Part II)

• Existential dialectics rejects any dualism (or dichotomy) as too rigid and instead allows the multiplicity of entities, to the extent that between the two opposites in each pair exist many other alternatives to choose from. They are named in that short form for aesthetic elegance, instead of listing all possible entities between the two opposites in the title.

• To be dialectic is to go beyond any rigid dichotomy and transcend into something different altogether in the long haul. One may be tempted to call the dialectic logic here with a different name like existential "multilectics" (instead of "dialectics"), but this naming is incorrect (or even misleading), for the two reasons aforestated.

• After all, each ontological pair come in all shapes and sizes (with different degrees) and do not exist by themselves but also interact with other entities unlike them. The virtue of revealing an ontological pair is to show how they relate within themselves (in multiple versions) and also interact with others unlike them (also in multiple versions).

• This will be clear in the table on "the dialectic constraints imposed by the principles."

Source: From Ch.1 of *FPHK*

Table 4.6. The Syntax of Existential Dialectics III: The Principles as Family Resemblances

• Each ontological principle is generic, with some other comparable ontological pairs to be put in the same family (like a *family resemblance*). Ludwig Wittgenstein (1953) once suggested the idea of "family resemblance" in explaining different games classified under the same family called *games*.

• Why should, for instance, playing football and chess as playing "games," when it is well understood that football is not the same as chess? The answer is that, although each game is different and has different rules, many of them (though not all) share, more or less, some commonalites (e.g., scoring as necessary for winning).

• And this is so, even though some games share more than some others in any given selection of critieria, and no two games are exactly identical. So, his point here is that there is no essential core which is common to all games, and the best that one can look for is some characteristics which are common to many (but not all) games. (A. Biletzki 2006)

• By the same logic—in the previous section on selection criteria, the flexibility-inflexibility pair can be put in the family resemblance of the simpleness-complicatedness principle, although the two pairs are not exactly identical. Likewise, the directness-indirectness pair can be put in the family resemblance of the preciseness-vagueness principle, although, again, the two pairs are not exactly identical.

• In this sense, which specific pair in a family should be used to designate the name of the family can be at times a bit arbitrary, but with good reason.

• For illustration, in the context of method, the partiality-totality principle can take the different form like individualisticness-holisticness, just as the explicability-inexplicability principle can take the different form like underlyingness-regularness—although each two pairs are not exactly identical and have slightly different meanings and usages. A more comprehensive listing of this family resemblance for all other principles is shown in the table on the syntax of existential dialectics.

Source: From Ch.1 of *FPHK*

Table 4.7. The Syntax of Existential Dialectics IV: The Dialectic Constraints Imposed by the Principles (Part I)

• **Co-Existent and Asymmetric**

—The principles, as they constitute the syntax of existential dialectics, are dialectic in character, such that, when they are applied, they impose dialectic constraints on how reality is to be understood. Consider, say, the symmetry-asymmetry principle as an illustration here, in order to summarize two main characters of the dialectic constraints in question.

—Firstly, to be dialectic here is to go beyond the narrow dichotomies (and, for that matter, any rigid multi-dimensional classificatory scheme), be they about "self" vs. "world," "freedom" vs. "unfreedom," "barbarity" vs. "civilization," "individuality" vs. "communality," and so on.

—One way to do so (to go beyond) is to consider them all in terms of co-existence (without favoring one over the rest). For instance, my theory of "post-civilization" (to be summarized later in the section on the pragmatics of existential dialectics) is to go beyond barbarity and civilization in terms of understanding barbarity and civilization as being co-existent. And the same logic can be said in relation to my theories of "post-democracy," "post-capitalsm," and others (also to be introduced later in the section on the pragmatics of existential dialectics), in regard to freedom vs. unfreedom, equality vs. inequality, communality vs. individuality, spirituality vs. materiality, and so on.

—But to consider them all (in the dichotomies—and, for that matter, in any rigid multi-dimensional classifactory schme) as co-existent is not the same as to imply that the opposites in any classificatory scheme are all equal, since, in accordance to the symmetry-asymmetry principle (as an illustration here), if they are equal in terms of being considered as co-existent, they are asymmetric in terms of being unequal in dialectic interaction (e.g., X can be more dominant than Y in case A, or Y is more dominant than X in case B).

(continued on next page)

Table 4.7. The Syntax of Existential Dialectics IV: The Dialectic Constraints Imposed by the Principles (Part II)

—For this reason, there are different versions of "post-democracy" and "post-capitalism" in my theories. As an illustration, in version I of the theory of post-democracy, freedom is more dominant than equality, whereas in version II of the theory of post-democracy, equality is more so than freedom.

—But this "X more than Y" has to be understood in the context of dialectic logic (not in conventional logic), in that both "X" and "Y" are important in post-democracy (in the context of dialectic logic), but in an asymmetry way. By contrast, in conventional logic, it often favors one over the other—be it in regard to privileging freedom over equality in Fascism, favoring freedom relatively more than equality in Liberal Democracy, or favoring equality relatively more than freedom in Socialist Democracy. In the latter two cases (about Liberal Democracy and Socialist Democracy), the difference between dialectic logic and conventional logic can be one in degree, not in kind—in this sense, albeit not in other senses.

—The same logic can be said about the relatonships between individuality and communality, between spirituality and materiality, and between formal legalism and informal legalism in the different versions of my theory of post-capitalism.

- **Transcendent**

—Secondly, to be dialectic is to go beyond the narrow dichotomies (and, for that matter, any rigid multi-dimensional classificatory scheme) in another way, this time, in a transcendent way, that is, in exploring other possibilities or even other issues not considered within the narrow confines of narrow dichotomies (and, for that matter, any rigid multi-dimensional classificatory scheme).

(continued on next page)

Table 4.7. The Syntax of Existential Dialectics IV: The Dialectic Constraints Imposed by the Principles (Part III)

—As an analogy, to go beyond the narrow color dichotomy of "black" and "white" is not just to choose both "black" and "white" (as in the first meaning) but also to explore other color options (e.g., "green," "purple," "blue," etc.—and, alternatively, "shade," "line," "curve," etc.). By the same logic, to go beyond "democracy" is to transcend democracy (as in version III of the theory of "post-democracy") and to explore other possibilties of lifeforms (e.g., floating consciousness, hyper-spatial consciousness, etc., to live beyond the narrow obsession with freedom and equality).

—This dialectic character of the principles in existential dialectics has important implications for the pragmatics of existential dialectics (as will be clear shortly, in the section on the pragmatics of existential dialectics).

Notes: The examples here are solely illustrative (not exhaustive). The comparison is also relative (not absolute), nor are they mutually exclusive. As generalities, they allow exceptions.
Sources: First explicitly stated in *ALD*. Also from all other books of mine.

Table 4.8. The Syntax of Existential Dialectics V: Further Clarifications (Part I)

- **Pioneering**
  —Firstly, the total number of ontological principles is unknown, to be discovered later, as our knowledge of the world becomes more advanced.
  —Consequently, the principles as introduced in my books are not exhaustive, with new ones being added, whenever more of them are discovered in later research.
  —At least, future generations can pick up where I leave off and continue the discovery.
  —In this sense, my work should be treated as a pioneering effort for the development of a systematic, comprehensive analysis of a new general ontology for the future of knowledge.

- **Flexible**
  —Secondly, the principles are not rigidly classified, as they can be reclassified in a different way.
  —For instance, the preciseness-vagueness principle is classified under the category of "structure" but can be reclassified under the category of "method," although in so doing, it has a different meaning in the context of method.
  —The same logic applies to the same-difference principle under the category of "outcome," which can be reclassified under the category of "structure," although in so doing, once more, it has a different meaning in the context of structure.

- **Mutually Constraining**
  —Thirdly, the principles are to be understood together, not that each prinicple is to be analyzed independently of others. After all, the principles are mutually constraining, in that they work together as a whole.

(continued on next page)

Table 4.8. The Syntax of Existential Dialectics V: Further Clarifications (Part II)

- **Selectively Useful**

  —Fourthly, the principles are relevant to all subject matters, but some principles are more useful to some subject matters than others—as implied in the symmetry-asymmetry principle. This is true, even if different studies of the same kind can yield different views about the degree of relevance for each ontological principle, depending on the specific nature of a research in question, needless to say.

- **Anti-Reductionistic**

  —And fifthly, the principles constitute only two levels of analysis, this time, at the ontological and methodological levels—while other levels of analysis (from the perspectives of the mind, nature, society, and culture) are also needed, in order to understand reality in its totality (as explained in the section on "sophisticated methodological holism," so as to avoid the dual dangers of reductionism and reverse-reductionism).

Notes: The examples here are solely illustrative (not exhaustive). The comparison is also relative (not absolute), nor are they mutually exclusive. As generalities, they allow exceptions.
Sources: From my previous books like *FHC*, *FCD*, *FPHC*, *BDPD*, *BCPC*, *BNN*, *FC*, *FAE*, and so on.

Table 4.9. The Syntax of Existential Dialectics VI: The Dilemma of Specific vs. General Ontology (Part I)

• Ontology is often more complicated and imposes some seemingly insurmountable difficulties concerning what constitutes a kind of ontology which can be valid enough to be accepted by the wider intellectual community in question. An excellent example concerns what I want to call *the dilemma of ontology* in relation to the relationship between "specific ontology" and "general ontology." (M. Bunge 1999; R. Corazzon 2007)

• On the one hand, "general ontology," as Mario Bunge (1999) put it, "studies all existents." But, on the other hand, "special ontology studies one genus of thing or process—physical, chemical, biological, social, etc." (M. Bunge 1999; R. Corazzon 2007)

• Consequently, the relationship between the former and the latter is that, as an illustration, "whereas general ontology studies the concepts of space, time, and event, the ontology of the social investigates such general sociological concepts as those of social system, social structure, and social change." (M. Bunge 1999; R. Corazzon 2007)

• With this formal definition in mind, the relationship between the two forms of ontology only poses *the dilemma of ontology*, in that general ontology is possible to the extent that its general contructs must be broad enough to serve as the "lowest common denominator" for all disciplines, but this character of the lowest common denominator is not only hard to find but also its usefulness is limited. (WK 2008b)

• But specific ontology fares no better either, in that its specific constructs primarily serve a specific discipline in question, although other disciplines can appropriate them for their own purposes, but in a different context (especially, though not exclusively, when used as a metaphor from one field to another). For instance, the Kantian category of understanding under the heading of "quantity" constitutes a type of specific ontology, which is specific to the discipline of mathematics, although other disciplines can make use of the categories in their own context.

(continued on next page)

Table 4.9. The Syntax of Existential Dialectics VI: The Dilemma of Specific vs. General Ontology (Part II)

• A solution lies in my proposal of "sophisticated methodological holism," which requires all levels of analysis in any subject matter. See the tables on sophisticated methodological holism, and on reductionism and reverse-reductionism, for more details.

Source: From *Sec. 1.4* of *FPHK*

Table 4.10. The Syntax of Existential Dialectics VII: Types of Inappropriate Family Resemblances

- **Inadequate Family Resemblances**
  —Ex: consistency-inconsistency: the formalness-informalenss principle
  —Ex: cognition-noncognition: the formalness-informalness principle

- **Compound Family Resemblances**
  —Ex: reality-nonreality: the formalness-informalness principles, the absoluteness-relativeness principle
  —Ex: normality-abnormality: the symmetry-asymmetry principle, the density-emptiness principle, the convention-novelty principle, the same-difference principle
  —Ex: structure-context: the simpleness-complicatedness principle, the theory-praxis principle
  —Ex: freedom-unfreedom: the symmetry-asymmetry principle, the expansion-contraction principle
  —Ex: morality-immorality: the denseness-emptiness principle, the symmetry-asymmetry, the same-difference principle
  —Ex: particularness-universalness: the preciseness-vagueness principle, the simpleness-complicatedness principle
  —Ex: pureness-mixedness: the absoluteness-relativeness principle, the simpleness-complicatedness principle, the same-difference principle
  —Ex: shortsightedness-foresightedness: the partiality-totality principle, the simpleness-complicatedness principle
  —Ex: shorttermness-longtermness: the partiality-totality principle, the simpleness-complicatedness principle, the slowness-quickness principle

Notes: The examples here are solely illustrative (not exhaustive). The comparison is also relative (not absolute), nor are they mutually exclusive. As generalities, they allow exceptions.
Source: First summarized in *BEPE* (and updated in later books)

Table 4.11. The Semantics of Existential Dialectics

- **Abstract Structure vs. Specific Meanings**
  - —The syntax of existential dialectics so understood in terms of ontological principles only gives us the structure of ontology in the world, in an abstract (general) sense. These principles by themselves do not tell us the specific meanings in a given context.
  - —In order to grasp the specific meanings of the principles in a given context, it is necessary to study the semantics of existential dialectics. And the analysis of the ontological principles as family resemblances in the section on syntax is only a starting point (and thus overlaps a bit with the semantics of existential dialectics here).
  - —With this caveat in mind, the reason that I have often gone in great lengths in my previous books on different subjects is to explain the specific meanings of the principles when applied in different contexts.

- **Specific Meanings in Specific Fields**
  - —For instance, in *FPHST*, I used the first three principles (i.e., the change-constancy principle, the regression-progression principle, and the symmetry-asymmetry principle) to propose "the perspectival theory of space-time," for a better way to understand space and time—especially, though not exclusively, in relation to future post-human history (as summarized in *Table 3.6*, *Table 3.7*, *Table 3.8*, and *Table 3.9*). In so doing, I had to introduce concepts and theories specific to the field of physics and other related fields (e.g., "absolute space" and "absolute time" in "classical mechanics" and "relative space-time" in "the theory of relativity").
  - —In *BNN*, I also exploited the three principles to propose the "transcendent" approach to the study of genes and memes as a new way to understand the interaction between nature and nurture. In so doing, I had to explore concepts and theories in the world of evolutionary theory (e.g., "mutation," "variation," "adaptation," "selection," and "inheritance" in Darwinian evolutionary theory) and neural biology (e.g., "chromosome," "gene," "DNA," "RNA," "protein," "neuron," "neural network," and "behavior").

Notes: The categories and examples in each are solely illustrative (not exhaustive). The comparison is also relative (not absolute), nor are they mutually exclusive. As generalities, they allow exceptions.
Sources: First explicitly stated in *ALD*. Also from all other books of mine.

Table 4.12. The Pragmatics of Existential Dialectics (Part I)

- **The Two-Way Street Connecting Theory and Meta-Theory**
  - —The pragmatics of existential dialectics is not a one-way street (that is, using the ontological principles for theoretical insights in praxis) but a two-way one, that is, (a) from meta-theory to theory, and (b) from theory to meta-theory.
  - —(a) On one side of the street, the ontological principles can inspire some theoretical insights in praxis, that is, in relation to some specific fields.
  - —(b) On the other (opposing) side of the street, however, the study of a subject matter in the specific fields in turn reveals some more hitherto unknown ontological principles to be discovered and identified. For this reason, three new principles were added in *FC*, one in *FAE*, and two in *ALD*, on top of the original three in *BCPC*—after some research on the specific subject matters.

- **Direct and Indirect Applications**
  - —*Direct*
    - The logic of existential dialectics can shed some theoretical insights on diverse phenomena in the world, and good instances are the pertinent use of the principles of existential dialectics for the theoretical insights on the freedom/unfreedom dialectics, the equality/inequality dialectics, and the wealth/poverty dialectics in my previous works.
    - My latest books like *FPHST* and *BNN* also use the principles to reveal some theoretical insights on the perspectives of space and time (as in *FPHST*) and of nature and nurture (as in *BNN*).
  - —*Indirect*
    - The theoretical insights can further be used to reveal other phenomena directly from them (viz., the theoretical insights) and therefore indirectly from the principles themselves. A good instance is the use of the theoretical insights on the freedom/unfreedom and equality/inequality dialectics for the understanding of the civilization/barbarity dialectics.

(continued on next page)

Table 4.12. The Pragmatics of Existential Dialectics (Part II)

- **Direct and Indirect Applications (*cont'd*)**
    - Even in indirect applications, however, a phenomenon under study can still be directly related back to the principles themselves. In the example as cited above, the civilization/barbarity dialectics can be directly related to the principles of existential dialectics without the intermediate role of the freedom/unfreedom and equality/inequality dialectics.

- **Multiple Levels of Application**

—The theoretical insights can be applied to different levels of analysis, even though in a given example, it may refer to one level only. For instance, in the example concerning the freedom/unfreedom dialectics, it can be used at the structural level (e.g., in relation to the theory of cyclical progression of hegemony), but it can be exploited as well for other levels (e.g., the theory of post-capitalism at the institutional level).

Notes: The categories and examples in each are solely illustrative (not exhaustive). The comparison is also relative (not absolute), nor are they mutually exclusive. As generalities, they allow exceptions.
Sources: From Ch.6 of *BCPC*. See also other books of mine.

Table 4.13. The Freedom-Unfreedom Dialectics (Part I)

• **On Having**

—*In Relation to the Technological*

(1) if freer from submission to Nature, then less free from ecological degradation (Deep and Social Ecology), even if in a hi-tech form

(2) if freer from technological inconvenience / backwardness, then less free from technological control and the loss of privacy

(3) if freer from technological (material) backwardness, then less free from the abusive (barbaric) maltreatment of the primitive Others

—*In Relation to the Everyday*

(1) if freer from abject poverty, then less free from artificial needs/discontents (Frankfurt School)

(2) if freer from sensual suppression, then less free from violent sublimation (Freud)

(3) if freer from the snobbishness of high culture, then less free from the shabbiness (leveling-off effect) of mass culture (Tocqueville)

(4) if freer from the inefficiency of traditional "compassionate economy," then less free from the bondage of a "ruthless [competitive] economy" (Keynes)

(5) if freer from anarchy in the state of nature (system fragmentation), then less free from government regulations and controls in system integration

---

(continued on next page)

Table 4.13. The Freedom-Unfreedom Dialectics (Part II)

- **On Belonging**
  - —*In Relation to the Good and the Just*
    - (1) if freer from disciplinary society, then less free from society of control (Foucault)
    - (2) if freer from the tyranny of one or a few, then less free from the tyranny of the majority (or sometimes, minority veto)
    - (3) if freer from elitist decision making, then less free from political gridlock/cleavage
    - (4) if freer from arbitrary (discretionary) administration, then less free from bureaucratic irrationality (Weber) and legal trickery (loopholes)

- **On Being**
  - —*In Relation to the True*
    - (1) if freer from unscientific dogmas, then less free from instrumental abyss (nihilism). Or conversely, if freer from meaninglessness, then less free from dogmas.
    - (2) if freer from the bondage of partiality/partisanship (e.g., prejudice, discrimination), then less free from the danger of impartiality and neutrality (e.g., opportunism, unrealisticness, lack of compassion, inaction)
    - (3) if freer from making generalizations, then less free from being unable to understand much of anything
  - —*In Relation to the Holy*
    - (1) if freer from collective conscience, then less free from social loneliness
    - (2) if freer from religious absoluteness, then less free from spiritual emptiness
  - —*In Relation to the Beautiful/Sublime*
    - (1) if freer from artistic non-autonomy, then less free from aesthetic disillusion (deconstruction)

(continued on next page)

Table 4.13. The Freedom-Unfreedom Dialectics (Part III)

Notes: The examples in each category are solely illustrative (not exhaustive), and the comparison is relative (not absolute), nor are they necessarily mutually exclusive. And some can be easily re-classified elsewhere. As generalities, they allow exceptions.
Sources: A reconstruction from Ch.10 of *FCD*, based on *FHC*

Table 4.14. The Equality-Inequality Dialectics (Part I)

- **On Having**
  - —*In Relation to the Technological*
    - (1) if more equal in treating Nature with spiritual unity, then less equal in suppressing the dominant drive to transcend it altogether
  - —*In Relation to the Everyday*
    - (1) if more equal in building social plurality, then less equal in leveling-off effects (e.g., the subsequent relative intolerance of high/intellectual ethos in mass culture industry)
    - (2) if more equal in socioeconomic distribution beyond a certain point, then less equal in efficiency (e.g. resentment, the erosion of work ethics)
    - (3) if more equal in urging an affirmative action program, then less equal in creating victim mentality (in oneself), stigma (from others), reverse discrimination (against the once privileged), and mediocracy (against the more able)

- **On Belonging**
  - —*In Relation to the Good and the Just*
    - (1) if more equal in banning monarchic/oligarchic exclusion, then less equal in producing "the tyranny of the majority" or of "minority veto"
    - (2) if more equal in encouraging participatory decision making, then less equal in inducing political divisiveness (gridlock/cleavage in power blocs) and organizational oligarchy
    - (3) if more equal in institutionalizing a decentralized bureaucracy, then less equal in falling into more territorial/turf politics (intrigues)

(continued on next page)

Table 4.14. The Equality-Inequality Dialectics (Part II)

• **On Being**

—In Relation to the Beautiful / Sublime

(1) if more equal in accepting diverse styles ("anything goes" mentality), then less equal in artistic good quality (in leveling-off effects against the best)

—In Relation to the True

(1) if more equal in tolerating multiple viewpoints (no matter how extreme), then less equal in epistemic standards

—In Relation to the Holy

(1) if more equal in celebrating any cults and sects (no matter how questionable), then less equal in spiritual depth and authenticity

---

Notes: The examples in each category are solely illustrative (not exhaustive), and the comparison is relative (not absolute), nor are they mutually exclusive. And some can be easily reclassified elsewhere. As generalities, they allow exceptions.

Sources: A reconstruction from Ch.10 of *FCD*, based on *FHC*

Table 4.15. The Duality of Oppression in Existential Dialectics: Oppression and Self-Oppression (Part I)

- **From the Same to the Others and Itself**
  —The Oppression by the Same against the Others
    - *By way of downgrading differences*
      –Ex: on judiciary caprice for corporate crimes (*Sec.2.2.1.2.1*)
      –Ex: on the deceptive politics of liberation (*Sec.3.5*)
      –Ex: on the humanitarian mystique (*Sec.4.4*)
      –Ex: on the fad of emotional intelligence (*Sec.5.3*)
    - *By way of accentuating differences*
      –Ex: on the legal sophistry of self-defense (*Sec.2.3*)
      –Ex: on the legal semantics of proportionality (*Sec.2.4*)
      –Ex: on the tricky politics of external threat (*Sec.3.4*)
      –Ex: on the appeal of the Far Right for democracy (*Sec.5.4*)
      –Ex: on the democratic axis of evil (*Sec.5.5*)
      –Ex: on the democratic way of brutality and revenge (*Sec.5.6*)
      –Ex: on democratic autocracy (*Sec.6.4*)

  —The Oppression by the Same against Itself
    - *By way of downgrading differences*
      –Ex: on the politics of fear (*Sec.2.2*)
      –Ex: on the trickery of compassionate conservatism (*Sec.3.2*)
      –Ex: on the deceptive politics of patriotism (*Sec.3.3*)
    - *By way of accentuating differences*
      –Ex: on the caprice of due process on domestic suspects (*Sec.2.2*)
      –Ex: on the false security/freedom dilemma (*Sec.6.5.2*)

(continued on next page)

Table 4.15. The Duality of Oppression in Existential Dialectics: Oppression and Self-Oppression (Part II)

- **From the Others to the Same and Themselves**

—The Oppression by the Others against the Same
- *By way of downgrading differences*
  - –Ex: on judiciary caprice in the reverse direction (*Sec.2.2.1.2.2*)
  - –Ex: on equal pay (*Sec.6.2.1.1*)
  - –Ex: on equal representation (*Sec.6.2.1.2*)
  - –Ex: on affirmative action program (*Sec.6.3.1.1*)
  - –Ex: on same-sex marriage (*Sec.6.3.1.2*)
- *By way of accentuating differences*
  - –Ex: on sexual harassment (*Sec.6.2.2.1*)
  - –Ex: on physical violence (*Sec.6.2.2.2*)
  - –Ex: on sexual exploitation (*Sec.6.2.2.3*)

—The Oppression by the Others against Themselves
- *By way of downgrading differences*
  - –Ex: on the reverse-class mystique (*Sec.4.2*)
  - –Ex: on the reverse-black mystique (*Sec.4.3*)
  - –Ex: on self-discrimination by downgrading (*Sec.6.3.2.2*)
- *By way of accentuating differences*
  - –Ex: on self-discrimination by accentuating (*Sec.6.3.2.1*)

Notes: The examples are solely illustrative (not exhaustive), nor are they mutually exclusive. As generalities, they allow exceptions. Also, both forms of oppression co-exist in all of the examples, so the listing of them are only meant in a relative, not absolute, sense.

Source: A summary of the sections (as cited) in Chs.2-6 of *BDPD*. See text for more info and references.

Table 4.16. The Structure of Existential Dialectics I: The Freedom/Unfreedom and Equality/Inequality Dialectics

• Each freedom and equality produces its own unfreedom and inequality, regardless of whether the pair occurs in political society (with the nation-state), in civil society (with some autonomy from the state), or elsewhere (e.g., in the private sphere of individual homes)—and regardless of whether freedom and equality are understood as "negative" or "positive."

• Oppression is dualistic, as much by the Same against the Others and itself, as by the Others against the Same and themselves.

• Both forms of oppression and self-oppression can be achieved by way of downgrading differences (between the Same and the Others) and of accentuating them.

• The relationships are relatively asymmetric between the Same and the Others and relatively symmetric within them. This is true, even when the Same can be relatively asymmetric towards itself in self-oppression, just as the Others can be likewise towards themselves.

• Symmetry and asymmetry change over time, with ever new players, new causes, and new forms.

Notes: The examples in each category are solely illustrative (not exhaustive) nor necessarily mutually exclusive, and the comparison is relative (not absolute). As generalities, they allow exceptions. "Negative" freedom is freedom "from" (e.g., freedom from poverty), whereas "positive" freedom is freedom "to" (e.g., freedom to the state of enlightenment). "Negative" equality is "procedural" equality (e.g., equality of opportunity), while "positive" equality is "substantive" equality (e.g., equality of outcome). Existential dialectics impose constraints on freedom and equality in democracy, non-democracy, and post-democracy. There is no utopia, in the end; even should there be one, dystopia would exist within it.
Sources: From *Table 1.5* of *BDPD*—and also from *FHC*, *FCD*, and *FPHC*

Table 4.17. The Structure of Existential Dialectics II: The Wealth/Poverty Dialectics

---

• There is no wealth without poverty, just as there is no poverty without wealth.

• The wealth/poverty dialectics occurs in the realms of having, belonging, and being, in relation to the material, relational, and spiritual.

• The wealth/poverty dialectics also expresses itself at the multiple levels of analysis in accordance to methodological holism, be they about the micro-physical, the chemical, the biological, the psychological, the organizational, the institutional, the structural, the systemic, the cultural, and the cosmological.

• The wealth/poverty dialectics is a different manifestation of existential dialectics in general, subject to the principles in its logic of ontology—just as the freedom/unfreedom and equality/inequality dialectics are likewise.

• There is no economic utopia, in the end; even should there be one, dystopia would exist within it.

Notes: The main points here are solely illustrative (not exhaustive) nor necessarily mutually exclusive, and the comparison is relative (not absolute). As generalities, they allow exceptions.
Sources: From *BCPC*. See also *FCD* and *FHC*.

Table 4.18. The Structure of Existential Dialectics III: The Civilization/Barbarity Dialectics

• There is no civilization without barbarity.

• The civilization/barbarity dialectics applies in the four civilizing processes (e.g., the rationalizing process, the pacifying process, the stewardizing process, and the subliming process).

• The civilization/barbarity dialectics is another (different) manifestation of existential dialectics in general, subject to the principles in its logic of ontology—just as the freedom/unfreedom and equality/inequality dialectics and the wealth/poverty dialectics are likewise.

• There is no utopia, in the end; even should there be one, dystopia would exist within it.

Notes: The main points here are solely illustrative (not exhaustive) nor necessarily mutually exclusive, and the comparison is relative (not absolute). As generalities, they allow exceptions.
Sources: From *BCIV*. See also *FCD*, *FHC*, and *BDPD*.

Table 4.19. The Double Sides of Virtual Organizations (Part I)

- **Psychological**
  —Ex: virtual psychosis
  —Ex: impersonality and loneliness in quaternary social relations

- **Organizational**
  —Ex: the race for power and interests
  —Ex: the world of unequal successes
  —Ex: the bureaucratic life of its own
  —Ex: lesser accountability and transparency

- **Economic**
  —Ex: the economic divides
  —Ex: the erosive impact of commercialization

- **Political**
  —Ex: the anti-authoritarian myth
  —Ex: different power struggles among groups (e.g., Conservative, Reformist, Radical)

- **Structural**
  —Ex: different localities (e.g., climates, scenery)
  —Ex: divided domains (e.g., core, peripheral, sub-peripheral)
  —Ex: substitution effect of social capital

- **Systemic**
  —Ex: uneven advances in transportation and communications, and geographical migration

- **Cultural**
  —Ex: conflicting civilizational fabrics (e.g., Confucian, Islamic, Western)
  —Ex: the bias of the Liberal Democratic agenda
  —Ex: the addiction to moral fanaticism

(continued on next page)

Table 4.19. The Double Sides of Virtual Organizations (Part II)

Notes: These categories and examples are solely illustrative (not exhaustive), and some of the items can be reclassified somewhere else. Nor are they mutually exclusive. Since they are generalities, exceptions are expected.
Source: From Ch.7 of *FCD*

Table 4.20. Beyond the World of Titans, and the Remaking of the World Order

- **Hyper-Empires**
  —Ex: The Chinese Union
  —Ex: The Indian Union

- **Meso-Empires**
  —Ex: The European Union
  —Ex: The North American Union

- **Micro-Empires**
  —Ex: The Latin American Union
  —Ex: The Middle Eastern Union

- **The Rest of the World**
  —Odd Powers
    - Ex: Japan
    - Ex: Russia

  —The Poor Club
    - Ex: The African Union

  —Ambivalent Regions
    - Ex: Southeast Asia
    - Ex: Oceania
    - Ex: South Asia
    - Ex: Central Asia
    - Ex: Southern/Eastern Europe
    - Ex: North Africa
    - Ex: Central America
    - Ex: Others (e.g., the Korean peninsula)

Source: A summary of Chs.2-5 (of *BWT*)

Table 4.21. The Origins of Authoritarian Liberal Democracy

- **The Geopower of Nature (Ch.4 of *ALD*)**
  —Ex: Power Character and Geographical Strategy
  —Ex: Living Space and Territorial Expansion
  —Ex: Strategic Heartland and Containment

- **The Biopsychology of the Mind (Ch.5 of *ALD*)**
  —Ex: The Bell Curve and Mass Intelligence
  —Ex: Group Analysis and Mass Knowledge
  —Ex: Groupthink and Elite (Mis)calculation

- **The Disciplinary Control of Society (Ch.6 of *ALD*)**
  —Ex: Social Organizations of Ruthlessness
  —Ex: Social Institutions of Greed
  —Ex: Social Structure of Exclusion
  —Ex: Social Systems of Violence

- **The Molding Force of Culture (Ch.7 of *ALD*)**
  —Ex: The Tradition of Conquest
  —Ex: The Rationalization of Unreason

Notes: The examples in each category are solely illustrative (not exhaustive), and they are also relative (not absolute), nor are they mutually exclusive. As generalities, they allow exceptions.
Source: A summary of Chs.4-7 of *ALD*

**Table 4.22. The Theory of Post-Democracy I:
The Priority of Freedom over Equality
(Part I)**

- **Differences**
  - *—For the aggressive Lions (the strong Elitists)*
    - •Setting up rank distinctions among unequals (e.g., between inferior humans and superior post-humans, or later among inferior post-humans and superior ones, relatively speaking)
    - •Yearning for being not only distinguished from unequals, but also the first among equals (the best of the very best)
    - •Soul-searching for a high spiritual culture (not the trashy one for the masses). Mass culture is a dirty joke for them.
  - *—For the manipulative Foxes (the weak Counter-Elitists)*
    - •Seeking a gentle hegemony by way of more communitarian concerns (for inferior humans and, later, inferior post-humans)
    - •Being more sympathetic to less formal-legalistic institutions and values

- **Similarities**
  - *—For both Lions and Foxes*
    - •Exploring different spheres of non-human consciousness in the cosmos (something vastly superior than the human one)
    - •Recognizing the democratic illusions (e.g., no freedom without unfreedom, no equality without inequality, or simply no justice without injustice, and vice versa)

(continued on next page)

Table 4.22. The Theory of Post-Democracy I: The Priority of Freedom over Equality (Part II)

Notes: The two callings and examples in each category are solely illustrative (not exhaustive), since there will be many different post-human value ideals in the distant future of post-human civilization. The comparison is also relative (not absolute) towards post-democracy, so this is not just a version of free-market democracy (nor Fascism/Nazism, as shown in the table later on democracy, non-democracy, and post-democracy). Nor are they mutually exclusive. As generalities, they allow exceptions. And the specific forms of post-human post-democratic ideals need to be further developed in future after-postmodern history, as they will be different from the ones we now know. The point here is to solely give an extremely rough picture of a small part of the world to come that we have never known.

Source: From Ch.10 of *FCD*. Refer to text for more info and references.

Table 4.23. The Theory of Post-Democracy II: The Priority of Equality over Freedom

- **Hybrid Versions of**
  —Ex: the Trans-Feminine Calling
  —Ex: the Trans-Sinitic Calling
  —Ex: the Trans-Islamic Calling
  —Ex: the Trans-Outerspace Calling

- **Qualifications**
  —These four versions of post-capitalist value ideals need not automatically be post-democratic, just as capitalism does not necessarily mean democracy. They are two different entities—though closely related.
  —But up to a certain threshold of elevating equality at the farther expense of freedom, the democratic ideals will be overcome and cease to exist.
  —The overcome will not be socialist or communist, but post-democratic with no freedom without unfreedom and no equality without inequality, subject to the constraints of existential dialectics.

Notes: The callings are solely illustrative (not exhaustive), since there will be many different post-human value ideals in the distant future of post-human lifeforms. The comparison is also relative (not absolute), nor are they mutually exclusive. As generalities, they allow exceptions. And the specific forms of post-human post-democratic ideals need to be further developed in future after-postmodern history, as they will be different from the ones we now know. The point here is to solely give an extremely rough picture of a small part of the world to come that we have never known.
Source: From Ch.10 of *FCD*. Refer to text for more info and references.

Table 4.24. The Theory of Post-Democracy III: The Transcendence of Freedom and Equality (Part I)

- **Transcending Freedom in Floating Existence**
  —*Freedom*: seeking an ultimate elimination of the body. Being without the body. The aim is to transcend freedom in the end into a metaphysical state (i.e., beyond the physique).
  —*Unfreedom*: yet facing difficult trade-offs. The sacrifice of bodily existence and its joyfulness. An eternal boredom in floating existence in dark deep space, though with alternative pleasures. There is no free lunch even in the state of transcending freedom.

- **Transcending Equality in the Rivalry of Cosmic Hegemony**
  —*Inequality*: competing to outlast other lifeforms in floating existence, or just marginalizing them for one's hegemonic expansiveness in the rest of the cosmos (and even beyond). Universalism is only for the mediocre.
  —*Equality*: accepting only those of one's rank as equal partners in the vast spacetime for cosmic supremacy. Even here, the aim is to transcend equality into a metaphysical state.

(continued on next page)

Table 4.24. The Theory of Post-Democracy III: The Transcendence of Freedom and Equality (Part II)

Notes: Do not confuse this transcendence of freedom and equality (as one version of post-democracy) with the naïve temptation to transcend the freedom/unfreedom and equality/inequality dialectics. Existential dialectics hold true for freedom and equality in all cultures and societies—past, present, or future (i.e., democracy, non-democracy, and post-democracy), regardless of whether freedom and equality are conventionally understood as "negative" or "positive."

Also, the two features and examples in each are solely illustrative (not exhaustive), since there will be many different post-human value ideals in the distant future of post-human lifeforms. The comparison is also relative (not absolute), nor are they mutually exclusive. As generalities, they allow exceptions. And the specific forms of post-human ideals even for these radically alien floating lifeforms (and others unknown to us) need to be further developed in future after-postmodern history, as they will likely be different from the ones herein illustrated. The point here is to solely give a very rough picture of a small part of the extremely alien world to come that we have never known.

Source: From Ch.10 of *FCD*. Refer to text for more info and references.

Table 4.25. Democracy, Non-Democracy, and Post-Democracy (Part I)

• **Democracy**

—*Theoretical Constructs*

•The pursuit of freedom and equality (in various degrees), regardless of whether freedom and equality can be understood as "negative" or "positive"

(1) more equality than freedom: The relative priority of the good over the right

(2) more freedom than equality: The relative priority of the right over the good

—*Types*

•Only (1): Different versions of communitarian moral universalism

•Only (2): Different versions of liberal moral universalism

•(1) or (2): Different versions of anarchic (non-nation-state) moral universalism

•(1) or (2): Different versions of postmodern moral localism

• **Non-Democracy**

—*Theoretical Constructs*

•The focus on (1') equality or (2') freedom, but not both, regardless of whether freedom and equality can be understood as "negative" or "positive"

—*Types*

•Only (1'): Different versions on the Far Left (e.g., Stalinism, Robespierrianism)

•Only (2'): Different versions on the Far Right (e.g., Nazism, absolute monarchism)

(continued on next page)

Table 4.25. Democracy, Non-Democracy, and Post-Democracy (Part II)

- **Post-Democracy**

  —*Theoretical Constructs*

  - The priority of (1'') equality over freedom, or (2'') freedom over equality, or (3'') the transcendence of freedom and equality, regardless of whether freedom and equality are "negative" or "positive." In degree, (1'') or (2'') is less than (1') or (2') but more than (1) or (2)—respectively.
  - Like democracy and non-democracy, post-democracy is also subject to the freedom/unfreedom and equality/inequality dialectics (or existential dialectics in general). Unlike them, post-democracy acknowledges the constraints of existential dialectics and no longer value freedom and equality as sacred virtues. There is no utopia, in the end; even were there one, dystopia would exist within it.

  —*Types*

  - (1''): Different versions of trans-Sinitic value ideals
  - (1''): Different versions of trans-feminine value ideals
  - (1''): Different versions of trans-Islamic value ideals
  - (1''): Different versions of trans-outerspace value ideals
  - (2''): Different versions of post-human elitist value ideals
  - (3''): Different versions of the value ideals of floating consciousness (etc.)

(continued on next page)

Table 4.25. Democracy, Non-Democracy, and Post-Democracy (Part III)

Notes: The examples are solely illustrative (not exhaustive), nor are they mutually exclusive. As generalities, they allow exceptions. "Negative" freedom is freedom "from" (e.g., freedom from poverty), whereas "positive" freedom is freedom "to" (e.g., freedom to the state of enlightenment). "Negative" equality is "procedural" equality (e.g., equality of opportunity), while "positive" equality is "substantive" equality (e.g., equality of outcome). Existential dialectics impose constraints on freedom and equality in democracy, non-democracy, and post-democracy, regardless of whether freedom and equality can be understood as "negative" or "positive" in conventional discourse. Therefore, do not confuse the transcendence of freedom and equality in (3'') with the naïve temptation to transcend existential dialectics. There is no utopia, in the end; even should there be one, it would not exist without dystopia embedded within it.
Sources: A summary, based on my previous works, especially Ch.5 of *FHC*, Chs.5-10 of *FCD*, Chs.2-4 of *FPHC*, and Chs.1 & 7 of *BDPD*. The reader should consult the books for more analysis, as this is only a summary here.

Table 4.26. Multiple Causes of the Emergence of Post-Democracy (Part I)

• **At the Micro-Physical Level**
—Ex: intelligent life without the human physical-chemical system
—Sources: Ch.7 of *FHC*; Chs.9-10 of *FCD*; Ch.1 of *FPHC*

• **At the Chemical Level**
—Ex: space radiation and toxins
—Sources: Ch.7 of *FHC*; Chs.9-10 of *FCD*

• **At the Bio-Psychological Level**
—Ex: exo-biological evolution in deep space
—Ex: genetic engineering of new beings
—Ex: limits of cognitive partiality
—Ex: illusions of emotional neutrality
—Ex: human biological inequality
—Ex: the rise of unfolding unconsciousness
—Sources: Ch.2 & Chs.9-10 of *FCD*; Ch.7 of *FHC*; Ch.4 of *BCPC*; *FPHU*

• **At the Institutional Level**
—Ex: the flawed logic of equality
—Ex: the conflicting nature of governance
—Sources: Ch. 5 of *FHC*; Chs. 6 & 10 of *FCD*; Ch. 3 of *FPHC*; Chs.2-5 of *BDPD*

• **At the Organizational Level**
—Ex: e-civic alienation
—Ex: the dark sides of formal-legalistic routines
—Sources: Ch.3 of *FHC*; Ch.7 of *FCD*; Ch.3 of *FPHC*

• **At the Structural Level**
—Ex: ever new forms of inequities, at home and abroad
—Ex: the emergence of China, women, and Islam as major actors
—Sources: Chs.5-6 of *FHC*; Chs.7, 9 & 10 of *FCD*; Chs.4-5 of *BDPD*

(continued on next page)

Table 4.26. Multiple Causes of the Emergence of Post-Democracy (Part II)

• **At the Cultural Level**
—Ex: freedom/unfreedom dialectics
—Ex: equality/inequality dialectics
—Ex: system fragmentation and integration
—Sources: Ch.5 of *FHC*; Chs. 3, 9 & 10 of *FCD*; Ch.4 of *FPHC*; Ch.1 of *BDPD*; Ch.4 of *BCPC*

• **At the Systemic Level**
—Ex: space habitats (in zero-gravity) and colonization
—Ex: ultra advanced future info systems
—Ex: qualitative demography
—Sources: Ch.7 of *FHC*; Chs.9 &10 of *FCD*

• **At the Cosmological Level**
—Ex: the colonization of multiverses
—Ex: the alteration of space-time and the creation of new matter-energy
—Ex: the expansion of floating consciousness
—Ex: the spread of hyper-spatial consciousness
—Sources: Ch.7 of *FHC*; Chs.9 &10 of *FCD*; Ch.4 of *FPHC*; *FPHST*

Notes: The examples in each category are solely illustrative (not exhaustive), and some of the items can be reclassified somewhere else. Nor are they always mutually exclusive. Since they are generalities, exceptions are expected.
Sources: Especially from *FHC*, *FCD*, *FPHC*, *BCPC*, *BDPD*, *FPHST*, and *FPHU*. See also other books and my perspectives on civilizational holism.

Table 4.27. Some Clarifications about Post-Capitalism and Post-Democracy (Part I)

• The prefix "trans-" in the first category of post-capitalism (with its four versions) refers to something "going beyond" (not "uniting" or "combining"). Ex: *Sec.10.3.3* of *FCD*; *Sec.2.4* & *Sec.4.4* of *FPHC*; *Sec.7.2* of *BCPC*

• Such terms like "post-democracy," "post-capitalism," "post-human elitist," "trans-feminine calling," and the like as used in my works are more for our current intellectual convenience than to the liking of future humans and post-humans, who will surely invent more tasteful neologisms to call their own eras, entities, and everything else, for that matter. But the didactic point here is to use the terms to foretell what the future might be like, not that its eras and entities must be called so exactly and permanently.
Ex: *Sec.11.1* of *FCD*; *Sec.7.2* of *BCPC*

• The four versions in the first category of post-capitalist value ideals need not automatically be post-democratic, just as capitalism does not necessarily mean democracy. They are two different entities—though closely related. But up to a certain threshold of elevating equality at the farther expense of freedom, the democratic ideals will be overcome and cease to exist. The same is true for the post-human elitist calling in the second category of post-capitalism in relation to post-democracy, depending on the extent to which freedom is elevated at the expense of equality.
Ex: *Sec.10.4.3.3* of *FCD*; *Table 3.9* of *FPHC*; *Table 7.6* of *BDPD*

• The comparison in each of the three realms of existence in all forms of post-capitalism is not absolute, but relative. Examples include "communal" vs. "individualistic," and the like.
Ex: Notes in *Table 10.8*, *Table 10.9*, *Table 10.10*, & *Table 10.11* of *FCD*; Chs.2-4 of *FPHC*; *Sec.7.2* of *BCPC*

(continued on next page)

Table 4.27. Some Clarifications about Post-Capitalism and Post-Democracy (Part II)

• The emergence of post-capitalism (and post-democracy, for that matter) has multiple causes (to not be reduced to one or only a few).
Ex: Ch.10 of *FCD*, Chs.2-4 of *FPHC*; *Sec.1.3* & *Sec.7.2* of *BCPC* (or *Table 1.8* & *Table 7.11*)

• The specific forms of post-capitalism (and post-democracy, for that matter) need to be further developed in future after-postmodern history, as they will be different from the ones we now know. The point here is to solely give an extremely rough sketch of a world to come that we have never known.
Ex: *Sec.10.3.3* & *Sec.10.4.3.3* of *FCD*; *Table 10.14* & *Table 10.15* of *FCD*; *Sec.7.2* of *BCPC*

• All forms of post-capitalism are not part of a "teleological law," but of "historical trends" only. The same is also true for all forms of post-democracy.
Ex: *Sec.7.1* of *FHC*; *Sec.9.5.3.2* & *Sec.10.3.4.2* of *FCD*; *Sec.7.2* of *BCPC*

• Post-capitalism is not better than capitalism in an "absolute" sense but only fits in better, on the basis of the historical contingency of culture, society, nature, and the mind in some future eras. The same is true for post-democracy in relation to democracy. The term "better" is historically relative.
Ex: *Sec.10.3.3* of *FCD*; *Sec.1.7* of *BDPD*; *Sec.1.5* of *BCPC*

(continued on next page)

Table 4.27. Some Clarifications about Post-Capitalism and Post-Democracy (Part III)

• All forms of post-capitalism and post-democracy are subject to the constraints of existential dialectics. In the process, the dialectic direction is to go beyond the conventional "either-or" dichotomies (e.g., freedom vs. unfreedom, equality vs. inequality, freedom vs. equality, individuality vs. communality, spirituality vs. materiality, formal legalism vs. informal legalism, etc.). As is true in post-civilization, to go beyond the dichotomies is to acknowledge the co-existence of both in each dichotomy, although the degree of scaling one over the over varies from case to case (e.g., the theory of post-capitalism I, the theory of post-capitalism II, the theory of post-democracy I, the theory of post-democracy II, etc.)—but is not to be extreme in largely favoring one over the other, *on average* (all things considered). There is no utopia to be had in the end; even should there be one, dystopia would exist within it.
Ex: Ch.5 of *FHC*; *Sec.10.4.4.2* of *FCD*; *Sec.1.5* of *BDPD*; *Sec.1.3* of *BCPC*; *BCIV*

• All forms of post-capitalism, however different from each other though they are, share one common feature, in that they all inspire for a higher spiritual culture. The same is also true for post-democracy.
Ex: *Sec.10.3*, *Sec.10.4* & *Sec.10.5* of *FCD*; *Chs.2-4* of *FPHC*; *Sec.7.2* of *BCPC*

• All forms of post-capitalism try to avoid the excess in capitalist consumerism by favoring more basic than artificial needs in having, but the quality and quantity of these "basic" needs will be measured by future standards, not by our current ones. Standards are historically relative.
Ex: *Sec.10.3*, *Sec.10.4* & *Sec.10.5* of *FCD*; Ch.2 of *FPHC*; *Sec.7.2* of *BCPC*

(continued on next page)

Table 4.27. Some Clarifications about Post-Capitalism and Post-Democracy (Part IV)

• All forms of post-capitalism make use of a different degree of political authority with advanced info systems in future history and strives for higher spiritual cultures (especially in the post-human age), while acknowledging the constraints of existential dialectics and no longer valuing free market (as in capitalism) and economic control (as in non-capitalism) as sacred virtues.
Ex: *Sec.10.3.4.2*, *Sec.10.3*, *Sec.10.4* & *Sec.10.5* of *FCD*; Chs.2-4 of *FPHC*; *Sec.1.5* of *BDPD*; *Sec.7.2* of *BCPC*

Notes:: The main points here are solely illustrative (not exhaustive) nor necessarily mutually exclusive, and the comparison is relative (not absolute). As generalities, they allow exceptions. The sections as cited are only illustrative (not exhaustive).
Sources: From *FHC*, *FCD*, *FPHC*, and *BDPD*

Table 4.28. The Theory of Post-Capitalism I.1: By Group— Ex: Spiritual/Communal in the Trans-Feminine Calling

- **More Communal Than Individual**
  —*Sharing*: learning from others, as different ideas mutually enrich
  —*Cooperative*: encouraging a sense of shared leadership and teamwork

- **More Informal-Legalistic Than Formal-Legalistic**
  —*Specific*: listening more from the heart than from the head, to know a person as a concrete, not as an abstract, unit
  —*Affective*: thinking and acting with others on a more affective tone. Business can mix with an emotional touch.
  —*Ascriptive*: hiring (or firing) can be done on the basis of merit (or lack of it), but deep solidarity (sisterhood) is important too.
  —*Particularistic*: making decisions on the basis of cost-benefit analysis, but a given group relationship is vital

- **More Spiritual Than Secular**
  —*Long-Term Looking*: sharing for a long-term relationship (e.g., love, friendship), not just for a short-term gain
  —*Loving/Caring*: showing compassion for the sufferings of others, without quickly blaming and pre-judging
  —*Respectful*: showing acceptance about others' feelings (and thoughts)

Notes: The categories and examples are solely illustrative, since there can be different versions, and the comparison is relative (not absolute), nor are they mutually exclusive. As generalities, they allow exceptions. The specific forms of the trans-feminine version need to be further developed in future after-postmodern history, as they will be different from the ones we now know, since the prefix "trans-" here means going beyond or deconstructing the feminine values, while using them as the inspirational point at the beginning.
Source: From Ch.10 of *FCD*. Refer to text for more info and references.

Table 4.29. The Theory of Post-Capitalism I.2: By Nation-State— Ex: Spiritual/Communal in the Trans-Sinitic Calling

• **More Communal Than Individualistic**
—*Centralized*: being more top-down in management
—*Collective*: encouraging more group cooperation
—*Social*: investing in trust and connection

• **More Informal-Legalistic Than Formal-Legalistic**
—*Specific*: knowing more of those related or connected
—*Affective*: behaving in a paternalistic, hierarchical way
—*Ascriptive*: favoring family members and those related
—*Particularistic*: building connection (guanxi) as imperative

• **More Spiritual Than Secular**
—*Expansionist*: diffusing civilizational values (e.g., the superiority complex of civilizationalism)
—*Holistic*: synthesizing things into a panoramic horizon
—*Historical*: learning from the lessons of the ancient past
—*Respectful*: deferential to elders and superiors

Notes: The categories and examples are solely illustrative, since there can be different versions, and the comparison is relative (not absolute), nor are they mutually exclusive. As generalities, they allow exceptions. The specific forms of the trans-Sinitic version need to be further developed in future after-postmodern history, as they will be different from the ones we now know, since the prefix "trans-" here means going beyond or deconstructing the Sinitic values, while using them as the inspirational point at the beginning.
Source: From Ch.10 of *FCD*. Refer to text for more info and references.

Table 4.30. The Theory of Post-Capitalism I.3: By Region— Ex: Spiritual/Communal in the Trans-Islamic Calling

- **More Communal Than Individualistic**
  —*Collective*: building the webs of relationships to bind individuals
  —*Sharing*: cultivating the established "wisdom" through common experience
  —*Cooperative*: stressing harmony, solidarity, and commonality

- **More Informal-Legalistic Than Formal-Legalistic**
  —*Specific*: making efforts to know well the participants (family and larger community) in matters of common concern
  —*Affective*: mixing work with language and ritual on explicit religious (Islamic) ideals, texts, stories, and examples
  —*Ascriptive*: privileging local history and custom on relationships among kinship groups
  —*Particularistic*: preferring an unbiased insider with ongoing connections to all parties

- **More Spiritual Than Secular**
  —*Historical*: learning from the lessons of the past as a source of stability and guidance
  —*Deferential*: showing respect for age, experience, status, and leadership in communal affairs
  —*Honorable*: emphasizing face, dignity, prestige, and fairness
  —*Compassionate*: giving mercy and charity ("Zahah") to others

Notes: The categories and examples are solely illustrative (not exhaustive), and the comparison is relative (not absolute), nor are they mutually exclusive. As generalities, they allow exceptions. The specific forms of the trans-Islamic version need to be further developed in future after-postmodern history, as they will be different from the ones we now know, since the prefix "trans-" here means going beyond or deconstructing the Islamic values, while using them as the inspirational point at the beginning.

Sources: From Ch.10 of *FCD*. Refer to text for more info and references, especially from the works by George Irani (2000) and C. Murphy (September 19, 2001).

Table 4.31. The Theory of Post-Capitalism I.4: By Universe— Ex: Spiritual/Communal in the Trans-Outerspace Calling

- **More Communal Than Individual**
  —*Cooperative*: requiring teamwork in small space habitats
  —*Sharing*: learning from, and enjoying being with, each other in a small group in outer space

- **More Informal-Legalistic Than Formal-Legalistic**
  —*Specific*: knowing more about each other to facilitate living and working together in space, both as fellow astronauts and space-mates
  —*Affective*: being friendly and social to each other as vital to working and living in small space quarters
  —*Ascriptive*: nurturing comaraderie among fellow astronauts as if they are family members over time
  —*Particularistic*: building work relationship with enduring memory in a space mission

- **More Spiritual Than Secular**
  —*Long-Term*: looking beyond selfish materialistic concerns in a precarious space environment with potential life or death
  —*Loving/Caring*: cultivating deep bondage for the success of a long term space mission
  —*Transcendent*: searching for life meaning in outer space

Notes: The calling and examples in each category are solely illustrative (not exhaustive), since there will be many different outer-space value ideals in the distant future of space colonization. The comparison is also relative (not absolute), nor are they mutually exclusive. As generalities, they allow exceptions. And the specific forms of trans-outer-space calling need to be further developed in future after-postmodern history, as they will be different from the ones we now know, since the prefix "trans-" here means going beyond or deconstructing the current outer-space values, while using them as the inspirational point at the beginning. The point here is to solely give an extremely rough picture of a small part of the world to come that we still do not know much about.
Source: From Ch.10 of *FCD*. Refer to text for more info and references.

Table 4.32. The Theory of Post-Capitalism II: Spiritual/Individualistic in the Post-Human Elitist Calling (Part I)

- **More Individualistic Than Communal**
  —Setting up rank distinctions among unequals (e.g., between inferior humans and superior post-humans, or later among inferior post-humans and superior ones, relatively speaking)
  —Yearning for being not only distinguished from unequals, but also the first among equals (the best of the very best)
  —Recognizing the constraints of equality/inequality dialectics (or existential dialectics in general)

- **More Spiritual Than Secular**
  —Soul-searching for a high spiritual culture (not the trashy one for the masses). Mass culture is a dirty joke for them.
  —Exploring different spheres of non-human consciousness in the cosmos (something vastly superior than the human one)
  —Recognizing the constraints of freedom/unfreedom dialectics (or existential dialectics in general)

- **Qualifications**
  —Although post-human elitist post-democracy is comparable to post-human elitist post-capitalism in some respects, the former does not necessarily imply the latter (post-human elitist post-capitalism), just as democracy does not have to entail capitalism. They are two different (though related) entities.
  —But up to a certain threshold of incorporating government intervention with advanced info systems in future civilizations for higher spiritual concerns at the expense of the free market and materialist pursuit, the capitalist ideal will be overcome.
  —The overcome will not be Fascist or feudalistic, but post-capitalist, subject to the constraints of existential dialectics.

(continued on next page)

Table 4.32. The Theory of Post-Capitalism II: Spiritual/Individualistic in the Post-Human Elitist Calling (Part II)

Notes: The calling and examples in each category are solely illustrative (not exhaustive). The comparison is also relative (not absolute), nor are they mutually exclusive. As generalities, they allow exceptions. And the specific forms of post-human elitist post-capitalism need to be further developed in future after-postmodern history, as they will be different from the ones we now know, while using them as the inspirational point at the beginning. The point here is to solely give an extremely rough picture of a small part of the world to come that we still do not know much about.
Sources: From Ch.10 of *FCD* (and also *FPHC*, *BDPD*, and *BCPC*). Refer to the text for more info and references.

Table 4.33. Capitalism, Non-Capitalism, and Post-Capitalism (Part I)

- **Capitalism**
  - —*Theoretical Constructs*
    - •Allocation of scarce resources among alternative wants largely by free market for competition (whose characteristics in its ideal form include, for instance, no barrier to entry or exit, homogeneity, perfect information, a large number of buyers/sellers, and perfect factor mobility)
    - •More formal-legalistic than informal-legalistic, more individualistic than communal, and more material (secular) than spiritual
    - •Either (1) minimal government or (2) relatively active government
  - —*Types*
    - •Only (1): Different versions of market capitalism (e.g., the U.S.)
    - •Only (2): Different versions of welfare capitalism (e.g., Sweden)

- **Non-Capitalism**
  - —*Theoretical Constructs*
    - •Allocation of scarce resources among alternative wants mainly by political authority for policies (which can be regulative, redistributive, symbolic, and participatory)
    - •More informal-legalistic than formal-legalistic
    - •Either (1') more individualistic (for the elites), often (though not always) for material (secular) concerns, or (2') more communal (for the masses), often (though not always) for spiritual concerns
  - —*Types*
    - •Only (1'): Different versions on the Right (e.g., Fascist corporate-state economy for the glory of the new Rome, medieval lord-vassal-serf economy for the power of the feudalistic order)
    - •Only (2'): Different versions on the Left (e.g., Soviet command economy for the creation of the New Socialist Man)

(continued on next page)

Table 4.33. Capitalism, Non-Capitalism, and Post-Capitalism (Part II)

- **Post-Capitalism**

—*Theoretical Constructs*

- •Allocation of scarce resources among alternative wants largely by political authority with advanced info systems in future civilizations, subject to existential dialectics. In degree of allocating by authority, post-capitalism is more than capitalism but less than non-capitalism.
- •More spiritual than secular (material)
- •Either (1'') more individualistic or (2") more communal
- •Like capitalism and non-capitalism, post-capitalism is also subject to the freedom/unfreedom and equality/inequality dialectics (or existential dialectics in general). There is no utopia, in the end; even were there one, dystopia would exist within it.
- •Unlike capitalism and non-capitalism, post-capitalism makes use of a different degree of political authority with advanced info systems in future civilizations and strives for higher-spiritual cultures (especially in the post-human age), while acknowledging the constraints of existential dialectics and no longer valuing free market (as in capitalism) and economic control (as in non-capitalism) as sacred virtues.

—*Types*

- •Only (1"): Different versions of post-human elitist value ideals
- •Only (2''): Different versions of trans-Sinitic value ideals
- •Only (2''): Different versions of trans-feminine value ideals
- •Only (2''): Different versions of trans-Islamic value ideals
- •Only (2''): Different versions of trans-outerspace value ideals

(continued on next page)

Table 4.33. Capitalism, Non-Capitalism, and Post-Capitalism (Part III)

Notes: The calling and examples in each category are solely illustrative (not exhaustive). The comparison is also relative (not absolute), nor are they mutually exclusive. As generalities, they allow exceptions. And the specific forms of each calling need to be further developed in future after-postmodern history, as they will be different from the ones we now know, while using them as the inspirational point at the beginning. The point here is to solely give an extremely rough picture of a small part of the world to come that we still do not know much about.

Source: From Ch.10 of *FCD*. Refer to the text for more info and references.

Table 4.34. Multiple Causes of the Emergence of Post-Capitalism (Part I)

- **At the Micro-Physical Level**
  —Ex: intelligent life without the human physical-chemical system
  —Ex: mastering of quantum mechanics, electromagnetism, and other fields for the understanding of a broad range of anomalous experiences and the application for artificial intelligence for spiritual quest
  —Sources: Ch.7 of *FHC*; Chs.9-10 of *FCD*; Ch.1 of *FPHC*

- **At the Chemical Level**
  —Ex: space radiation and toxins
  —Sources: Ch.7 of *FHC*; Chs.9-10 of *FCD*

- **At the Bio-Psychological Level**
  —Ex: exo-biological evolution in deep space
  —Ex: genetic engineering of new beings
  —Ex: limits of human cognition
  —Ex: the rise of unfolding unconsciousness
  —Sources: Ch.2 & Chs.9-10 of *FCD*; Ch.7 of *FHC*; *FPHU*

- **At the Institutional Level**
  —Ex: the flawed logic of the free market
  —Ex: the need of a post-autistic economics
  —Sources: Ch.10 of *FCD*

- **At the Organizational Level**
  —Ex: the dark sides of formal-legalistic routines
  —Sources: Ch.3 of *FHC*; Ch.7 of *FCD*; Ch.3 of *FPHC*

- **At the Structural Level**
  —Ex: ever new forms of inequities, at home and abroad
  —Ex: the emergence of China, women, and Islam as major actors
  —Sources: Chs.5-6 of *FHC*; Chs.7, 9 & 10 of *FCD*; Chs.4-5 of *BDPD*

(continued on next page)

Table 4.34. Multiple Causes of the Emergence of Post-Capitalism (Part II)

• **At the Cultural Level**
—Ex: freedom/unfreedom dialectics
—Ex: equality/inequality dialectics
—Sources: Ch.5 of *FHC*; Chs.3 & 10 of *FCD*; Ch.4 of *FPHC*; Ch.1 of *BDPD*

• **At the Systemic Level**
—Ex: space habitats (in zero-gravity) and colonization
—Ex: ultra advanced future info systems
—Ex: qualitative demography
—Sources: Ch.7 of *FHC*; Chs. 9 & 10 of *FCD*

• **At the Cosmological Level**
—Ex: the colonization of multiverses
—Ex: the alteration of space-time and the creation of new matter-energy
—Ex: the expansion of floating consciousness
—Ex: the spread of hyper-spatial consciousness
—Sources: Ch.7 of *FHC*; Chs. 9 & 10 of *FCD*; Ch.4 of *FPHC*; *FPHST*

Notes: The examples in each category are solely illustrative (not exhaustive), and some of the items can be reclassified somewhere else. Nor are they always mutually exclusive. Since they are generalities, exceptions are expected.
Sources: Especially from *FHC*, *FCD*, *FPHC*, *BCPC*, *BDPD*, *FPHST*, and *FPHU*. See also other books and my perspectives on civilizational holism.

Table 4.35. The Theoretical Debate on Civilization

- **The Progressive Theory of Civilization**
  —*Thesis*: The "civilizing" process is "good," as opposed to the "barbarizing" process as something "bad," relatively speaking.
  —*Discourse*: Especially, though not exclusively, in the Enlightenment era and a bit before. Example: Thomas Hobbes—in that the tribes in primitive societies were "savages."

- **The Romantic Theory of Barbarity**
  —*Thesis*: The "civilizing" process is "bad," as opposed to the "barbarizing" process as something "good," relatively speaking.
  —*Discourse*: Especially, though not exclusively, in the Counter-Enlightenment circle. Example: Jean-Jacques Rousseau—in that civilization "corrupts" men, and the "savages" are in fact "noble."

- **The Moderate Theory of Civilization**
  —*Thesis*: The "civilizing" process is "good," but there is a price to pay, especially in systematic (compulsive) self-control.
  —*Discourse*: Especially, though not exclusively, in some late modern and postmodern circles. Example: Norbert Elias—in that social manners become more refined in the civilizing process, but self-control also becomes more systematic.

- **The Theory of Post-Civilization**
  —*Thesis*: The civilizing process is as evil and good as barbarity, and each cannot exist without the other, to be eventually superseded by post-civilization unto the post-human age.
  —*Discourse*: Proposed by Peter Baofu. See the rest of *BCIV* for more analysis.

Source: From *BCIV* on the theoretical debate

Table 4.36. No Freedom Without Unfreedom in the Civilizing Processes (Part I)

• **The Rationalizing Process (at the Level of Culture)**
—if freer from the dominance of unreason (as in barbarism) in the civilizing process, then less free from the rationalizing process (be it in the form of the principle of either transcendence or immanence)
—if freer from the principle of immanence in the rationalizing process, then less free from the inclination to commit terror in the name of reason and the relative underdevelopment of non-reason (e.g., in relation to yoga and meditation)
—if freer from the principle of transcendence in the rationalizing process, then less free from the relative underdevelopment of reason (e.g., in relation to systematic methodology) and the occurrence of oppression in the name of non-reason

• **The Pacifying Process (at the Level of Society)**
—if freer from the dominance of pillage (as in savagery) in the civilizing process, then less free from the pacifying process (be it in the form of external control or self-control)
—if freer from self-control in the pacifying process, then less free from the temptation of expansionist oppression and rebellious mindset in external control
—if freer from external control in the pacifying process, then less free from the gruesome psychological self-torture and conformism in self-control

• **The Stewardizing Process (at the Level of Nature)**
—if freer from the dominance of nature (as in the state of nature) in the civilizing process, then less free from the stewardizing process (be it in the form of the stewardship of creation or the covenant with nature)
—if freer from the stewardship of creation in the stewardizing process, then less free from material underdevelopment, relatively speaking, and spiritual exclusion in the covenant with nature
—if freer from the covenant with nature in the stewardizing process, then less free from ecological degradation and spiritual disconnection from nature in the stewardship of creation

---

(continued on next page)

Table 4.36. No Freedom Without Unfreedom in the Civilizing Processes (Part II)

- **The Subliming Process (at the Level of the Mind)**
  —if freer from the dominance of spontaneity (as in the wild state of the mind) in the civilizing process, then less free from the subliming process, be it in the form of (cyclical-centric) self-refinement or (linear-centric) self-discipline
  —if freer from (cyclical-centric) self-refinement in the subliming process, then less free from the (linear-centric) self-regimen (as a form of neurosis)
  —if freer from (linear-centric) self-discipline in the subliming process, then less free from the (cyclical-centric) self-torture (equally as a form of neurosis)

Notes: The examples in each category are solely illustrative (not exhaustive), and the comparison is relative (not absolute), nor are they necessarily mutually exclusive. And some can be easily re-classified elsewhere. As generalities, they allow exceptions.
Sources: From *BCIV*. See also *FHC*, *FCD*, *FPHC*, *BDPD*, and *BCPC*.

Table 4.37. No Equality Without Inequality in the Civilizing Processes (Part I)

- **The Rationalizing Process (at the Level of Culture)**
  —if more equal for the role of rationalization in the rationalizing process (of civilizational making), then less equal for that of mythicization (as in barbarism)
  —if more equal for the principle of transcendence in (linear-centric) rationalizing process, then less equal for the principle of immanence
  —if more equal for the principle of immanence in (cyclical-centric) rationalizing process, then less equal for the principle of transcendence

- **The Pacifying Process (at the Level of Society)**
  —if more equal for pacification in civilizational making, then less equal for the institution of pillaging and others (as in savagery)
  —if more equal for external control, relatively speaking, in pacifying process, then less equal for self-control
  —if more equal for self-control, relatively speaking, in pacifying process, then less equal for external-control

- **The Stewardizing Process (at the Level of Nature)**
  —if more equal for stewardship in the stewardizing process (of civilizational making), then less equal for reverent (submissive) existence (as in barbarism)
  —if more equal for the stewardship of creation in (linear- centric) stewardizing process, then less equal for the (cyclical-centric) covenant with nature for harmonious co-existence
  —if more equal for the (cyclical-centric) covenant with nature in the stewardizing process, then less equal for the (linear-centric) stewardship of nature for domination

---

(continued on next page)

Table 4.37. No Equality Without Inequality in the Civilizing Processes (Part II)

- **The Subliming Process (at the Level of the Mind)**
  —if more equal for the role of reason in the subliming process, then less equal for that of unreason (as in the natural state of wildness)
  —if more equal for the primacy of reason in (linear-centric) subliming process, then less equal for other faculties (e.g., intuition, existential feelings, and analogous thinking) in cyclical-centric one
  —if more equal for the exercise of other faculties (e.g., intuition, existential feelings, and analogous thinking) in cyclical-centric subliming process, then less equal for the role of reason in linear-centric counterpart

Notes: The examples in each category are solely illustrative (not exhaustive), and the comparison is relative (not absolute), nor are they mutually exclusive. And some can be easily reclassified else-where. As generalities, they allow exceptions.
Sources: From *BCIV*. See also *FHC*, *FCD*, *FPHC*, *BDPD*, and *BCPC*.

Table 4.38. Five Theses on Post-Civilization

• Post-civilization no longer treats civilization as good and barbarity as evil (relatively speaking), nor does it nostalgically regard barbarity as good and civilization as evil (relatively speaking again). Civilization is as evil and good as barbarity.

• Post-civilization also no longer accepts the dichotomy between civilization and barbarity. Civilization cannot exist without barbarity. It is no longer necessary to preserve civilization, any more than it is imperative to destroy barbarity. To go beyond civilization and barbarity is to acknowledge the co-existence of both, although the degree of scaling one over the over varies from case to case—but is not to be extreme in largely favoring one over the other, *on average* (subject to the constraints of existential dialectics).

• Post-civilization is thus subject to the constraints of existential dialectics. There is no freedom without unfreedom, and no equality without inequality, for instance. There will be no utopia; even should there be one, there would be dystopia embedded within it.

• Post-civilization will eventually replace civilization (as a form of life settlement), to be dominated by post-capitalist and post-democratic lifeforms here on earth and in deep space (besides other alien lifeforms that we have never known), unto the post-human age in multiverses. Those few post-humans who keep civilization will live in a "post-human civilization," while the rest (the majority), who choose post-civilization, will evolve towards the state of "post-human post-civilization." One therefore should not confuse "post-human civilization" with "post-human post-civilization," as the two are not the same.

• Post-civilization will confront psychosis as a primary problem in the culture of virtuality unto the post-human age, just as civilization has neurosis as a primary one of its own (although both neurosis and psychosis are major problems in both).

Notes: The comparison in each category is relative (not absolute), nor are they necessarily mutually exclusive. And some can be easily re-classified elsewhere. As generalities, they allow exceptions.
Sources: From *BCIV*. See also *FHC*, *FCD*, *FPHC*, *BDPD*, and *BCPC*.

Table 4.39. Barbarity, Civilization, and Post-Civilization

- **The Rationalizing Process (at the Level of Culture)**
  —*Barbarity*
    - More mythicizing than rationalizing, relatively speaking
  —*Civilization*
    - More rationalizing than mythicizing, relatively speaking
  —*Post-Civilization*
    - Beyond the dichotomy, subject to existential dialectics

- **The Pacifying Process (at the Level of Society)**
  —*Barbarity*
    - More pillaging than pacifying, relatively speaking
  —*Civilization*
    - More pacifying than pillaging, relatively speaking
  —*Post-Civilization*
    - Beyond the dichotomy, subject to existential dialectics

- **The Stewardizing Process (at the Level of Nature)**
  —*Barbarity*
    - More revering than stewardizing, relatively speaking
  —*Civilization*
    - More stewardizing than revering, relatively speaking
  —*Post-Civilization*
    - Beyond the dichotomy, subject to existential dialectics

- **The Subliming Process (at the Level of the Mind)**
  —*Barbarity*
    - More impulsing than subliming, relatively speaking
  —*Civilization*
    - More subliming than impulsing, relatively speaking
  —*Post-Civilization*
    - Beyond the dichotomy, subject to existential dialectics

Notes: The comparison in each category is relative (not absolute), nor are they necessarily mutually exclusive. And some can be easily re-classified elsewhere. As generalities, they allow exceptions.
Sources: From *BCIV*. See also *FHC*, *FCD*, *FPHC*, *BDPD*, and *BCPC*.

Table 4.40. Types of Super Civilization in the Cosmos (Part I)

• **Type I**

—a civilization which gains control of and uses the total energy output "falling on its planet from its sun for interstellar communication" (or, in general, space colonization). For N. Kardashev, who proposed the first three types, human civilization is currently Type Zero (Type O), which is below even Type I, since its present energy consumption for all purposes, let alone for interstellar communication, is still 10,000 times less.

• **Type II**

—a civilization which gains control of and uses directly the total energy output of its sun for interstellar communication (or, in general, space colonization).

• **Type III**

—a civilization which gains control of and uses the total energy output of its galaxy for interstellar communication (or, in general, space colonization).

• **Type IV**

—a civilization which gains control of and uses the total energy output of its cluster of galaxies for interstellar communication (or, in general, space colonization).

• **Type V**

—a civilization which gains control of and uses the total energy output of its supercluster of galaxies for interstellar communica-tion (or, in general, space colonization).

(continued on next page)

Table 4.40. Types of Super Civilization in the Cosmos (Part II)

- **Type…n**
  —So continues the series in what I call the cyclical progression of hegemony in the cosmos and beyond.

Notes: The Russian astrophysicist Nikolai Kardashev proposed the first three types of super civilization in terms of total energy out-put for interstellar communication. (CSM 1979) I extend his argument further to propose Type IV, Type V, Type VI, and Type…n, in the context of my claim about the cyclical progression of he-gemony in the cosmos and beyond.
Sources: From *Table 9.4* of *FCD*. See *FHC*, *FCD*, and *FPHC* for more info.

Table 4.41. The Civilizational Project from Pre-Modernity to After-Postmodernity (Part I)

	Pre-Modern	*Modern*	*Postmodern*	*After-Postmodern*
Main narratives	•Sacralness •Courtliness •Vitalism •Animism	•Freedom •Equality •Fraternity	•Multiplicity •Hybridization	•Naked contingency •Cyclical progres-sion of hegemony
Main institutions	•Monarchy •Aristocracy •Feudalism •Holy order •Primitivism	•Capitalism •Liberalism •Socialism •Nazism •Fascism	•Capitalism •Liberalism •Postmodern politics of difference	•Post-Capitalism •Post-Democracy •Others
Main techno-logical and economic revolutions	•Agricultural	•Service •Industrial	•Informational	•Biological •Material •Energy •Space •Others

(continued on next page)

Table 4.41. The Civilizational Project from Pre-Modernity to After-Postmodernity (Part II)

	Pre-Modern	*Modern*	*Postmodern*	*After-Postmodern*
Main agents	•Males •Upper strata •Mini-states	•Males •Upper strata •Whites •Empires	•Males •Upper strata •Whites •Others •Supra-states •IO's	•Post-humans •Humans •Others
Main impacts	•Local	•Inter-national	•Global	•Outer-space •Multiverse
Main outcomes	•Towards moderntiy •Rise of linear- & cyclical-centric civiliza-tions	•Towards post-modernity •Dominance of linear-centric civilization	•Towards after-post-moderntiy •Linear-centric civiliza-tion in crisis	•Towards human (& maybe post-human) extinction •Rise of post-civiliza-tion, especially in post-human forms of space-time

Notes: The examples in each category are solely illustrative (not exhaustive) nor necessarily mutually exclusive, and the comparison is relative (not absolute). As generalities, they allow exceptions.

Sources: From *Table 10.16* of *FCD*—and also from *BCIV* on post-civilization (and *FPHST*)

Table 4.42. Civilizational Holism
(Part I)

• **At the Micro-Physical Theoretical Level**
—Ex: Mastering of quantum mechanics, electromagnetism, and other fields for the understanding of a broad range of anomalous experiences and theapplication for artificial intelligence (*Sec.1.4.1* of *FPHC*)

• **At the Chemical Theoretical Level**
—Ex: Unprecedented expansion of (and violence to) the mind through ever new forms of drugs (and virtual technologies, for that matter) (Ch.9 of *FCD*)

• **At the Biological Theoretical Level**
—Ex: Humans are not biologically equal, on the basis of race, gender, ethnicity, age, and whatnot. (*Sec.2.6* & Ch.10 of *FCD*; *BNN*) And post-humans will experience the same fate, in an even more amazing way.

• **At the Psychological Theoretical Level**
—Ex: Human cognitive impartiality and emotional neutrality are quite limited. (*Secs.2.4-2.5* of *FCD*)
—Ex: Rise of Floating Consciousness (Ch.10 of *FCD*; Chs.1 & 4 of *FPHC*) and Unfolding Unconsciousness (*FPHU*)

• **At the Organizational Theoretical Level**
—Ex: Administrative colonization of deep space, with less legal-formalism in some corners. (Chs.9-10 of *FCD*)

• **At the Institutional Theoretical Level**
—Ex: Both capitalism and democracy will not last, to be superseded by different versions of post-capitalism and post-democracy in after-postmodernity. (Ch.10 of *FCD*)

(continued on next page)

**Table 4.42. Civilizational Holism
(Part II)**

- **At the Structural Theoretical Level**
  —Ex: Social stratification reappears in ever new forms, also with new causes and new players in the cyclical progression of hegemony. (Chs.8-10 of *FCD*)
  —Ex: The world of hyper-empires, and the union of the unions (*BWT*)

- **At the Systemic Theoretical Level**
  —Ex: Outerspace expansion: local → regional → global → solar → galactic → clustery → multiversal (Ch.9 of *FCD*)
  —Ex: Demographic transition: human extinction, and the rise of post-humans (e.g., cyborgs, thinking machines, thinking robots, genetically altered superior beings, floating consciousness, hyper-spatial consciousness) (Ch.4 of *FPHC*; Ch.10 of *FCD*; & Ch.7 of *FHC*)
  —Ex: New technological forces in material sciences, electronic and communication sciences, energy sciences, biosciences, manufacturing and engineering sciences, and space sciences (Ch.10 of *FCD* & Ch.7 of *FHC*)
  —Ex: Systematic dominance towards nature for space colonization (Chs.9-10 of *FCD*; Chs.2 & 7 of *FHC*)

- **At the Cultural Theoretical Level**
  —Ex: The post-human transcendence of freedom and equality (Ch.10 of *FCD*)
  —Ex: Methodological Holism (Ch.1 of *FCD*; Ch.1 of *FPHC*; *Sec.2.1* & *Sec.2.5* of *BCPC*)
  —Ex: The Evolution from Barbarity to Post-Civilization (*BCIV*)

- **At the Cosmological Theoretical Level**
  —Ex: Mastering of dark matter and dark energy, and the exploration of multiverses (Ch.4 of *FPHC*; Ch.10 of *FCD*; & Ch.7 of *FHC*)
  —Ex: Alternation of space-time (*FPHST*)
  —Ex: The emergence of hyper-spatial consciousness (*FPHC*)

(continued on next page)

**Table 4.42. Civilizational Holism
(Part III)**

- **At Other Levels**
  —Ex: Historical: pre-modernity → modernity → postmodernity → after-postmodernity (human distinction, and the rise of post-humans, including floating consciousness, hyper-spatial consciousness, and unfolding unconsciousness) (Ch.7 of *FHC*; Ch.10 of *FCD*; *FPHC*; *FPHU*)

Notes: These examples are solely illustrative (not exhaustive), and some of the items can be reclassified somewhere else. Nor are they always mutually exclusive. Since they are generalities, exceptions are expected. And the comparison is relative, not absolute.
Sources: From *Table 5.1* of *FPHC*—with details from *FHC*, *FCD*, and the rest of my books. Not every aspect in each category as discussed in all my books are presented here, since there are too many issues. For more info, also consult the table on theories on civilizational holism and, of course, the books themselves.

Table 4.43. Theories on Civilizational Holism (Part I)

I. Theories in Relation to Nature

—*At the Macro-Physical (Cosmological) Theoretical Level*

- 43. Theory of Post-Cosmology (Peter Baofu) (*BCOS*)
- 42. Theory of Hyper-Spatial Consciousness (Peter Baofu) (Ch.4 of *FPHC*; *FPHG*)
- (• 40). Selective Theory of Geometry (Peter Baofu) (*FPHG*)
- (• 39). Perspectival Theory of Space-Time (Peter Baofu) (*FPHST*)
- (• 38). Dialectic Theory of Complexity (Peter Baofu) (*FC*)
- (• 9). Theory of Floating Consciousness (Peter Baofu) (Ch.10 of *FCD*; Chs.1 & 4 of *FPHC*)
- 41. Theory of the Geopower of Nature (Peter Baofu) (Ch.4 of *ALD*)

—*At the Micro-Physical Theoretical Leve*

- 40. Selective Theory of Geometry (Peter Baofu) (*FPHG*)
- 39. Perspectival Theory of Space-Time (Peter Baofu) (*FPHST*)
- 38. Dialectic Theory of Complexity (Peter Baofu) (*FC*)

(continued on next page)

Table 4.43. Theories on Civilizational Holism (Part II)

II. Theories in Relation to Culture

—At the Cultural Theoretical Level

- • 37. Theory of Post-Ethics (Peter Baofu)
  (*BEPE*)
- • 36. Dualistic Theory of Mass Culture (Peter Baofu)
  (Ch.2 of *FHC*)
- • 35. Comparative Theory of Religion (Peter Baofu)
  (Ch.3 of *FHC*; Ch.9 of *FCD*; Ch.1 of *FPHK*)
- • 34. Theory of Post-Civilization (Peter Baofu)
  (*BCIV*)
- • 33. Theory of the Trinity of Modernity to Its After-Postmodern Counterpart (Peter Baofu)
  (*FHC*; Ch.10 of *FCD*)
- • 32. Transformative Theory of Aesthetic Experience (Peter Baofu)
  (*FAE*)
- (• 17). Theory of Post-Capitalism (Peter Baofu)
  (Ch.10 of *FCD*; Chs.2 & 4 of *FPHC*; *BCPC*)
- (• 16). Theory of Post-Democracy (Peter Baofu)
  (Ch.10 of *FCD*; Chs.3 & 4 of *FPHC*; *BDPD*)
- (• 5). Theory of Existential Dialectics, or the Holistic Theory of Knowledge (Peter Baofu)
  (*FHC*; *FCD*; *FPHC*; *BDPD*; *FC*; *FAE*; *ALD; FIA; FPHK*; etc,)
- (• 4). Contrastive Theory of Rationality (Peter Baofu)
  (*FPHML*)
- (• 2). Theory of Methodological Holism (Peter Baofu)
  (Ch.1 of *FCD*; Ch.1of *FPHC*; *Sec.2.1* & *Sec.2.5* of *BCPC*; *FC*; *FPHK*; etc.)

(continued on next page)

Table 4.43. Theories on Civilizational Holism (Part III)

III. Theories in Relation to the Mind

—*At the Biological Theoretical Level*

- 31. Theory of Contrastive Advantages (Peter Baofu) (*Sec.2.6* & Ch.10 of *FCD; BNN*)
- (• 24). Theory of Floating Consciousness (Peter Baofu) (Ch.10 of *FCD*; Chs.1 & 4 of *FPHC*)

—*At the Psychological Theoretical Level*

- 30. Contrarian Theory of Personality (Peter Baofu) (*FPHP*)
- 29. Theory of Virtual Sexuality (Peter Baofu) (*FPHS*)
- 28. Expansive-Contractive Theory of Martial Arts (Peter Baofu) (*FPHMA*)
- 27. Multilogical Theory of Learning (Peter Baofu) (*FPHL*)
- 26. Comprehensive Theory of Creative Thinking (Peter Baofu) (*FPHCT*)
- 25. Theory of Unfolding Unconsciousness—also known as the Unfolding Theory of Anomalous Experience (Peter Baofu) (*FPHU*)
- 24. Theory of Floating Consciousness (Peter Baofu) (Ch.10 of *FCD*; Chs.1 & 4 of *FPHC*)
- 23. Theory of Cognitive Partiality (Peter Baofu) (*Sec.2.4* of *FCD*; *Sec.4.5.1.1* of *BCPC*)
- 22. Theory of Emotional Non-Neutrality (Peter Baofu) (*Sec.2.5* of *FCD*; *Sec.4.5.2* of of *BCPC*)
- 21. Theory of Behavioral Alteration (Peter Baofu) (*Sec.4.5.3* of *BCPC*)

Table 4.43. Theories on Civilizational Holism (Part IV)

IV. Theories in Relation to Society

—*At the Organizational Theoretical Level*

- 20. Theory of E-Civic Alienation (Peter Baofu) (Ch.7 of *FCD*)
- 19. Combinational Theory of Organization (Peter Baofu) (*FPHO*; Ch.6 of *ALD*)

—*At the Institutional Theoretical Level*

- 18. Reconstruction of Theory of Law (Peter Baofu) (*FPHLAW*)
- 17. Theory of Post-Capitalism (Peter Baofu) (Ch.10 of *FCD*; Chs.2 & 4 of *FPHC*; *BCPC*)
- 16. Theory of Post-Democracy (Peter Baofu) (Ch.10 of *FCD*; Chs.3 & 4 of *FPHC*; *BDPD*)
- 15. Dynamic Theory of Comparative Political Systems (Peter Baofu) (*ALD*)

—*At the Systemic Theoretical Level*

- 14. Contingent Theory of Urban Planning (Peter Baofu) (*FPHUP*)
- 13. Totalistic Theory of Communication (Peter Baofu) (*FPHMM*; *FCD*; *FHC*)
- 12. Ambivalent Theory of Technology (Peter Baofu) (*FPHE*; *FCD*; *FHC*)
- 11. Multifaceted Theory of War and Peace (Peter Baofu) (Ch.9 of *FCD*; Ch.1 of *FPHK*)
- 10. Theory of Post-Humanity (Peter Baofu) (Ch.7 of *FHC*; Chs.3, & 10 of *FCD*; Chs.1, 3 & 4 of *FPHC*; and other books of mine)
- 9. Theory of the Cyclical Progression of System Integration and Fragmentation (Peter Baofu) (Chs.9-10 of *FCD*)
- 8. Synthetic Theory of Information Architecture (Peter Baofu) (*FIA*)

(continued on next page)

Table 4.43. Theories on Civilizational Holism (Part V)

IV. Theories in Relation to Society (*cont'd*)

—At the Structural Theoretical Level

- 7. Theory of the Cyclical Progression of Hegemony (Peter Baofu) (Chs.9-10 of *FCD*; Chs.1, 3 & 4 of *FPHC*; *BDPD*)
- 6. Theory of the Cyclical Progression of Empire-Building (Peter Baofu) (*BWT*)

V. Meta-Theories (in Relation to Theories)

—At the Ontological Meta-Theoretical Level

- 5. Theory of Existential Dialectics, or the Holistic Theory of Knowledge (Peter Baofu) (*FHC*; *FCD*; *FPHC*; *BDPD*; *FC*; *FAE*; *ALD; FIA; FPHK*; etc,)
- 4. Contrastive Theory of Rationality (Peter Baofu) (*FPHML*)

—At the Methodological Meta-Theoretical Level

- 3. Critical-Dialectic Theory of Formal Science (Peter Baofu) (*FPHFS*)
- 2. Theory of Methodological Holism (Peter Baofu) (Ch.1 of *FCD*; Ch.1of *FPHC*; *Sec.2.1* & *Sec.2.5* of *BCPC*; *FC*; *FPHK*; etc.)

VI. Theories in Relation to the Rest

—At Other Levels (Historical)

- 1. Theory of the Evolution from Pre-Modernity to After-Postmodernity (Peter Baofu) (*FHC*; Ch.9-10 of *FCD*; *FPHC*)

Notes: All these theories are my constructions, as some of the main contributions of my grant project on civilization and its future. These examples are solely illustrative (not exhaustive), and some of the items can be reclassified somewhere else. Nor are they always mutually exclusive. Since they are generalities, exceptions are expected.

Sources: From my previous books.

BIBLIOGRAPHY

Adams, Fred; & Gregory Laughlin. 1999. *The Five Ages of the Universe*. New York: The Free Press.

_____.& Gregory Laughlin. 1997. "A Dying Universe: The Long-Term Fate and Evolution of Astrophysical Bbjects," *Reviews of Modern Physics*, 69 (2): 337–372. <http://rmp.aps.org/abstract/RMP/v69/i2/p337_1>.

Alpher, R. A.; & G. Gamow. 1948. "The Origin of Chemical Elements." *Physical Review*, 73: 803. <http://adsabs.harvard.edu/abs/1948PhRv...73..803A>.

_____.1948a. "Evolution of the Universe." *Nature,* 162: 774. <doi:10.1045/march2004-featured.collection>.

American Association for the Advancement of Science, the (AAAS). 2006. "Statement on the Teaching of Evolution" (February 16). <http://www.aaas.org/news/releases/2006/pdf/0219boardstatement.pdf>.

Aguirre, Anthony; & Steven Gratton. 2003. "Inflation without a Beginning: A Null Boundary Proposal." *Phys. Rev.* D67 083515. <http://arxiv.org/abs/gr-qc/0301042>.

_____. & S. Gratton. 2002. "Steady-State Eternal Inflation." *Phys. Rev.* D65 083507. <http://arxiv.org/abs/astro-ph/0111191>.

Arany-Prado, Lilia I.; Eduardo F. del Peloso; Licio da Silva; & Gustavo F. Porto de Mello. 2005. "The Age of the Galactic Thin Disk from Th/Eu Nucleocosmochronology: Extended Sample." *Proceedings of the International Astronomical Union*, 1: 485-486. Cambridge: Cambridge University Press.

Aurobindo, Sri. 1977. *The Life Divine*. Sri Aurobindo Ashram Trust.

Baofu, Peter. 2010. *Beyond Ethics to Post-Ethics: A Preface to a New Theory of Morality and Immorality*. Charlotte, NC: Information Age Publishing.

_____.2010a. *The Future of Post-Human Personaltiy: A Preface to a New Theory of Normality and Abnormality*. Cambridge, England: Cambridge International Science Publishing, Ltd.

_____.2010b. *The Future of Post-Human War and Peace: A Preface to a New Theory of Aggression and Pacificity*. Cambridge, England: Cambridge Scholars Publishing.

_____.2010c. *The Future of Post-Human Formal Science: A Preface to a New*

Theory of Abstraction and Application. Cambridge, England: Cambridge Scholars Publishing.

_____.2010d. *The Future of Post-Human Law: A Preface to a New Theory of Necessity, Contingency, and Justice*. Cambridge, England: Cambridge Scholars Publishing.

_____.2010e. *The Future of Post-Human Sexuality: A Preface to a New Theory of the Body and Spirit of Love Makers*. Cambridge, England: Cambridge Scholars Publishing.

_____,2009. *The Future of Post-Human Martial Arts: A Preface to a New Theory of the Body and Spirit of Warriors*. Cambridge, England: Cambridge Scholars Publishing.

_____.2009a. *The Future of Post-Human Organization: A Preface to a New Theory of Communication, Decision-Making, and Leadership*. Cambridge, England: Cambridge Scholars Publishing.

_____.2009b. *The Future of Post-Human Language: A Preface to a New Theory of Structure, Context, and Learning*. Cambridge, England: Cambridge Scholars Publishing.

_____.2009c. *The Future of Post-Human Urban Planning: A Preface to a New Theory of Density, Void, and Sustainability*. Cambridge, England: Cambridge Scholars Publishing.

_____.2009d. *The Future of Post-Human Geometry: A Preface to a New Theory of Infinity, Symmetry, and Dimensionality*. Cambridge, England: Cambridge Scholars Publishing.

_____.2009e. *The Future of Post-Human Creative Thinking: A Preface to a New Theory of Invention and Innovation*. Cambridge, England: Cambridge Scholars Publishing.

_____.2009f. *The Future of Post-Human Mass Media: A Preface to a New Theory of Communication*. Cambridge, England: Cambridge Scholars Publishing.

_____.2009g. *The Future of Post-Human Engineering: A Preface to a New Theory of Technology*. Cambridge, England: Cambridge Scholars Publishing.

_____.2008. *The Future of Post-Human Mathematical Logic: A Preface to a New Theory of Rationality*. Cambridge, England: Cambridge Scholars Publishing.

_____.2008a. *The Future of Post-Human Knowledge: A Preface to a New Theory of Methodology and Ontology*. Oxford, United Kingdom: Chandos Publishing (Oxford) Ltd.

_____.2008b. *The Future of Post-Human Unconsciousness: A Preface to a New Theory of Anomalous Experience*. Cambridge, England: Cambridge Scholars Publishing.

_____.2008c. *The Future of Information Architecture: Conceiving a New Way to Understand Taxonomy, Network, and Intelligence*. Oxford, United Kingdom:

Chandos Publishing (Oxford) Ltd.

_____.2007. *The Rise of Authoritarian Liberal Democracy: A Preface to a New Theory of Comparative Political Systems*. Cambridge, England: Cambridge Scholars Publishing.

_____.2007a. *The Future of Aesthetic Experience: Conceiving a Better Way to Beauty, Ugliness, and the Rest.* Cambridge, England: Cambridge Scholars Publishing.

_____.2007b. *The Future of Complexity: Conceiving a Better Way to Understand Order and Chaos.* London, United Kingdom: World Scientific Publishing Co.

_____.2007c. *Beyond the World of Titans, and the Remaking of World Order: A Preface to a New Logic of Empire-Building.* Cambridge, England: Cambridge Scholars Publishing.

_____.2006. *Beyond Nature and Nurture: Conceiving a Better Way to Understand Genes and Memes.* Cambridge, England: Cambridge Scholars Publishing.

_____.2006a. *The Future of Post-Human Space-Time: Conceiving a Better Way to Understand Space and Time.* New York: Peter Lang Publishing, Inc.

_____.2006b. *Beyond Civilization to Post-Civilization: Conceiving a Better Model of Life Settlement to Supersede Civilization.* New York: Peter Lang Publishing, Inc.

_____.2005. *Beyond Capitalism to Post-Capitalism: Conceiving a Better Model of Wealth Acquisition to Supersede Capitalism.* New York: The Edwin Mellen Press.

_____.2004. *The Future of Post-Human Consciousness.* New York: The Edwin Mellen Press.

_____.2004a. *Beyond Democracy to Post-Democracy: Conceiving a Better Model of Governance to Supersede Democracy.* Volume 1. New York: The Edwin Mellen Press.

_____.2004b. *Beyond Democracy to Post-Democracy: Conceiving a Better Model of Governance to Supersede Democracy.* Volume 2. New York: The Edwin Mellen Press.

_____.2002. *The Future of Capitalism and Democracy.* Maryland: The University Press of America.

_____.2000. *The Future of Human Civilization.* Volume 1. New York: The Edwin Mellen Press.

_____.2000a. *The Future of Human Civilization.* Volume 2. New York: The Edwin Mellen Press.

Barrow, John D. 1998. *Impossibility: The Limits of Science and the Science of Limits.* Oxford: Oxford University Press.

BBC News (BBC). 2000. "Quiet Cult's Doomsday Deaths" (March 29).

<http://news.bbc.co.uk/2/hi/africa/683813.stm>.

Beretta, Francesco. 2005. "Galileo, Urban VIII, and the Prosecution of Natural Philosophers." *The Church and Galileo*. Ernan McMullin, ed. Notre Dame, IN: University of Notre Dame Press.

Biletzki, Anat; & Anat Matar. 2006. "Ludwig Wittgenstein" (Nov. 17). *Stanford Encyclopedia of Philosophy*. <http://plato.stanford.edu/entries/wittgenstein/>.

Blackwell, Richard J. 1991. *Galileo, Bellarmine, and the Bible*. Notre Dame, IN: University of Notre Dame Press.

Blavatsky, Helena P. 1977. *The Secret Doctrine: The Synthesis of Science, Religion, and Philosophy*. Theosophical University Press.

Boggess, N.W., et al. 1992. "The COBE Mission: Its Design and Performance Two Years after the launch." *Astrophysical Journal*, 397: 420. <http://adsabs.harvard.edu/doi/10.1086/171797>.

Borzello, Anna. 2000. "A Party, Prayers, Then Mass Suicide" (March 20). *The Guardian*. <http://www.guardian.co.uk/world/2000/mar/20/3>.

_____.2000a. "Mass Graves Found in Sect House" (March 25). *The* Guardian. <http://www.guardian.co.uk/world/2000/mar/25/1>.

Bunge, Mario. 1999. *Dictionary of philosophy*. Amherst, MA: Prometheus Books.

Callan, C.; & S. Coleman. 1977. "Fate of the False Vacuum. II. First Quantum Corrections." *Phys. Rev*., D16: 1762–68.

Cambridge Cosmology (CC). 2010. "The Four Pillars of the Standard Cosmology" (February 27). <http://www.damtp.cam.ac.uk/research/gr/public/bb_pillars.html>.

Carroll, Rory. 2000. "Pope Says Sorry for Sins of Church" (March 13). *The Guardian*. <http://www.guardian.co.uk/world/2000/mar/13/catholicism.religion>.

Caston, Victor. 2006. "Aristotle's Psychology." *A Companion to Ancient Philosophy*. Mary Gill and Pierre Pellegrin, eds. Hoboken: Wiley-Blackwell.

_____.1999. "Aristotle's Two Intellects: A Modest Proposal." *Phronesis*, 44.3: 199-227.

Cauvin, Henri. 2000. "Fateful Meeting Led to Founding of Cult in Uganda" (March 27). *The New York Times*. <http://www.nytimes.com/2000/03/27/world/fateful-meeting-led-to-founding-of-cult-in-uganda.html?pagewanted=1>.

_____.2000a. "Evidence Indicates Uganda Cult Held an Eerie Prelude to Fire" (March 26). *The New York Times*. <http://www.rickross.com/reference/tencommandments/tencommandments66.html>.

Chase, Scott. 1993. "Tachyons" (March). *Physics FAQ*. <http://math.ucr.edu/home/baez/physics/ParticleAndNuclear/tachyons.html>.

Christianson, E. 1995. *Edwin Hubble: Mariner of the Nebulae*. New York: Farrar, Straus and Giroux.

Cline, Austin. 2008. "Cosmological Argument: Does the Universe Require a First Cause?" (June 20). *About.com: Agnosticism/Atheism.* <http://atheism.about.com/od/argumentsforgod/a/cosmological.htm>.

Coleman, S. 1977. "Fate of the False Vacuum: Semiclassical Theory." *Phys. Rev.*, D15: 2929–36.

Collinson, Diane; & Robert Wilkinson. 1994. *Thirty-five Oriental Philosophers.* NY: Routledge.

Corazzon, Raul. 2007. "Definitions of Ontology: From Nicolai Hartmann to the Present Time." <http://www.formalontology.it/section_4_two.htm>.

Corbett, Dan. 2008. "Introduction to String Theory" (November 23). *ThinkQuest.* <http://library.thinkquest.org/27930/stringtheory1.htm>.

Craig, William L. 2008. "Initial Arguments: A Defense of the Cosmological Argument for the Existence of God" (June 20). <http://www.leaderu.com/offices/billcraig/docs/craig-smith1.html>.

Cross, Richard. 1998. *The Physics of Duns Scotus.* Oxford: Oxford University Press.

Deutsch, Eliot. 1980. *Advaita Vedanta: A Philosophical Reconstruction.* Hawaii: University of Hawaii Press.

Drake, Stillman. 1978. *Galileo At Work.* Chicago, IL: University of Chicago Press.

Dyson, Freeman J. 1979. "Time Without End: Physics and Biology in an Open Universe." *Reviews of Modern Physics*, 51: 447. <http://www.think-aboutit.com/Misc/time_without_end.htm>.

Ellis, George; U. Kirchner; & W. R. Stoeger. 2004. "Multiverses and Physical Cosmology." *Monthly Notices of the Royal Astronomical Society*, 347: 921–936. <http://arxiv.org/abs/astro-ph/0305292>.

Evans, Robert. 2010. "Big Bang Experiment May Reveal Dark Universe: CERN" (March 08). *Yahoo! News.* <http://news.yahoo.com/s/nm/20100308/sc_nm/us_science_cern>.

Everett, Hugh. 1957. "'Relative State' Formulation of Quantum Mechanics" (July). *Reviews of Modern Physics*, vol. 29, no.3. <http://www.univer.omsk.su/omsk/Sci/Everett/paper1957.html>.

Fahlbusch, Erwin; Geoffrey William Bromiley; & David B. Barrett. 1999. "Panentheism." *The encyclopedia of Christianity*, 4: 21. Wm. B. Eerdmans Publishing.

Fantoli, Annibale. 2005. "The Disputed Injunction and Its Role in Galileo's Trial." *The Church and Galileo.* Ernan McMullin, ed. Notre Dame, IN: University of Notre Dame Press.

Feldman, Fred. 2005. "The Open Question Argument: What It Isn't; and What It Is." *Philosophical Issues*, 15. <http://www3.interscience.wiley.com/journal/120710700/abstract?CRETRY=1&SRETRY=0>.

Feuerbacher, B.; & R. Scranton. 2006. "Evidence for the Big Bang" (January 25). *TalkOrigins.* <http://www.talkorigins.org/faqs/astronomy/bigbang.html#evid

ence>.

Feynman, Richard P. & A. R. Hibbs. 1965. *Quantum Physics and Path Integrals.* New York: McGraw–Hill.

Finocchiaro, Maurice. 1989. *The Galileo Affair: A Documentary History.* Berkeley, CA: The University of California Press.

Firehammer, Reginald. 2005. "Basic Principles of Ontology" (January 05). <http://usabig.com/autonomist/philosophy/ontology2.html>.

Fisher, Ian. 2000. "Uganda Survivor Tells of Questions When World Didn't End" (April 03). *The New York Times.* <http://www.nytimes.com/2000/04/03/world/uganda-survivor-tells-of-questions-when-world-didn-t-end.html?pagewanted=1>.

_____.2000a. "Uganda Cult's Mystique Finally Turned Deadly" (April 02). *The New York Times.* <http://www.nytimes.com/2000/04/02/world/uganda-cult-s-mystique-finally-turned-deadly.html?pagewanted=all>.

Frampton, P. H. 1977. "Consequences of Vacuum Instability in Quantum Field Theory." *Phys. Rev.*, D15: 2922–28.

_____.1976. "Vacuum Instability and Higgs Scalar Mass." *Phys. Rev. Lett.*, 37: 1378–1380.

Friedman, A. 1999. "On the Curvature of Space." *General Relativity and Gravitation*, 31: 1991–2000. <http://adsabs.harvard.edu/abs/1999GReGr..31.1991F>.

Gladders, M.D.; H. K. C. Yee; Subhabrata Majumdar; L. Felipe Barrientos; Henk Hoekstra; Patrick B. Hall; & Leopoldo Infante. 2007. "Cosmological Constraints from the Red-Sequence Cluster Survey." *The Astrophysical Journal*, 655 (1): 128–134. <http://adsabs.harvard.edu/abs/2007ApJ...655..128G>.

Gott, J. Richard. 2002. *Time Travel in Einstein's Universe.* Boston : Houghton Mifflin.

Gribbin, John. 2010. "Inflation for Beginners" (February 28). <http://www.lifesci.sussex.ac.uk/home/John_Gribbin/cosmo.htm>.

Guth, Alan. 1997. *The Inflationary Universe: The Quest for a New Theory Of Cosmic Origins.* New York: Helix Books.

Harrison, Paul. 1999. *Elements of Pantheism.* Boston, MA: Element Books.

Hawking, Stephen; & Leonard Mlodinow. 2010. *The Grand Design.* NY: Bantam Books.

_____.2005. "Gödel and the End of Physics." <http://www.damtp.cam.ac.uk/strings02/dirac/hawking/>.

_____.1988. *A Brief History of Time: From the Big Bang to Black Hole.* New York: Bantam Books.

_____.1973. *The Large Scale Structure of Space-Time.* Cambridge, UK: Cambridge University Press.

Heindel, Max. 1909. "The Rosicrucian Cosmo-Conception." <http://www.rosicrucian.com/rcc/rcceng14.htm#par67>.

Hexham, Irving; & Karla Poewe. 1997. "UFO Religion—Making Sense of the Heaven's Gate Suicides" (May 07). *Christian Century*: 439–440. <http://www.ucalgary.ca/~nurelweb/papers/irving/HGCC.html>.

Hinshaw, G.; et al. 2009. "Five-Year Wilkinson Microwave Anisotropy Probe (WMAP) Observations: Data Processing, Sky Maps, and Basic Results." *Astrophysical Journal Supplement*, 180: 225–245. <http://lambda.gsfc.nasa.gov/product/map/dr3/pub_papers/fiveyear/basic_results/wmap5basic.pdf>.

Hoyle, F. 1948. "A New Model for the Expanding Universe." *Monthly Notices of the Royal Astronomical Society*, 108: 372. <http://adsabs.harvard.edu/abs/1948MNRAS.108..372H>.

Hubble, E. 1929. "A Relation Between Distance and Radial Velocity Among Extra-Galactic Nebulae." *Proceedings of the National Academy of Sciences*, 15 (3): 168–73. <http://antwrp.gsfc.nasa.gov/debate/1996/hub_1929.html>.

Interacademy Panel, the (IAP). 2006. "IAP Statement on the Teaching of Evolution" (June 21). <http://www.interacademies.net/Object.File/Master/6/150/Evolution%20statement.pdf>.

Irion, Robert; Neil de Grasse Tyson; & Charles Tsun-Chu Liu. 2000. *One Universe: At Home in the Cosmos*. Washington, D.C.: Joseph Henry Press.

Irwin, Terence. 1990. *Aristotle's First Principles*. Oxford, United Kindgom: Oxford University Press.

Joyce, George Hayward. 1922. *Principles of Natural Theology*. NY: Longmans Green.

Kaku, Michio. 1994. *Hyperspace: A Scientific Odyssey Through Parallel Universes, Time Warps, and The Tenth Dimension*. Illustrations by Robert O'Keefe. Oxford: Oxford University Press.

Kenny, Anthony. 1993. *Aquinas on Mind*. London, United Kingdom: Routledge.

Khoury1, J.; B. A. Ovrut; N. Seiberg; P. Steinhardt; & N. Turok. 2002. "From Big Crunch to Big Bang." *Physical Review D*, 65 (8). <http://prd.aps.org/abstract/PRD/v65/i8/e086007>.

Kolb, Edward; & Michael Turner. 1988. *The Early Universe*. Reading, MA: Addison–Wesley.

Komatsu, E., et al. 2009. "Five-Year Wilkinson Microwave Anisotropy Probe Observations: Cosmological Interpretation." *Astrophysical Journal Supplement*, 180: 330. <http://www.iop.org/EJ/abstract/0067-0049/180/2/330/>.

Kragh, H. 1996. *Cosmology and Controversy*. Princeton, NJ: Princeton University Press.

Krauss, Lawrence, M.: & Glenn D. Starkman. 2000. "Life, the Universe, and Nothing: Life and Death in an Ever-expanding Universe." *Astrophysical*

Journal, 531: 22–30. <http://iopscience.iop.org/0004-637X/531/1/22/>.

Lemaître, G. 1931. "A Homogeneous Universe of Constant Mass and Growing Radius Accounting for the Radial Velocity of Extragalactic Nebulae." *Monthly Notices of the Royal Astronomical Society*, 91: 483–490. <http://adsabs.harvard.edu/abs/1931MNRAS..91..483L>.

_____.1931a. "The Evolution of the Universe: Discussion." *Nature*, 128: 699–701. <http://www.nature.com/doifinder/10.1038/128704a0>.

Lewis, David. 1983. "New Work for a Theory of Universals." *Australasian Journal of Philosophy*.

Linde, Andre. 2007. "Sinks in the Landscape, Boltzmann Brains and the Cosmological Constant Problem." *Journal of Cosmology and Astroparticle Physics*: 022. <http://www.iop.org/EJ/abstract/1475-7516/2007/01/022>.

_____.1986. "Eternal Chaotic Inflation." *Modern Physics Letters*, A1: 81.

_____.1986a. "Eternally Existing Self-Reproducing Chaotic Inflationary Universe." *Physics Letters B*, 175: 395–400.

Loux, Michael J. 2001. "The Problem of Universals." *Metaphysics: Contemporary Readings*. Michael J. Loux, ed. New York: Routledge.

MacLeod, M.; & E. Rubenstein. 2006. "Universals." *The Internet Encyclopedia of Philosophy*. J. Fieser & B. Dowden, eds. <http://www.iep.utm.edu/universa/>.

Macmillan Encyclopedia of Philosophy (MEP). 1967. "Cosmological Argument for the Existence of God." Vol. 2: 233ff.

Maitreya.org (MAI). 2010. "He Comes as…The Maitreya" (March 16). <http://www.maitreya.org/English/PBuddhism.htm>.

McAleer, Neil. 1987. *The Mind-Boggling Universe*. New York: Doubleday & Co.

McEvilley, Thomas. 2002. *The Shape of Ancient Thought*. New York: Allworth.

McMullin, Ernan, ed. 2005. "The Church's Ban on Copernicanism, 1616." *The Church and Galileo*. Ernan McMullin, ed. Notre Dame, IN: University of Notre Dame Press.

Mendell, Henry. 2004. "Aristotle and Mathematics" (March 26). *Stanford Encyclopedia of Philosophy*. Stanford, CA: Stanford University. <http://plato.stanford.edu/entries/aristotle-mathematics/>.

Menegoni, Eloisa et al. 2009. "New Constraints on Variations of the Fine Structure Constant from CMB Anisotropies." *Physical Review D*, 80 (8). <http://arxiv.org/abs/0909.3584>.

Merriam-Webster's Collegiate Dictionary Online (MWD). 2010. "Esoteric" (February 24). <http://www.merriam-webster.com/dictionary/esoteric>.

_____.2007. "Agency" (March 20). <http://www.m-w.com/dictionary/agency>.

Moss, I. G.; & S. W. Hawking. 1982. "Supercooled Phase Transitions in the Very Early Universe." *Physics Letters*, B110: 35–8.

Munitz, M. K. 1959. "One Universe or Many?" *Journal of the History of* Ideas, 12: 231–255. <http://links.jstor.org/sici?sici=0022-5037(195104)12%3A2%3C2

31%3AOUOM%3E2.0.CO%3B2-F>.

Musser, George; & J. Minkel. 2002. "A Recycled Universe: Crashing Branes and Cosmic Acceleration May Power an Infinite Cycle in which Our Universe Is but a Phase" (February 11). *Scientific American*. <http://www.sciam.com/article.cfm?id=a-recycled-universe>.

National Center for Science Education (NCSE). 2008. "Statements from Scientific and Scholarly Organizations" (August 28). <http://ncse.com/media/voices/science

New Scientist (NS). 2007. "The Cosmos—Before the Big Bang" (April 28). <http://www.astro.caltech.edu/~george/ay21/readings/beforethebigbang.pdf>.

_____.2004. "An Open Letter to the Scientific Community" (May 22). <http://www.cosmologystatement.org/>.

New York Times (NYT). 2000. "Cult in Uganda Poisoned Many, Police Say" (July 28). <http://www.nytimes.com/2000/07/28/world/cult-in-uganda-poisoned-many-police-say.html?pagewanted=1>.

_____.1997. "Families Learning of 39 Cultists Who Died Willingly" (March 29). <http://query.nytimes.com/gst/fullpage.html?res=9400E7DB133AF93AA15750C0A961958260&sec=health&spon=&pagewanted=all>.

_____.1997a. "Ex-Cultist Dies In Suicide Pact; 2d Is 'Critical'" (May 7). <http://query.nytimes.com/gst/fullpage.html?res=9E06E7DD1E30F934A35756C0A961958260>.

Normore, Calvin. 2007. "The Matter of Thought." *Representation and Objects of Thought in Medieval Philosophy*. Henrik Lagerlund, ed. Hampshire: Ashgate.

Nova Online (NO). 2000. "Time Can Vary." <http://www.pbs.org/wgbh/nova/time/thinktime.html>.

Numbers, Ronald L. 2010. "Creationism History: Topic Index" (February 24). *Counterbalance Meta-Library*. <http://www.counterbalance.net/history/intro-frame.html>.

Owen, H. P. 1971. *Concepts of Deity*. London: Macmillan.

Oxford English Dictionary, the Compact Edition of the (OED). 1971. "Universe." Volume II: 3518. Oxford: Oxford University Press.

Page, Don. 1976. "Particle Emission Rates from a Black Hole: Massless Particles from an Uncharged, Nonrotating Hole." *Physical Review*, D 13: 198–206. <http://prd.aps.org/abstract/PRD/v13/i2/p198_1>.

Pasnau, Robert. 2001. *Thomas Aquinas on Human Nature*. Cambridge, United Kingdom: Cambridge University Press.

Partridge, Christopher. 2005. *Introduction to World Religions*. Fortress.

Peebles, P. J. E. 2003. "The Cosmological Constant and Dark Energy." *Reviews of Modern Physics*, 75: 559–606. <http://rmp.aps.org/abstract/RMP/v75/i2/p

559_1>.

Penzias, A. A.; & R. Wilson. 1965. "A Measurement of Excess Antenna Temperature at 4080 Mc/s." *Astrophysical Journal*, 142: 419. <http://adsabs.harvard.edu/abs/1965ApJ...142..419P>.

PBS.org (PBS). 1998. "People and Discoveries: Big Bang Theory." <http://www.pbs.org/wgbh/aso/databank/entries/dp27bi.html>.

Price, H. H. 1953. "Universals and Resemblance." *Thinking and Experience*. London: Hutchinson.

Quine, W. V. O. 1961. "On What There is." *From a Logical Point of View*. N.Y: Harper and Row.

Raman, Shankar. 1997. Lectures on Literary Theory. Cambridge, MA: MIT (spring).

Ramsland, Katherine. 2006. "Death Mansion" (September 20). *All about Heaven's Gate cult* CourtTV Crime Library). <http://www.crimelibrary.com/notorious_murders/mass/heavens_gate/5.html>.

Ratzinger, Joseph Cardinal. 1994. *Turning Point for Europe? The Church in the Modern World—Assessment and Forecast*. Brian McNeil, tr. San Francisco, CA: Ignatius Press.

Rees, M. 2000. *Just Six Numbers*. London: Orion Books.

Reichenbach, Bruce. 2006. "Cosmological Argument." *The Stanford Encyclopedia of Philosophy*. Edward N. Zalta, ed. <http://plato.stanford.edu/archives/fall2006/entries/cosmological-argument/>.

Ross, Rich. 1999. "'Heaven's Gate' Suicides" (October). The Rick A. Ross Institute. <http://www.culteducation.com/hgate.html>.

Russell, Bertrand. 1912. *The Problems of Philosophy*. New York: H. Holt.

Sackmann, I. J.; A. I. Boothroyd; & K. E. Kraemer. 1993. "Our Sun. III. Present and Future." *Astrophysical Journal*, 418: 457. <http://adsabs.harvard.edu/cgi-bin/nph-bib_query?bibcode=1993ApJ...418..457S>.

San Diego Herald Tribute (SDHT). 2007. "Heaven's Gate: A Timeline" (March 18). <http://legacy.signonsandiego.com/uniontrib/20070318/news_lz1n18 time lin.html>.

Schröder, K. P.; & Robert Connon Smith. 2008. "Distant Future of the Sun and Earth Revisited." *Monthly Notices of the Royal Astronomical Society*, 386: 155. <http://www3.interscience.wiley.com/journal/119396337/abstract?CRETRY=1&SRETRY=0>.

Sharratt, Michael. 1994. *Galileo: Decisive Innovator*. Cambride, UK: Cambridge University Press.

Shields, Christopher. 2007. *Aristotle*. London, United Kingdom: Routledge.

_____.1993. "Some Recent Approaches to Aristotle's De Anima." *De Anima: Books II and* III (With Passages From Book I). W. D. Hamlyn, tr. Oxford:

Clarendon.

_____.1988. "Soul as Subject in Aristotle's De Anima." *Classical Quarterly*, 38 (1): 140-49.

Shlain, Leonard. 1991. *Art and Physics: Parallel Visions in Space,Time, and Light*. New York: William Morrow and Co., Inc.

Singh, S. 2010. "Big Bang" (February 26). <http://www.simonsingh.net/Big_Bang.html>.

Sivananda, Swami. 2010. "Brahma Sutras" (February 25). <http://www.swami-krishnananda.org/bs_2/bs_2.1.03.html>.

SLAC Virtual Visitor Center (SLAC). 2008. "Theory: Decays." <http://www2.slac.stanford.edu/vvc/theory/decays.html>.

Slipher, V. M. 1915. "Spectrographic Observations of Nebulae." *Popular Astronomy*, 23: 21–24. <http://adsabs.harvard.edu/abs/1915PA.....23Q..21S>.

_____.1913. "The Radial Velocity of the Andromeda Nebula." *Lowell Observatory Bulletin*, 1: 56–57. <http://adsabs.harvard.edu/abs/1913LowOB ...2...56S>.

Spergel, D. N., et al. 2006. "Wilkinson Microwave Anisotropy Probe (WMAP) Three Year Results: Implications for Cosmology." *Astrophysics*. <http://arxiv.org/abs/astro-ph/0603449v2>.

Spielvogel, Jackson; & David Redles. 1986. "Hitler's Racial Ideology: Content and Occult Sources." *Simon Wiesenthal Center Annual*, 3. <http://motlc.wiesenthal.com/site/pp.asp?c=gvKVLcMVIuG&b=395043>.

Steinhardt, Paul J.; & Neil Turok. 2002. "Cosmic Evolution in a Cyclic Universe." *Physical Review D*, 65 (12): 126003. <http://link.aps.org/doi/10.1103/PhysRevD.65.126003>.

_____. & Neil Turok. 2001. "A Cyclic Model of the Universe." <http://arxiv.org/abs/hep-th/0111098>.

Stenger, Victor J. 2010. "Is the Universe Fine-Tuned for Us?" (February 27). <http://www.colorado.edu/philosophy/vstenger/Cosmo/FineTune.pdf>.

Stone, M. 1977. "Semiclassical Methods for Unstable States." *Phys.Lett.*, B 67: 186–183.

_____.1976. "Lifetime and Decay of Excited Vacuum States." *Phys. Rev.*, D 14: 3568–3573.

Tipler, Frank. 1994. *The Physics of Immortality*. New York: Doubleday.

Vaas, Rudiger, 2003. "Time before Time: Classifications of Universes in Contemporary Cosmology, and How to Avoid the Antinomy of the Beginning and Eternity of the World." <http://philsci-archive.pitt.edu/archive/00001910/01/VAASTIME.PDF>.

Vallee, Jacques. 1979. *Messengers of Deception: UFO Contacts and Cults*. Ronin.

Vella, John. 2008. *Aristotle: A Guide for the Perplexed*. New York: Continuum.

Wang, Aihe. 2000. *Cosmology and Political Culture in Early China*. Cambridge,

UK: Cambridge University Press.

Weber, Max. 1930. *The Protestant Ethic and the Spirit of Capitalism.* Talcott Parsons, tr. New York: Scribner.

Weinberg, Steven. 1993. *The First Three Minutes: A Modern View of the Origin of the Universe.* New York: Basic Books.

Wikipedia (WK). 2010. "Cosmology" (February 23). <http://en.wikipedia.org/wiki/Cosmology>.

_____.2010a. "Cosmogony" (February 23). <http://en.wikipedia.org/wiki/Cosmogony>.

_____.2010b. "Universe" (February 23). <http://en.wikipedia.org/wiki/Universe>.

_____.2010c. "Geocentric Model" (February 23). <http://en.wikipedia.org/wiki/Geocentric>.

_____.2010d. "Heliocentric Model" (February 23). <http://en.wikipedia.org/wiki/Heliocentric>.

_____.2010e. "General Relativity" (February 23). <http://en.wikipedia.org/wiki/General_relativity>.

_____.2010f. "Cosmological Constant" (February 23). <http://en.wikipedia.org/wiki/Cosmological_constant>.

_____.2010g. "Critical Density" (February 23). <http://en.wikipedia.org/wiki/Critical_density>.

_____.2010h. "Non Standard Cosmology" (February 23). <http://en.wikipedia.org/wiki/Non-standard_cosmology>.

_____.2010i. "Monism" (February 24). <http://en.wikipedia.org/wiki/Monism>.

_____.2010j. "Pantheism" (February 24). <http://en.wikipedia.org/wiki/Pantheism>.

_____.2010k. "Panentheism" (February 24). <http://en.wikipedia.org/wiki/Panentheism>.

_____.2010l. "Creationism" (February 24). <http://en.wikipedia.org/wiki/Creationism>.

_____.2010m. "Religious Cosmology" (February 24). <http://en.wikipedia.org/wiki/Religious_cosmology>.

_____.2010n. "Mechanism (Philosophy)" (February 24). <http://en.wikipedia.org/wiki/Mechanism_(philosophy)>.

_____.2010o. "Dynamism" (February 24). <http://en.wikipedia.org/wiki/Dynamism>.

_____.2010p. "Atomism" (February 24). <http://en.wikipedia.org/wiki/Atomism>.

_____.2010q. "Esoteric Cosmology" (February 24). <http://en.wikipedia.org/wiki/Esoteric_cosmology>.

_____.2010r. "Emanation" (February 24). <http://en.wikipedia.org/wiki/

Emanation>.

_____.2010s. "Involution (Esoterism" (February 24). <http://en.wikipedia.org/wiki/Involution_(esoterism)>.

_____.2010t. "Planes (Esotericism)" (February 24). <http://en.wikipedia.org/wiki/Planes_(esotericism)>.

_____.2010u. "Advaita Vedanta" (February 25). <http://en.wikipedia.org/wiki/Advaita>.

_____.2010v. "Acosmism" (February 25). <http://en.wikipedia.org/wiki/Acosmism>.

_____.2010w. "Idealism" (February 26). <http://en.wikipedia.org/wiki/Idealism>.

_____.2010x. "Universal (metaphysics)" (February 26). <http://en.wikipedia.org/wiki/Universal_(metaphysics)>.

_____.2010y. "Abstract Objects" (February 26). <http://en.wikipedia.org/wiki/Abstract_objects)>.

_____.2010z. "Problem of Universals" (February 26). <http://en.wikipedia.org/wiki/Problem_of_universals>.

_____.2010aa. "Big Bang" (February 26). <http://en.wikipedia.org/wiki/Big_bang>.

_____.2010bb. "Timeline of the Big Bang" (February 26). <http://en.wikipedia.org/wiki/Timeline_of_the_Big_Bang>.

_____.2010cc. "Brain Cosmology" (February 27). <http://en.wikipedia.org/wiki/Brane_cosmology>.

_____.2010dd. "Ekpyrotic Universe" (February 27). <http://en.wikipedia.org/wiki/Ekpyrotic>.

_____.2010ee. "Cyclic Model" (February 27). <http://en.wikipedia.org/wiki/Cyclic_model>.

_____.2010ff. "Chaotic Inflation" (February 27). <http://en.wikipedia.org/wiki/Chaotic_inflation>.

_____.2010gg. "Galileo Affair" (March 01). <http://en.wikipedia.org/wiki/Galileo_affair>.

_____.2010hh. "The Cosmological Argument" (March 10). <http://en.wikipedia.org/wiki/Cosmological_argument>.

_____.2010ii. "Hylomorphism" (March 10). <http://en.wikipedia.org/wiki/Hylomorphism>.

_____.2010jj. "Substantial Form" (March 10). <http://en.wikipedia.org/wiki/Substantial_form>.

_____.2010kk. "Involution" (March 11). <http://en.wikipedia.org/wiki/Involution>.

_____.2010ll. "Satchitananda" (March 11). <http://en.wikipedia.org/wiki/

Satchitananda>.

_____.2010mm. "Spiritual Evolution" (March 11). <http://en.wikipedia.org/wiki/Spiritual_evolution>.

_____.2010nn. "Theosophy" (March 12). <http://en.wikipedia.org/wiki/Theosophy>.

_____.2010oo. "The Secret Doctrine" (March 12). <http://en.wikipedia.org/wiki/Root_Race>.

_____.2010pp. "The Anthroposophy" (March 12). <http://en.wikipedia.org/wiki/Anthroposophy>.

_____.2010qq. "The Ultimate Fate of the Universe" (March 12). <http://en.wikipedia.org/wiki/Ultimate_fate_of_the_universe>.

_____.2010rr. "Friedmann Equations" (March 13). <http://en.wikipedia.org/wiki/Friedmann_equations>.

_____.2010ss. "Heaven's Gate" (March 14). <http://en.wikipedia.org/wiki/Heaven's_Gate_(cult)>.

_____.2010tt. "1 E 19 s and More" (March 14). <http://en.wikipedia.org/wiki/1_E19_s_and_more>.

_____.2010uu. "Iron-56" (March 14). <http://en.wikipedia.org/wiki/Iron-56>.

_____.2010vv. "Eschatology" (March 16). <http://en.wikipedia.org/wiki/Eschatology>.

_____.2010ww. "Apocalypse" (March 16). <http://en.wikipedia.org/wiki/Apocalpyse>.

_____.2010xx. "End Time" (March 16). <http://en.wikipedia.org/wiki/End_time>.

_____.2010yy. "Movement for the Restoration of the Ten Commandments of God" (March 17). <http://en.wikipedia.org/wiki/Movement_for_the_Restoration_of_the_Ten_Commandments_of_God>.

_____.2008. "Posthumanism" (October 20). <http://en.wikipedia.org/wiki/Posthumanism>.

_____.2008a. "Transhumanism" (October 20). <http://en.wikipedia.org/wiki/Transhumanism>.

_____.2008b. "Lowest Common Denominator" (March 26). <http://en.wikipedia.org/ wiki/Least_common_denominator>.

_____.2008c. "Infinity" (November 11). <http://en.wikipedia.org/wiki/Infinity>.

_____.2008d. "Transfinite Number" (November 10). <http://en.wikipedia.org/wiki/Transfinite_ number>.

_____.2008e. "Quantum Mechanics" (November 26). <http://en.wikipedia.org/wiki/Quantum_mechanics>.

_____.2008f. "Copenhagen Interpretation" (November 26).

<http://en.wikipedia.org/wiki/Copenhagen_interpretation >.

_____.2008g. “Many Worlds Interpretation” (November 26). <http://en.wikipedia.org/wiki/Many_Worlds_Interpretation>.

_____.2008h. “Hyperspace (Science Fiction)” (November 23). <http://en.wikipedia.org/wiki/Hyperspace_(science_fiction)>.

_____.2008i. “Time Travel” (November 26). <http://en.wikipedia.org/wiki/Time_travel#Time_dilation>.

_____.2008j. “Wormhole” (November 25). <http://en.wikipedia.org/wiki/Wormhole>.

_____.2007. “Linguistics” (March 04). <http://en.wikipedia.org/wiki/Linguistics>.

Wikiversity (WKV). 2008. “Why 10 Dimensions” (November 23). <ttp://en.wikiversity.org/wiki/Why_10_dimensions>

Wilczek, Frank. 2008. “The Lightness of Being: Mass, Ether, and the Unification of Forces: Anticipating a New Golden Age” (October 16). *M.I.T. World*. <http://mitworld.mit.edu/video/618>.

Williams, D. C. 1953. “On the Elements of Being.” *Review of Metaphysics*, 17.

Wittgenstein, Ludwig.1953. *Philosophical Investigations*. G. E. M. Anscombe and R. Rhees, eds. G.E.M. Anscombe, tr. Oxford: Blackwell.

Wojik, Daniel. 2003. “Apoaaacalyptic and Millenarian Aspects of American UFOism.” *UFO Religions*. Chistopher Partridge, ed. Routledge.

Wright, E. L. 2009. “What Is the Evidence for the Big Bang?” (May 09). *Frequently Asked Questions in Cosmology*. UCLA, Division of Astronomy and Astrophysics. <http://www.astro.ucla.edu/~wright/cosmology_faq.html#BBevidence>.

_____.2009a. “Cosmology and Religion.” *Ned Wright's Cosmology Tutorial*. <http://www.astro.ucla.edu/~wright/cosmo-religion.html>.

_____.2004. “Big Bang Nucleosynthesis.” <http://www.astro.ucla.edu/~wright/BBNS.html>.

Yao, W. M.; et al. 2006. “Review of Particle Physics.” *Journal of Physics G*, 33: 1–1232. <http://pdg.lbl.gov/2006/reviews/darkmatrpp.pdf>.

INDEX

•A•

•B•

•C•

•D•

•E•

•F•

•G•

•H•

•I•

•J•

•K•

•L•

•M•

•N•

•O•

•P•

•Q•

•R•

•S•

•T•

•U•

•V•

•W•

•X•

•Y•

•Z•